Iterative Methods for
Large Linear Systems

Iterative Methods for Large Linear Systems

Edited by

David R. Kincaid and Linda J. Hayes

Center for Numerical Analysis
The University of Texas at Austin
Austin, Texas

ACADEMIC PRESS, INC.

Harcourt Brace Jovanovich, Publishers

Boston San Diego New York
Berkeley London Sydney
Tokyo Toronto

ACADEMIC PRESS, INC.
1250 Sixth Avenue, San Diego, CA 92101

United Kingdom Edition published by
ACADEMIC PRESS LIMITED
24-28 Oval Road, London NW1 7DX

Library of Congress Cataloging-in-Publication Data

Iterative methods for large linear systems / edited by David R.
  Kincaid and Linda J. Hayes.
     p.   cm.
  Papers from a conference held Oct. 19–21, 1988, at the Center for
Numerical Analysis of the University of Texas at Austin.
  Includes bibliographical references.
  ISBN 0-12-407475-8 (alk. paper)
  1. Iterative methods (Mathematics) — Congresses. 2. Vector
processing (Computer science) — Congresses. 3. Parallel processing
(Electronic computers) — Congresses. I. Kincaid, David (David
Ronald II. Hayes, Linda J.
QA432.I84   1989
515'.72 — dc20   89-28453
                                                            CIP

Printed in the United States of America
89 90 91 92   9 8 7 6 5 4 3 2 1

Preface

Recently, the use of iterative methods for solving linear systems has experienced a resurgence of activity as scientists attack extremely complicated three-dimensional problems using vector and parallel supercomputers. This book contains a wide spectrum of research topics related to iterative methods such as searching for optimum parameters, using hierarchical basis preconditioners, utilizing software as a research tool, and developing algorithms for vector and parallel computers. Current research on a variety of different iterative algorithms is presented including the successive overrelaxation (SOR) method, the symmetric and unsymmetric SOR methods, a local (ad-hoc) SOR scheme, the alternating direction implicit (ADI) method, block iterative methods, asynchronous iterative procedures, multilevel methods, adaptive algorithms, and domain decomposition algorithms. The applications involve partial differential equations (elliptic problems, Navier-Stokes equations, etc.) with various ways of approximating boundary conditions and with different types of associated linear systems (symmetric, nonsymmetric, indefinite symmetric Toeplitz, etc.). The analysis utilized ranges over finite differences, finite elements, semi-implicit discretizations, and spectral analysis, for example.

This book is a direct outgrowth of a conference on *Iterative Methods for Large Linear Systems*, held October 19–21, 1988, hosted by the Center for Numerical Analysis of The University Texas at Austin. This book contains 17 papers from the conference with the objective of providing an overview of the state of the art in the use of iterative methods for solving sparse linear systems. The emphasis is on identifying current and future research directions in the mainstream of modern scientific computing with an eye to contributions of the past, present, and future. Many research advances in the development of iterative methods for high-speed computers over the past forty years were discussed with current research as the primary focus. Forty-six papers were presented during the conference with approximately 150 participants, among them many of the world's most distinguished mathematicians and computer scientists.

The highlight of the conference was a banquet honoring David M. Young, Jr., for his many contributions to the field of computational mathematics and on the occasion of his 65th birthday. Many friends and colleagues of Professor Young from academia, industry, and government told of their association with him over a career of almost forty years that parallels the development of high-speed computing. The banquet and conference offered the unique opportunity to hear from many of those who, along with Young, had firsthand involvement in the development of scientific computing and modern numerical analysis.

Thanks goes to all of the speakers at the conference. Their names and the titles of the papers given are listed herein. Associate Dean James W. Vick of the College of Natural Sciences gave the opening remarks, and the following individuals chaired sessions: George D. Byrne, John R. Cannon, Graham F. Carey, E. Ward Cheney, Len Colgan, Linda J. Hayes, Olin G. Johnson, David R. Kincaid, Esmond G. Ng, J. Tinsley Oden, L. Duane Pyle, Robert van de Geijn, Kamy Sepehrnoori, and Charles H. Warlick. The banquet was enriched by comments by the following people: Loyce Adams, John Cannon, Graham Carey, Garrett Birkhoff, Lou Ehrlich, Anne Elster, Gene Golub, Margaret Gregory, Anne Greenbaum, Bob Greenwood, Linda Hayes, Cecilia Jea, David Kincaid, Tom Manteuffel, Tom Nance, Tinsley Oden, Paul Saylor, Mark Seager, Dick Varga, Gene Wachspress, Mary Wheeler, John Whiteman, and Charles Warlick.

Special guests at the banquet were Mrs. Robert T. Gregory, Charles and Suzanne Warlick, Tinsley and Barbara Oden, Al and Nell Dale, Efraim and Edna Armendariz, and members of the Young family: William and Linda Young, Arthur Young, Carolyn Young, Christine Sorenson, and David Sorenson.

The conference was organized by David R. Kincaid, Linda J. Hayes, Graham F. Carey, and E. Ward Cheney, all of The University of Texas at Austin. The assistance of Katy Burrell, Sheri Brice, and Adolfo Sanchez in providing much of the support work for the conference is gratefully acknowledged. Also making the conference a success with special activities were Celeste Hamman and Julie Burrell.

Lisa Laguna is be to commended for reworking all of the papers in this book. Putting them into a uniform format using LaTeXwas no small task by any means. Kay Nettle was quite helpful in assisting us to overcome technical difficulties. Sheri Brice contributed to the production of this book in numerous ways as well.

Appreciation goes to the Computer Sciences Department and the Computation Center of The University of Texas at Austin for the use of their excellent computing facilities.

The Conference on Iterative Methods for Large Linear Systems was co-sponsored by the Society of Industrial and Applied Mathematics (SIAM) Activity Groups for Linear Algebra and Supercomputing. Support for the conference was provided, in part, by the Office of Naval Research, the Department of Energy, the National Science Foundation, the Air Force Office of Scientific Research, and The University of Texas at Austin.

David R. Kincaid
Linda J. Hayes
September 1989

Authors of Chapters

Loyce M. Adams, Department of Applied Mathematics, University of Washington, Seattle, WA 98195

Owe Axelsson, Department of Mathematics, University of Nijmegen, Toernooiveld 6525 ED Nijmegen, The Netherlands

Garrett Birkhoff, Department of Mathematics, Harvard University, Cambridge, MA 02138

Paul Concus, Lawrence Berkeley Laboratory, University of California, Berkeley, CA 94720

John E. de Pillis, Department of Mathematics, University of California, Riverside, CA 92521-0135

Maksymilian Dryja, Institute of Informatics, Warsaw University, Warsaw, Poland

Louis W. Ehrlich, Johns Hopkins University, Applied Physics Laboratory, Laurel, MD 20707-6099

Howard C. Elman, Department of Computer Science, University of Maryland, College Park, MD 20742

Gene H. Golub, Computer Science Department, Stanford University, Stanford, CA 94305

Louis A. Hageman, Westinghouse Electric Corporation, Bettis Atomic Power Laboratory, West Mufflin, PA 15122-0079

David L. Harrar II, Applied Mathematics & Computer Sciences Department, University of Virginia, Charlottesville, VA 22903

Wayne D. Joubert, Center for Numerical Analysis, University of Texas at Austin, Austin, TX 78713-8510

Xiezhang Li, Institute for Computational Mathematics, Kent State University, Kent, OH 44242-4001

Robert E. Lynch, Department of Computer Science, Purdue University, West Lafayette, IN 47907

Tsun-Zee Mai, Department of Mathematics, University of Alabama, Tuscaloosa, AL 35487

Thomas A. Manteuffel, Computational Mathematics Group, University of Colorado, Denver, CO 80202

Dan C. Marinescu, Department of Computer Science, Purdue University, W. Lafayette, IN 47907

James M. Ortega, Applied Mathematics & Computer Sciences Department, University of Virginia, Charlottesville, VA 22903

John R. Rice, Department of Computer Science, Purdue University, W. Lafayette, IN 47907

Paul E. Saylor, Department of Computer Science, University of Illinois at Urbana-Champaign, Urbana, IL 61801

Richard S. Varga, Institute for Computational Mathematics, Kent State University, Kent, OH 44242-4001

Eugene L. Wachspress, Department of Mathematics, University of Tennessee, Knoxville, TN 37996-1300

Olof B. Widlund, Courant Institute of Mathematical Sciences, Department of Computer Science, New York University, New York, NY 10012

David M. Young, Center for Numerical Analysis, University of Texas at Austin, Austin, TX 78713-8510

Papers Presented at Conference

Loyce M. Adams* (University of Washington) "Fourier Analysis of Two-Level Hierarchical Basis Preconditioners"

David V. Anderson* & Alice E. Koniges (Lawrence Livermore National Laboratory) "The Solution of Large Striped Matrix Systems Derived from Multiple Coupled 3D PDEs"

Steven F. Ashby* (Lawrence Livermore National Laboratory) "Polynomial Preconditioning for Conjugate Gradient Methods"

Owe Axelsson* (University of Nijmegen, The Netherlands) "An Algebraic Framework for Hierarchical Basis Functions Mutilevel Methods or the Search for 'Optimal' Preconditoners"

Garrett Birkhoff* (Harvard University) & Robert E. Lynch (Purdue University) "ELLPACK and ITPACK as Research Tools for Solving Elliptic Problems"

Randall B. Bramley* (University of Illinois at Urbana-Champaign) "A Projection Method for Large Sparse Linear Systems"

Graham F. Carey*, David R. Kincaid, Kamy Sepehrnoori, & David M. Young (University of Texas at Austin) "Vector and Parallel Iterative Solution Experiments"

Paul Concus* (Lawrence Berkeley Laboratory) & Paul E. Saylor (University of Illinois) "Preconditioned Iterative Methods for Indefinite Symmetric Toeplitz Systems"

Jerome Dancis* (University of Maryland) "Diagonalizing the Adaptive SOR Iteration Method"

Seungsoo Lee, George S. Dulikravich* & Daniel J. Dorney (Pennsylvania State University) "Distributed Minimal Residual (DMR) Method for Explicit Algorithms Applied to Nonlinear Systems"

Louis W. Ehrlich* (John Hopkins Applied Physics Laboratory) "A Local Relaxation Scheme (Ad-Hoc SOR) Applied to Nine Point and Block Difference Equations"

---

*speaker

ix

Howard C. Elman* (University of Maryland) & Gene H. Golub (Stanford University) "Block Iterative Methods for Cyclically Reduced Non-self-adjoint Elliptic Problems"

Anne C. Elster* (Cornell University), Hungwen Li (IBM Almaden Research Center, San Jose) & Michael M.C. Sheng (National Chiao-Tung University, Taiwan, R.O. China) "Parallel Operations for Iterative Methods: A Polymorphic View"

David J. Evans* & C. Li (Loughborough University of Technology, England) "$D^{1/2}$-Norms of the SOR and Related Method for a Class of Nonsymmetric Matrices"

Bernd Fischer* (Stanford University & University of Hamburg, W. Germany) & Lothar Reichel (Bergen Scientific Centre, Bergen, Norway & University of Kentucky) "A Stable Richardson Iteration Method for Complex Linear Systems"

Gene H. Golub* (Stanford University) & John de Pillis (University of California, Riverside) "Toward an Effective Two-Parameter SOR Method"

Anne Greenbaum* (New York University) "Predicting the Behavior of Finite Precision Lanczos and Conjugate Gradient Computations"

S. Galanis, Apostolos Hadjidimos* & Dimitrois Noutsos (University of Ioannina, Greece, & Purdue University) "On an SSOR Matrix Relationship and Its Consequences"

Louis A. Hageman* (Westinghouse-Bettis Laboratory) "Relaxation Parameters for the IQE Iterative Procedure for Solving Semi-Implicit Navier-Stokes Difference Equations"

Martin Hanke* (Universitat Karlsruhe, W. Germany) "On Kaczmarz' Method for Inconsistent Linear Systems"

Kaibing Hwang* & Jinru Chen (Nanjing Normal University, P.R. China) "A New Class of Methods for Solving Nonsymmetric Systems of Linear Equations — Constructing and Realizing Symmetrizable Iterative Methods"

Kang C. Jea* (Fu Jen University, Taiwan, R.O. China) & David M. Young (University of Texas at Austin) "On The Effectiveness of Adaptive Chebyshev Acceleration for Solving Systems of Linear Equations"

David R. Kincaid* (University of Texas at Austin) "A Status Report on the ITPACK Project"

C.-C. Jay Kuo* & Tony F. Chan (University of California at Los Angeles) "Two-Color Fourier Analysis of Iterative Methods for Elliptic Problems with Red-Black Ordering"

Avi Lin* (Temple University) "Asynchronous Parallel Iterative Methods"

Robert E. Lynch* (Purdue University) "Hodie Approximation of Boundary Conditions"

Tsun-Zee Mai* (University of Alabama) & David M. Young (University of Texas at Austin) "A Dual Adaptive Procedure for the Automatic Determination of Iterative Parameters for Chebyshev Acceleration"

Thomas A. Manteuffel* (University of Colorado, Denver & Los Alamos National Laboratories) & Wayne D. Joubert (University of Texas at Austin) "Iterative Methods for Nonsymmetric Linear Systems"

Craig Douglas, J. Mandel & Willard L. Miranker* (IBM Watson Research Center, Yorktown Heights) "Fast Hybrid Solution of Algebraic Systems"

Michael I. Navon* & H.-I. Lu (Florida State University) "A Benchmark Comparison of the ITPACK Package on ETA-10 and Cyber-205 Supercomputers"

Apostolos Hadjidimos (Purdue University & University of Ioannina, Greece) & Michael Neumann* (University of Connecticut) "Convergence Domains and Inequalities for the Symmetric SOR Method"

Thomas C. Oppe* (University of Texas at Austin) "Experiments with a Parallel Iterative Package"

David L. Harrar & James M. Ortega* (University of Virginia) "Solution of Three-Dimensional Generalized Poisson Equations on Vector Computers"

M. G. Petkov* (Academy of Sciences, Bulgaria) "On the Matrix Geometric Progression and the Jordan Canonical Form"

M. Dryja (University of Warsaw, Poland) & Wlodck Proskurowski* (University of Southern California) "Composition Method for Solving Elliptic Problems"

Dan C. Marinescu & John R. Rice* (Purdue University) "Multi-level Asynchronous Iterations for PDEs"

Paul J. Lanzkron, Donald J. Rose* & Daniel B. Szyld (Duke University) "Convergence of Nested Iterative Methods for Linear Systems"

Paul E. Saylor* (University of Illinois) "An Adaptive Algorithm for Richardson's Method"

Robert van de Geijn* (University of Texas at Austin) "Machine Independent Parallel Numerical Algorithms "

Richard S. Varga* & Xiezhang Li (Kent State University) "A Note on the SSOR and USSOR Iterative Methods Applied to p-Cyclic Matrices"

A. Haegemans & J. Verbeke* (Katholieke Universiteit Leuven, Belgium) "The Symmetric Generalized Accelerated Overrelaxation (GSAOR) Method"

Eugene L. Wachspress* (University of Tennessee) "The ADI Minimax Problem for Complex Spectra"

Mary F. Wheeler* (University of Houston) "Domain Decomposition—Multigrid Algorithms for Mixed Finite Element Methods for Elliptic PDE's"

John R. Whiteman* (Brunel University, England) "Finite Element Treatment of Singularities in Elliptic Boundary Value Problems"

Maksymilian Dryja (Warsaw University) & Olof B. Widlund* (New York University) "Some Domain Decomposition Algorithms for Elliptic Problems"

Robert E. Wyatt* (University of Texas at Austin) "Iterative Methods in Molecular Collision Theory"

David M. Young* (University of Texas at Austin) & Tsun-Zee Mai (University of Alabama) "The Search for Omega"

Portrait of David M. Young, Jr., drawn by his sister Christine Sorenson
(credit: Christine Sorenson)

Professor David M. Young, Jr.

Professor David M. Young received his Ph.D. in Mathematics from Harvard University. Following a year of post-doctoral research at Harvard he went to the Computing Laboratory of the Aberdeen Proving Ground. A year later he moved to the University of Maryland where he held the rank of Associate Professor of Mathematics. Three years later he joined the staff of the Ramo-Wooldridge Computation in Los Angeles where he was Head of the Mathematics and Computation Department.

In 1958, Dr. Young joined the faculty of The University of Texas as Professor of Mathematics and Director of the Computation Center. While Director of the Computation Center he was involved in the acquisition of two large computer systems—the Control Data 1604 computer in 1961 and the Control Data 6600 computer in 1966. In 1970, Professor Young left the Computation Center to become Director of the Center for Numerical Analysis. He currently holds that position and has the title of Ashbel Smith Professor of Mathematics and Computer Sciences at The University of Texas at Austin. He is also a member of the Texas Institute for Computational Mechanics.

Professor Young has worked extensively on the numerical solution of partial differential equations with emphasis on the use of iterative methods for solving large sparse systems of linear algebraic equations. His early work on iterative methods involved the analysis of the successive overrelaxation (SOR) method for consistently ordered matrices. He was also one of the first to use Chebyshev polynomials for the solution of linear systems. He has written over 90 technical papers and is the author of three books. One of these books, which appeared in 1971, is devoted to the basic theory of iterative methods. A second book, with the late Robert Todd Gregory, is a two volume survey of numerical mathematics that first appeared in 1972 and was reprinted in 1988. The third book, with Louis A. Hageman, appeared in 1981 and is devoted to iterative algorithms, including automatic procedures for choosing iteration parameters, as well as accurate procedures for deciding when to stop the iterative process.

Some topics Professor Young has worked on include the following: alternating direction implicit methods, adaptive procedures for determining iteration parameters, conjugate gradient methods with emphasis on generalizations for nonsymmetric systems, high-order discretization methods for partial differential equations, norms of matrices for iterative methods, procedures for estimating the accuracy of an approximate solution of a linear system, and the symmetric successive overrelaxation (SSOR) method.

Since the mid-1970s, Professor Young has collaborated with David Kincaid and others on the ITPACK project. This project is concerned with research on iterative algorithms and on the development of research-oriented mathematical software. Several software packages have been developed with the capability of handling non-symmetric, as well as symmetric, systems. The emphasis of the work has been increasingly on the use of advanced computer architectures.

Books

- *Iterative Solution of Large Linear Systems*, Academic Press, New York, 1971.

- (with Robert Todd Gregory) *A Survey of Numerical Mathematics*, Vols. I and II, Addison-Wesley, Reading, Massachusetts. (reprinted by Dover, 1988.)

- (with Louis A. Hageman) *Applied Iterative Methods*, Academic Press, 1981.

Ph.D. Theses Supervised
at The University of Texas at Austin

Tsun-Zee Mai [1986] "Adaptive Iterative Algorithms for Large Sparse Linear Systems"

Cecilia Kang Chang Jea [1982] "Generalized Conjugate Gradient Acceleration of Iterative Methods"

Linda J. Hayes [1977] "Generalization of Galerkin Alternating-Direction Methods to Non-Rectangular Regions Using Isoparametric Elements"

Vitalius J. Benokraitis [1974] "On the Adaptive Acceleration of Symmetric Successive Overrelaxation"

David R. Kincaid [1971] "An Analysis of a Class of Norms of Iterative Methods for Systems of Linear Equations"

Jerry C. Webb [1970] "Finite-Difference Methods for a Harmonic Mixed Boundary Value Problem"

Alvis E. McDonald [1970] "A Multiplicity-Independent Global Iteration for Meromorphic Functions"

Thurman G. Frank [1966] "Error Bounds on Numerical Solutions of Dirichlet Problems for Quasilinear Elliptic Equations"

Louis W. Ehrlich [1963] "The Block Symmetric Successive Overrelaxation Method"

Master's Theses Supervised
at The University of Texas at Austin

Florian Jarré [1986] "Multigrid Methods"

Vona Bi Roubolo [1985] "Finite Difference Solutions of Elliptic and Parabolic
Partial Differential Equations on Rectangle Regions: Programs ELIPBVP and
IVPROBM"

Reza Oraiee Abbasian [1984] "Lanczos Algorithms for the Acceleration of Nonsymmetrizable Iterative Methods"

Ru Huang [1983] "On the Determination of Iteration Parameters for Complex SOR
and Chebyshev Methods"

Janis Anderson [1983] "On Iterative Methods Based on Incomplete Factorization
Methods"

Linda J. Hayes [1974] "Comparative Analysis of Iterative Techniques for Solving
Laplace's Equation on the Unit Square on a Parallel Processor"

Tran Phien [1972] "An Application of Semi-Iterative and Second-Degree Symmetric Successive Overrelaxation Iterative Methods"

Belinda M. Wilkinson [1969] "A Polyalgorithm for Finding Roots of Polynomial
Equations"

Harold D. Eidson [1969] "The Convergence of Richardson's Finite-Difference Analogue for the Heat Equation"

Alkis Mouradoglou [1967] "Numerical Studies on the Convergence of the Peaceman-
Rachford Alternating Direction Implicit Method"

James A. Downing [1966] "The Automatic Construction of Contour Plots with
Application to Numerical Analysis"

W.G. Poole [1965] "Numerical Experiments with Several Iterative Methods for
Solving Partial Differential Equations"

W.P. Champagne [1964] "On Finding Roots of Polynomials by Hook-or-by-Crook"

Charles A. Shoemake [1964] "A Technique for Extracting Multiple Roots from
Polynomial Equations"

Mary B. Fanett [1963] "Application of the Remes Algorithm to a Problem in
Rational Approximation"

M.C. Scully [1962] "A Study of Linear Programming Using Digital Computer
Techniques"

W.P. Cash [1961] "The Determination of Peaceman-Rachford Iteration Parameters"

Jessee J. Stephens [1961] "A Consideration of Computational Stability for Various Time Difference Analogues Used in Numerical Weather Prediction"

Robert A. Sibley [1959] "Intelligent Activity on the Part of Digital Computers"

Master's Theses Supervised
at The University of Maryland, College Park

Louis W. Ehrlich [1956] "The Use of the Monte Carlo Method for Solving Boundary Value Problems Involving the Difference Analogue of $u_{xx} + u_{yy} + k(u_y/y) = 0$"

Charles H. Warlick [1955] "Convergence Rates of Numerical Methods for Solving $\partial^2 u/\partial x^2 + (k/\rho)(\partial u/\partial \rho) + \partial^2 u/\partial \rho^2 = 0$"

Photographs from Conference

Group photograph taken at the Iterative Methods Conference (October 19, 1988, Austin, TX)
(photograph by Steve Armstrong)

David Young (seated) and some of his Ph.D. and Master's degree students, left to right: David Kincaid, Louis Ehrlich, Vona Bi Roubolo, Cecilia Jea, Mary Wheeler, Tsun-Zee Mai, Linda Hayes, Tom Oppe, and Wayne Joubert. (photograph by Steve Armstrong)

David Young showing the audience at the banquet the personalized license plate presented to him. David Kincaid and Mildred Young also in picture. (photograph by Carole Kincaid)

David Young shown with plaque presented to him at the banquet inscribed:

You're the Omega$_{best}$

DAVID M. YOUNG

From Your Colleagues, Students, and Friends
Celebrating Your 65th Birthday

October 20, 1988

Also in the picture are David Kincaid and Mildred Young. (photograph by Carole Kincaid)

Seated at the family table are left to right: Arthur Young, Linda Young, William Young, Christine Sorenson, David Sorenson, and Carolyn Young.
(photograph by Carole Kincaid)

Gene Golub addressing the audience at the banquet. Also in the picture in the foreground are, left to right, Nell and Al Dale, Suzanne and Charles Warlick, Edna and Efraim Armendariz and in the background are David Kincaid, Mildred and David Young, and Linda Hayes.
(photograph by Carole Kincaid)

Some of those seated at the head-table are, left to right, Graham Carey, Mary Wheeler, Gene Golub, Carole and David Kincaid.
(photograph by Steve Armstrong)

David Kincaid addressing the audience at the banquet. Seated at table in the foregound are, left to right, Barbara and Tinsley Oden, Nell Dale, Al Dale, Suzanne Warlick, Charles Warlick, Edna and Efraim Armendariz. Seated, at the table in the background from left to right, are Mildred and David Young, Linda Hayes, Garrett Birkhoff, and Margaret Gregory.
(photograph by Carole Kincaid)

A unique feature of the conference were the afternoon stretch sessions conducted by Celeste Hamman. Shown in the middle of the picture limbering-up are Garrett Birkhoff and John Whiteman, among others.
(photograph by David Kincaid)

Another unique feature of the conference was an All-Conference Tennis Tournament. In the back row from left to right, are David Young, Gene Wachspress, William Young, Robert Dopson, David Kincaid and, in the middle row, are Katy Burrell, Wlodek Proskurowski, and, in the front row, are Tom Nance, Raphaele Herbin, Anne Elster, and Anne Greenbaum.
(photograph by Carole Kincaid)

Contents

4 Preconditioned Iterative Methods for Indefinite Symmetric Toeplitz Systems   **65**

Paul Concus and Paul E. Saylor

5 A Local Relaxation Scheme (Ad-Hoc SOR) Applied to Nine Point and Block Difference Equations   **81**

L.W. Ehrlich

6 Block Iterative Methods for Cyclically Reduced Non-Self-Adjoint Elliptic Problems   **91**

Howard C. Elman and Gene H. Golub

Chapter 1

Fourier Analysis of Two-Level Hierarchical Basis Preconditioners

LOYCE M. ADAMS
University of Washington

Dedicated to David M. Young, Jr., on the occasion of his sixty-fifth birthday.

Abstract

We use Fourier analysis techniques to find the condition number of one, two, and three dimensional model problems that are preconditioned with a two-level hierarchical basis matrix. The results show that all these problems have constant condition numbers as the mesh size becomes small provided a block diagonal scaling is applied. Limitations and extensions of the Fourier procedure are discussed.

1 Introduction

We consider the second order symmetric, positive and self-adjoint elliptic problem,

$$(1.1) \qquad \begin{aligned} Lu &= f \quad \in \ \Omega \\ u &= 0 \ \text{on} \ \partial\Omega \end{aligned}$$

with variational formulation,

$$(1.2) \qquad a(u, v) = (f, v) \ \forall v \in H_0^1$$

where v is the test function belonging to Hilbert space H_0^1. The finite element method looks for a solution u^h in the finite dimensional subspace $S_h \subset H_0^1$ that is obtained by discretizing Ω with the standard finite elements (e.g., linear, bilinear, trilinear). The computational attractiveness of the finite element method comes

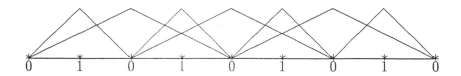

Figure 1.1 Hierarchical basis function — 2 levels.

from choosing a simple set of basis functions to span S_h. Typically, these are the *nodal basis functions*, ϕ_j, which satisfy

$$\phi_j = \begin{cases} 1 & \text{at node } j \\ 0 & \text{at other nodes.} \end{cases}$$

If we express the approximation, u^h as

(1.3)
$$u^h = \sum_{j=1}^{N} q_j \phi_j$$

and substitute into (1.2) we obtain the linear system,

(1.4)
$$\mathbf{Aq} = \mathbf{b}$$

where

$$A_{ij} = a(\phi_i, \phi_j) \quad i, j = 1, 2, ..., N$$
$$b_i = (f, \phi_i) \quad i = 1, 2, ..., N$$
$$q = \text{coefficient vector in the nodal basis.}$$

Another set of basis functions, $\bar{\phi}_j$, called the *hierarchical basis functions*, can be obtained by refining the domain Ω in a hierarchical fashion, Yserentant [4]. To obtain these basis functions, Ω is discretized with an initial coarse grid (level 0) and the standard nodal basis functions on this grid are used. The grid is then further refined, thereby introducing new basis functions (standard nodal) associated with the newly introduced nodes (level 1), while keeping the basis functions associated with the level 0 nodes. This is illustrated in Fig. 1.1 for a one dimensional problem and linear elements.

If we now express the approximation, u^h, as

(1.5)
$$u^h = \sum_{j=1}^{N} \bar{q}_j \bar{\phi}_j$$

and substitute into (1.2) we get the system

(1.6) $$\bar{\mathbf{A}}\bar{\mathbf{q}} = \bar{\mathbf{b}}$$

where

$$\bar{A}_{ij} = a(\bar{\phi}_i, \bar{\phi}_j) \quad i,j = 1, 2, ..., N$$
$$\bar{b}_i = a(f, \bar{\phi}_i)$$

and $\bar{\mathbf{q}}$ is the coefficient vector in the hierarchical basis.

In general, the matrix $\bar{\mathbf{A}}$ in (1.6) is not as sparse as $\mathbf{A}$ in (1.4) because the level 0 hierarchical basis functions have support across more elements than the level 0 standard nodal basis functions for the refined grid. The motivation for using this change of basis is that the matrix $\bar{\mathbf{A}}$ can be simply scaled to have lower condition number than matrix $\mathbf{A}$ and hence iterative methods like conjugate gradient applied to scaled (1.6) will converge faster than those applied to (1.4). In fact, we can view (1.6) as a preconditioning of (1.4) by relating $\mathbf{A}$ and $\bar{\mathbf{A}}$. If we evaluate (1.3) at node i we obtain q_i. If we evaluate (1.5) at node i we get

(1.7) $$q_i = \sum_{j=1}^{N} \bar{q}_j \bar{\phi}_j(\text{node } i).$$

Hence, we have

(1.8) $$\mathbf{q} = \mathbf{S}\bar{\mathbf{q}}$$

where

$$S_{ij} = \bar{\phi}_j(\text{node } i),$$

and

(1.9) $$\bar{\mathbf{A}} = \mathbf{S}^T \mathbf{A} \mathbf{S}$$
$$\bar{\mathbf{b}} = \mathbf{S}^T \mathbf{b}.$$

The condition number of a matrix obtained by block diagonally scaling (1.9) for one dimensional problems has been shown by Zienkiewicz, et. al. [5] to be bounded independent of the number of refinement levels. For two dimensional problems, Yserentant [4] has shown the condition number to be $O(j^2)$ where j is the number of refinement levels. Recently, Ong [3] has shown for three dimensional problems with uniform grids, the analogous result is $O(2^j)$.

The purpose of this paper is to show how to exactly calculate the condition number of a two-level model problem with Dirichlet boundary conditions in one, two, and three dimensions using Fourier analysis techniques. In Sec. 2, we examine the one dimensional Laplacian with linear elements. In Sec. 3, we consider the two dimensional Laplacian with bilinear elements. In Sec. 4, the matrix $\mathbf{S}$ of Sec. 3 is used to precondition the 5-pt approximation to the Laplacian. In Sec. 5, the three dimensional Laplacian with trilinear S and 7-pt approximation is considered. We conclude the paper by summarizing the results and discussing the extension and limitations of the procedure.

Figure 1.2 **S** for 1D — linear element.

Figure 1.3 **A** for 1D — nodal basis.

2   1D, Linear S

In this section, we illustrate the Fourier analysis technique by finding the eigenvalues of

(2.1) $$\mathbf{S}^T\mathbf{A}\mathbf{S}$$

and

(2.2) $$\mathbf{D}_0^{-1}(\mathbf{S}^T\mathbf{A}\mathbf{S})$$

where **A** is the standard 3-point approximation to

$$-u_{xx}=0, \quad 0<x<1$$
(2.3)
$$u(0)=0$$
$$u(1)=0$$

and **S** is the hierarchical matrix for linear elements. The matrix $\mathbf{D}_0$ is a block diagonal matrix used for scaling. This matrix will be determined by the Fourier analysis.

We begin by introducing an initial grid of level 0 nodes and refine this grid with level 1 nodes. These nodes and their associated hierarchical basis functions are shown in Fig. 1.1. From Fig. 1.1, it can be seen that the level 0 basis functions have the value $\frac{1}{2}$ at the level 1 nodes to their left and right. The matrix **S**, therefore, can be described by the two stencils below, for the level 0 and level 1 nodes, respectively. In Fig. 1.2, the number above the node indicates its type and the number below indicates the stencil weight. The matrix **A**, arising from the standard nodal basis, can be represented by the stencils shown in Fig. 1.3. To see how $\mathbf{S}^T\mathbf{A}\mathbf{S}$ amplifies a vector **u**, we let the components of **u** for level 0 and level 1 nodes be represented as

$$u_{j,0} = \sum_{\xi} \hat{u}_{\xi,0}\sin\xi\pi x_j$$

(2.4)
$$u_{j,1} = \sum_{\xi} \hat{u}_{\xi,1} \sin \xi \pi x_j.$$

From Fig. 1.2 and (2.4) we find that $\mathbf{Su} = \mathbf{w}$ gives the following 2×2 matrix in Fourier space,

(2.5)
$$\begin{bmatrix} \hat{w}_{\xi,0} \\ \hat{w}_{\xi,1} \end{bmatrix} = \begin{bmatrix} 1 & 0 \\ \cos \xi \pi h & 1 \end{bmatrix} \begin{bmatrix} \hat{u}_{\xi,0} \\ \hat{u}_{\xi,1} \end{bmatrix}$$

where h is the grid spacing between level 0 and level 1 nodes and $\mathbf{w}$ has a similar expansion as $\mathbf{u}$ in (2.4). We denote the matrix in (2.5) as $\widehat{\mathbf{S}}$.

Likewise, $\mathbf{q} = \mathbf{Aw}$ has the Fourier representation, $\hat{\mathbf{q}} = \widehat{\mathbf{A}}\hat{\mathbf{w}}$,

(2.6)
$$\begin{bmatrix} \hat{q}_{\xi,0} \\ \hat{q}_{\xi,1} \end{bmatrix} = \begin{bmatrix} 2 & -2\cos \xi \pi h \\ -2\cos \xi \pi h & 2 \end{bmatrix} \begin{bmatrix} \hat{w}_{\xi,0} \\ \hat{w}_{\xi,1} \end{bmatrix}$$

and $\mathbf{y} = \mathbf{S}^T \mathbf{q}$ has the representation, $\hat{\mathbf{y}} = \widehat{\mathbf{S}}^T \hat{\mathbf{q}}$,

(2.7)
$$\begin{bmatrix} \hat{y}_{\xi,0} \\ \hat{y}_{\xi,1} \end{bmatrix} = \begin{bmatrix} 1 & \cos \xi \pi h \\ 0 & 1 \end{bmatrix} \begin{bmatrix} \hat{q}_{\xi,0} \\ \hat{q}_{\xi,1} \end{bmatrix}.$$

Therefore, $\mathbf{y} = \mathbf{S}^T \mathbf{ASu}$ becomes $\hat{\mathbf{y}} = \widehat{\mathbf{A}}\hat{\mathbf{u}}$,

(2.8)
$$\begin{bmatrix} \hat{y}_{\xi,0} \\ \hat{y}_{\xi,1} \end{bmatrix} = \begin{bmatrix} 2 - 2\cos^2 \xi \pi h & 0 \\ 0 & 2 \end{bmatrix} \begin{bmatrix} \hat{u}_{\xi,0} \\ \hat{u}_{\xi,1} \end{bmatrix}.$$

If $h = \frac{1}{N+1}$, then the frequency, ξ, ranges from $\xi = 1, 2, ..., N$ for the N interior grid points. The eigenvalues 2 and $2 - 2\cos^2 \xi \pi h$ of $\widehat{\mathbf{A}}$ remain unchanged if we replace ξ by $(N + 1) - \xi$. Hence, if we let ξ range from $\xi = 1, 2, ..., \frac{N}{2}$, we get exactly N eigenvalues for $\mathbf{S}^T \mathbf{AS}$, the correct number.

The smallest and largest eigenvalues of $\bar{\mathbf{A}} = \mathbf{S}^T \mathbf{AS}$ can be seen from (2.8) to be

(2.9)
$$\lambda_{\min} = 2 - 2\cos^2 \pi h$$
$$\lambda_{\max} = 2$$

when $\xi = 1$. The condition number of $\bar{\mathbf{A}}$ is therefore,

(2.10)
$$\kappa(\bar{\mathbf{A}}) = \frac{1}{1 - \cos^2 \pi h}.$$

If $h = \frac{1}{8}$ as shown in Fig. 1.1, (2.10) yields $\kappa(\bar{\mathbf{A}}) = 6.83$. If we start with a finer and finer initial grid, (2.10) shows that

(2.11)
$$\kappa(\bar{\mathbf{A}}) \to O(h^{-2})$$

as $h \to 0$.

Figure 1.4 Level 2-node parents.

$$\begin{matrix} (0) & (0) & (0) \\ \bullet & \bullet & \bullet \\ -1/2^J & 2/2^J & -1/2^J \end{matrix} \qquad\qquad \begin{matrix} (j) \\ \bullet \\ 2/2^{J-j} \end{matrix}$$

Figure 1.5 Level J stencil for $\bar{\mathbf{A}}$. $(j > 0)$

The Fourier matrix in (2.8) is diagonal. This shows that as $h \to 0$, a perfectly conditioned matrix $\mathbf{D}_0^{-1}\mathbf{S}^T\mathbf{A}\mathbf{S}$ can be obtained by choosing $\mathbf{D}_0$ to be block diagonal with the Laplacian operator for the level 0 nodes and $2\mathbf{I}$ for the level 1 nodes. So, we essentially precondition $\mathbf{S}^T\mathbf{A}\mathbf{S}$ by the initial grid operator. In Fourier space, if

$$(2.12) \qquad \widehat{\mathbf{D}}_0^{-1} = \begin{bmatrix} \frac{1}{2-2\cos^2\xi\pi h} & 0 \\ 0 & \frac{1}{2} \end{bmatrix}$$

then $\widehat{\mathbf{D}}_0^{-1}\widehat{\mathbf{S}}^T\widehat{\mathbf{A}}\widehat{\mathbf{S}} = \mathbf{I}$.

For multiple levels, a matrix $\widehat{\mathbf{S}}$, independent of x_j, can not be determined as for two levels. This is because a level 2 node (there are two kinds) has both level 0 and level 1 parents as shown in Fig. 1.4. Since the parents are of different types, the $\sin\xi\pi x_j$ components do not combine to yield a quantity independent of x_j. However, multiple levels in 1D can be analyzed easily by directly computing $\bar{\mathbf{A}} = \mathbf{S}^T\mathbf{A}\mathbf{S}$ in the space domain and then applying the Fourier technique to the result. The component $\bar{A}_{ij}$ is gotten from the bilinear form of (2.3) using the hierarchical basis functions $\hat{\phi}_i$. That is,

$$(2.13) \qquad \bar{A}_{ij} = \int_0^1 \hat{\phi}_{i,x}\hat{\phi}_{j,x}dx.$$

$\bar{\mathbf{A}}$ for levels $j = 0, 1, 2, ..., J$ is quickly computed to have the stencil representation in Fig. 1.5. Note that the stencils in Fig. 1.5 show that a block diagonal matrix is obtained in the physical domain. The Fourier matrix, for multiple levels, will be diagonal, since there is no interaction between levels. That is,

$$(2.14) \qquad \begin{aligned} \widehat{\bar{A}}_{00} &= \frac{1}{2^{J-1}}(1 - \cos\xi\pi 2^J h) \text{ for level 0 nodes} \\ \\ \widehat{\bar{A}}_{ii} &= \frac{2}{2^{J-j}} \text{ for level } j \text{ nodes, } j = 1, 2, ..., J \end{aligned}$$

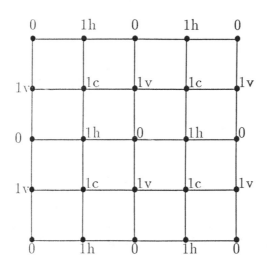

Figure 1.6 Node types — 2 levels.

where h is the grid spacing on level J. Even if we precondition by the initial grid operator, the condition number will be $O(2^{-J}) = O(h^{-1})$. But, the easy fix is to also diagonally scale the other grid levels by the appropriate scalar indicated in (2.14). The result of this block diagonal scaling is a perfectly conditioned matrix $\mathbf{D}_0^{-1}\mathbf{S}^T\mathbf{A}\mathbf{S}$, and is known in the literature ([4, 5]).

3   2D, Bilinear S, Bilinear A

In this section, we examine the condition number of (2.1) where the two-level $\mathbf{S}$ comes from the bilinear element and $\mathbf{A}$ is the standard nodal bilinear approximation to the two dimensional negative Laplacian. We then precondition (2.1) by an initial grid operator and find the resulting condition number of (2.2).

As in Adams, LeVeque, Young [1], the first task is to determine the different types of nodes so that each type has the same stencil for $\mathbf{S}$ and $\mathbf{A}$. Figure 1.6 shows this can be done with four types of nodes: type 0 (level 0), type 1c (level 1), type 1v (level 1), and type 1h (level 1). Notice that type 1h and 1v nodes are gotten by bisecting the horizontal and vertical sides of the level 0 boxes. Type 1c nodes are the centers of level 0 boxes. This categorization of nodes is different from the two-color Fourier analysis used by Kuo and Chan [2] to analyze various iterative methods, including multigrid. In their approach, two types of nodes and two types of frequencies are used, thereby resulting in 4 × 4 Fourier matrices.

The matrix $\mathbf{S}$ that arises from bilinear elements has the stencil representation

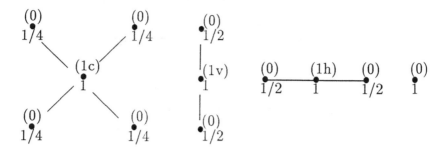

Figure 1.7 Bilinear **S** — 2 levels.

shown in Fig. 1.7 for the four node types. Likewise, the bilinear nodal basis matrix **A** is given in Fig. 1.8. Following the procedure of the last section, we define

$$(3.1) \qquad u_{ij,0} = \sum_{\xi,\eta} \hat{u}_{\xi\eta,0} \sin \xi\pi x_i \sin \eta\pi y_j$$

$$u_{ij,1c} = \sum_{\xi,\eta} \hat{u}_{\xi\eta,1c} \sin \xi\pi x_i \sin \eta\pi y_j$$

$$u_{ij,1v} = \sum_{\xi,\eta} \hat{u}_{\xi\eta,1v} \sin \xi\pi x_i \sin \eta\pi y_j$$

$$u_{ij,1h} = \sum_{\xi,\eta} \hat{u}_{\xi\eta,1h} \sin \xi\pi x_i \sin \eta\pi y_j$$

and see how $\mathbf{S}^T \mathbf{A} \mathbf{S}$ amplifies the vector **u**.

From Fig. 1.7 and (3.1), the Fourier matrix $\hat{\mathbf{S}}$ is easily computed to be

$$(3.2) \qquad \hat{\mathbf{S}} = \begin{bmatrix} 1 & 0 & 0 & 0 \\ \cos\xi\pi h \cos\eta\pi h & 1 & 0 & 0 \\ \cos\eta\pi h & 0 & 1 & 0 \\ \cos\xi\pi h & 0 & 0 & 1 \end{bmatrix}$$

and from Fig. 1.8 and (3.1), the Fourier matrix $\hat{\mathbf{A}}$ is computed to be

$$(3.3) \qquad \hat{\mathbf{A}} = \begin{bmatrix} \frac{8}{3} & a & b & c \\ a & \frac{8}{3} & c & b \\ b & c & \frac{8}{3} & a \\ c & b & a & \frac{8}{3} \end{bmatrix}$$

where $a = -\frac{4}{3} \cos\xi\pi h \cos\eta\pi h$, $b = -\frac{2}{3} \cos\eta\pi h$, and $c = -\frac{2}{3} \cos\xi\pi h$. Forming

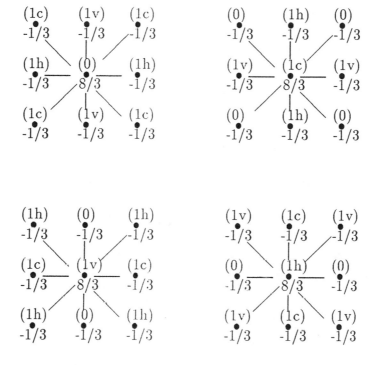

Figure 1.8 Bilinear **A** — 2 levels.

$\widehat{\bar{\mathbf{A}}} = \widehat{\mathbf{S}}^T \widehat{\mathbf{A}} \widehat{\mathbf{S}}$ we get

(3.4)
$$\widehat{\bar{\mathbf{A}}} = \begin{bmatrix} a & 0 & b & c \\ 0 & \frac{8}{3} & d & e \\ b & d & \frac{8}{3} & f \\ c & e & f & \frac{8}{3} \end{bmatrix}$$

where

$$a = \frac{1}{3}(8 - 2\cos 2\eta\pi h - 2\cos 2\xi\pi h - 4\cos 2\xi\pi h \cos 2\eta\pi h)$$

$$b = -\frac{1}{6}\cos \eta\pi h - \cos \eta\pi h \cos 2\xi\pi h$$

$$c = -\frac{1}{6}\cos \xi\pi h - \cos \xi\pi h \cos 2\eta\pi h$$

$$d = -\frac{2}{3}\cos \xi\pi h$$

$$e = -\frac{2}{3}\cos \eta\pi h$$

$$f = -\frac{4}{3}\cos \eta\pi h \cos \xi\pi h.$$

We could derive $\widehat{\bar{\mathbf{A}}}$ directly by performing the integration,

(3.5)
$$\bar{A}_{ij} = \int_0^1 \int_0^1 \left(\hat{\phi}_{i,x}\hat{\phi}_{j,x} + \hat{\phi}_{i,y}\hat{\phi}_{j,y} \right) dx\,dy$$

where the $\hat{\phi}_i$ are the hierarchical basis functions. Then, $\widehat{\bar{\mathbf{A}}}$ can be obtained by applying the Fourier analysis to $\bar{\mathbf{A}}$.

Since $\widehat{\bar{\mathbf{A}}}$ is a symmetric positive definite matrix, the elements along the diagonal must lie between its smallest and largest eigenvalues. Since $a \to 0$ as $O(h^2)$ as $h \to 0$, we know for this case that

(3.6)
$$\kappa(\widehat{\bar{\mathbf{A}}}) = \kappa(\mathbf{S}^T\mathbf{A}\mathbf{S}) \geq O(h^{-2}).$$

If we precondition (3.4) in Fourier space by the diagonal of $\widehat{\bar{\mathbf{A}}}$ in (3.4), we get

(3.7)
$$\widehat{\mathbf{D}}_0^{-1}\widehat{\bar{\mathbf{A}}} = \begin{bmatrix} 1 & 0 & a & b \\ 0 & 1 & c & d \\ e & c & 1 & g \\ f & d & g & 1 \end{bmatrix}$$

where

$$r = \frac{8}{3} - \frac{16}{3}\cos^2 \eta\pi h \cos^2 \xi\pi h + \frac{4}{3}\cos^2 \eta\pi h + \frac{4}{3}\cos^2 \xi\pi h$$

$$a = \frac{2\cos \eta\pi h(1 - \cos^2 \xi\pi h)}{r}$$

$$b = \frac{2\cos\xi\pi h(1 - \cos^2\eta\pi h)}{r}$$

$$c = -\frac{1}{4}\cos\xi\pi h$$

$$d = -\frac{1}{4}\cos\eta\pi h$$

$$e = \frac{3}{4}\cos\eta\pi h - \frac{3}{4}\cos^2\xi\pi h\cos\eta\pi h$$

$$f = \frac{3}{4}\cos\xi\pi h - \frac{3}{4}\cos^2\eta\pi h\cos\xi\pi h$$

$$g = -\frac{1}{2}\cos\xi\pi h\cos\eta\pi h$$

In the case $\xi = \eta$, we get $\lambda = 1$ and the cubic equation

$$(3.8) \qquad (1-\lambda)^3 - (2c^2 + 2ae + g^2)(1-\lambda) + 2g(c^2 + ae) = 0$$

which has roots

$$\lambda = 1 - g$$

$$\lambda = 1 + \frac{g}{2} \pm \frac{1}{2}\sqrt{g^2 + 8(c^2 + ae)}.$$

The largest condition number is obtained as $h \to 0$. This gives, for $\xi = \eta$,

$$(3.9) \qquad \kappa(\mathbf{D}_0^{-1}\mathbf{S}^T\mathbf{A}\mathbf{S}) = 3 + \sqrt{3}.$$

4 2D, Bilinear, 5-Point A

In the last section, we analyzed the two level $\bar{\mathbf{A}}$ that results from using the hierarchical bilinear basis functions in (3.5). This was viewed as preconditioning the matrix $\mathbf{A}$ that is obtained from the nodal bilinear basis functions. We can also consider using the same $\mathbf{S}$ to precondition the matrix $\mathbf{A}_5$ that results from the standard 5-point discretization of the two-dimensional negative Laplacian. Figure 1.9 shows the stencil representation of $\mathbf{A}_5$. In Fourier space, $\mathbf{A}_5$ is represented as

$$(4.1) \qquad \bar{\mathbf{A}}_5 = \begin{bmatrix} 4 & 0 & -2\cos\eta\pi h & -2\cos\xi\pi h \\ 0 & 4 & -2\cos\xi\pi h & -2\cos\eta\pi h \\ -2\cos\eta\pi h & -2\cos\xi\pi h & 4 & 0 \\ -2\cos\xi\pi h & -2\cos\eta\pi h & 0 & 4 \end{bmatrix}$$

and $\widehat{\mathbf{A}} = \widehat{\mathbf{S}}^T\widehat{\mathbf{A}}_5\widehat{\mathbf{S}}$ is calculated from (3.2) and (4.1) to be

$$\widehat{\mathbf{A}} = \begin{bmatrix} 4 - 4\cos^2\xi\pi h\cos^2\eta\pi h & 0 & 2\cos\eta\pi h\sin^2\xi\pi h & 2\cos\xi\pi h\sin^2\eta\pi h \\ 0 & 4 & -2\cos\xi\pi h & -2\cos\eta\pi h \\ 2\cos\eta\pi h\sin^2\xi\pi h & -2\cos\xi\pi h & 4 & 0 \\ 2\cos\xi\pi h\sin^2\eta\pi h & -2\cos\eta\pi h & 0 & 4 \end{bmatrix}$$

$$(4.2)$$

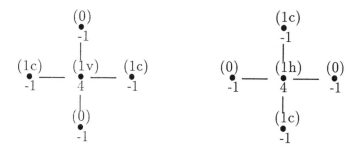

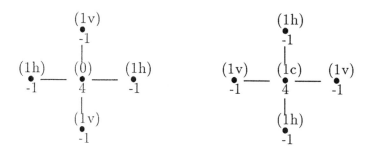

Figure 1.9 5-point **A** — 2 levels.

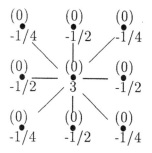

Figure 1.10 Coarse grid preconditioner.

As in Sec. 3, $\widehat{\mathbf{A}}$ has condition number $O(h^{-2})$ as $h \to 0$ since $4 - 4\cos^2 \xi\pi h \cos^2 \eta\pi h$ approaches zero as $O(h^2)$. So we consider preconditioning (4.2) by $\widehat{\mathbf{D}}_0^{-1}$, the diagonal of $\widehat{\mathbf{A}}$. In the space domain, this corresponds to $\mathbf{D}_0^{-1}\mathbf{S}^T\mathbf{A}\mathbf{S}$ where

$$(4.3) \qquad \mathbf{D}_0^{-1} = \begin{bmatrix} \mathbf{D}_{00}^{-1} & & & \\ & \frac{1}{4}\mathbf{I}_{1c} & & \\ & & \frac{1}{4}\mathbf{I}_{1v} & \\ & & & \frac{1}{4}\mathbf{I}_{1h} \end{bmatrix}$$

and $\mathbf{D}_{00}$ is the stencil on the level 0 nodes shown in Fig. 1.10.

In Fourier space, the preconditioned matrix becomes

$$(4.4) \qquad \widehat{\mathbf{D}}_0^{-1}\widehat{\mathbf{A}} = \begin{bmatrix} 1 & 0 & a & b \\ 0 & 1 & c & d \\ e & c & 1 & 0 \\ f & d & 0 & 1 \end{bmatrix}$$

where

$$a = \frac{\cos \eta\pi h \sin^2 \xi\pi h}{2(1 - \cos^2 \xi\pi h \cos^2 \eta\pi h)}$$

$$b = \frac{\cos \xi\pi h \sin^2 \eta\pi h}{2(1 - \cos^2 \xi\pi h \cos^2 \eta\pi h)}$$

$$c = -\frac{1}{2} \cos \xi\pi h$$

$$d = -\frac{1}{2} \cos \eta\pi h$$

$$e = \frac{1}{2} \cos \eta\pi h \sin^2 \xi\pi h$$

$$f = \frac{1}{2} \cos \xi\pi h \sin^2 \eta\pi h.$$

A quick inspection of (4.4) shows the eigenvalues, λ, satisfy

$$(1 - \lambda)^2 = \bar{\lambda}$$

where $\bar{\lambda}$ are eigenvalues of

$$(4.5) \qquad \begin{bmatrix} e & c \\ f & d \end{bmatrix} \begin{bmatrix} a & b \\ c & d \end{bmatrix}.$$

A quick calculation of the eigenvalues of (4.5) when $\xi = \eta$ gives

$$\bar{\lambda} = 0$$

$$\bar{\lambda} = \frac{\cos^2 \xi\pi h}{1 + \cos^2 \xi\pi h}$$

from which we obtain

(4.6)
$$\lambda = 1, 1, 1 \pm \frac{\cos \xi \pi h}{\sqrt{1 + \cos^2 \xi \pi h}}.$$

The largest condition number occurs when $\xi = 1$ and equals

(4.7)
$$\kappa(\widehat{\mathbf{D}}_0^{-1}\widehat{\mathbf{A}}) = \frac{1 + \frac{\cos \pi h}{\sqrt{1+\cos^2 \pi h}}}{1 - \frac{\cos \pi h}{\sqrt{1+\cos^2 \pi h}}}.$$

As $h \to 0$, κ becomes larger and we get

(4.8)
$$\kappa(\widehat{\mathbf{D}}_0^{-1}\widehat{\mathbf{A}}) = 3 + 2\sqrt{2}$$

which is only slightly larger than the $3 + \sqrt{3}$ obtained for this preconditioner used with the bilinear nodal $\mathbf{A}$ of Sec. 3.

5 3D, Trilinear S, 7-Point A

In this section, we consider the natural extension of the preconditioner in Sec. 4 to three dimensions for two levels. We first divide a cubical domain with a level zero grid into smaller cubes with the introduction of level 1 nodes. Each of the level 0 cubes are divided into eight smaller cubes by bisecting each of its edges. This introduces seven new types of nodes: 1x (x-direction edge midpoints), 1y (y-direction edge midpoints), 1z (z-direction edge midpoints), 1xy (x-y face centers), 1yz (y-z face centers), 1xz (x-z face centers) and 1c (original cube centroid). The stencils for matrix $\mathbf{S}$ for each node type all have one at the center. Types 1x, 1y, and 1z have weights of $\frac{1}{2}$ corresponding to the two level 0 nodes sharing the same edge. Types 1xy, 1yz, and 1xz have weights $\frac{1}{4}$ corresponding to the four level 0 nodes on the same face. Type 1c nodes have weights $\frac{1}{8}$ corresponding to the eight level 0 nodes on the corners of the level 0 box.

The matrix $\widehat{\mathbf{S}}$ has the form,

(5.1)
$$\begin{bmatrix} 1 & \mathbf{0} \\ \mathbf{q} & \mathbf{I} \end{bmatrix}$$

where

(5.2)
$$\mathbf{q} = \begin{bmatrix} \cos \xi \pi h \cos \eta \pi h \cos \zeta \pi h \\ \cos \xi \pi h \\ \cos \eta \pi h \\ \cos \zeta \pi h \\ \cos \xi \pi h \cos \eta \pi h \\ \cos \eta \pi h \cos \zeta \pi h \\ \cos \xi \pi h \cos \zeta \pi h \end{bmatrix}.$$

The matrix $\mathbf{A}_7$ comes from the 7-point approximation to the three-dimensional negative Laplacian. In Fourier space, $\widehat{\mathbf{A}}_7$ has the form

(5.3)
$$\widehat{\mathbf{A}}_7 = \begin{bmatrix} 6 & \mathbf{r}^T \\ \mathbf{r} & \mathbf{A}_{22} \end{bmatrix}$$

where $\mathbf{r}^T = (0, -2\cos\xi\pi h, -2\cos\eta\pi h, -2\cos\zeta\pi h, 0, 0, 0)$ and

(5.4)
$$\mathbf{A}_{22} = \begin{bmatrix} 6 & 0 & 0 & 0 & c & a & b \\ 0 & 6 & 0 & 0 & b & 0 & c \\ 0 & 0 & 6 & 0 & a & c & 0 \\ 0 & 0 & 0 & 6 & 0 & b & a \\ c & b & a & 0 & 6 & 0 & 0 \\ a & 0 & c & b & 0 & 6 & 0 \\ b & c & 0 & a & 0 & 0 & 6 \end{bmatrix}$$

where $a = -2\cos\xi\pi h$, $b = -2\cos\eta\pi h$, and $c = -2\cos\zeta\pi h$.

Using (5.1) and (5.3) we get

(5.5)
$$\widehat{\mathbf{S}}^T\widehat{\mathbf{A}}_7\widehat{\mathbf{S}} = \begin{bmatrix} 6 + 2\mathbf{q}^T\mathbf{r} + \mathbf{q}^T\mathbf{A}_{22}\mathbf{q} & \mathbf{r}^T + \mathbf{q}^T\mathbf{A}_{22} \\ \mathbf{r} + \mathbf{A}_{22}\mathbf{q} & \mathbf{A}_{22} \end{bmatrix}.$$

We now analyze the limiting case when $\cos\xi\pi h$, $\cos\eta\pi h$, and $\cos\zeta\pi h$ approach one as $h \to 0$. We get

$$\mathbf{q}^T \to [1,1,1,1,1,1,1]$$
$$\mathbf{r} + \mathbf{A}_{22}\mathbf{q} \to [0,0,0,0,0,0,0]$$
$$6 + 2\mathbf{q}^T\mathbf{r} + \mathbf{q}^T\mathbf{A}_{22}\mathbf{q} \to 0.$$

This suggests that if we scale $\widehat{\mathbf{S}}^T\widehat{\mathbf{A}}_7\widehat{\mathbf{S}}^T$ by

(5.6)
$$\widehat{\mathbf{D}}_0^{-1} = \begin{bmatrix} (6 + 2\mathbf{q}^T\mathbf{r} + \mathbf{q}^T\mathbf{A}_{22}\mathbf{q})^{-1} & \mathbf{0} \\ \mathbf{0} & \frac{1}{6}\mathbf{I} \end{bmatrix},$$

the eigenvalues of $\widehat{\mathbf{D}}_0^{-1}\widehat{\mathbf{S}}^T\widehat{\mathbf{A}}_7\widehat{\mathbf{S}}$ include 1 and the eigenvalues of $\frac{1}{6}\mathbf{A}_{22}$. The eigenvalues of $\mathbf{A}_{22}$ are easily calculated to be $8, 8, 4, 4, 6, 6 \pm 2\sqrt{7}$. The condition number of this limiting case is therefore,

(5.7)
$$\kappa(\widehat{\mathbf{D}}_0^{-1}\widehat{\mathbf{A}}) = \frac{1}{2}(3 + \sqrt{7})^2.$$

6   Concluding Remarks

We have shown how to analyze matrices $\bar{\mathbf{A}}$ that result from preconditioning $\mathbf{A}$ by $\mathbf{S}$ when two levels of nodes are used. This strategy depends on isolating different types of nodes such that each type has the same computational stencil for $\mathbf{S}$ and $\mathbf{A}$. The results for two levels in one, two, and three dimensions show that the condition number of $\bar{\mathbf{A}}$ grows as $O(h^{-2})$ as the coarse grid stepsize is allowed to approach zero. However, if the coarse grid nodes are preconditioned as we have shown, the condition number for the one, two, and three dimensional model problems are all constant as the coarse grid stepsize approaches zero.

Extension of these ideas to more levels in the case of two and three dimensional problems is not obvious. Two problems arise. First, too many types of nodes must be used as the number of levels increase, thereby complicating the analysis. But more importantly, the Fourier matrices for $\mathbf{A}$ and $\mathbf{S}$ can not be found that are independent of the spatial variables. Perhaps after $\widehat{\mathbf{S}}^T \widehat{\mathbf{A}} \widehat{\mathbf{S}}$ is formed, the spatial dependence disappears.

Acknowledgements

This research was supported by U.S. Air Force Office of Scientific Research Grant No. 86-0154. This paper was originally typed by the author in LaTeX and reformatted in LaTeX by Lisa Laguna.

References

[1] Adams, L.M., R.J. LeVeque, and D.M. Young. "Analysis of the SOR Iteration for the 9-Point Laplacian," *SIAM Journal Numerical Analysis*, **25** (1988) 1156–1180.

[2] Kuo, C.C.J., and T.F. Chan. "Analysis of Iterative Algorithms for Elliptic Problems with Red/Black Ordering," CAM Report 88-15, Department of Mathematics, UCLA, May 1988.

[3] Ong, M.E.G. "Hierarchical Basis Preconditioners for Elliptic Problems in Three Dimensions," Ph.D. Thesis, Department of Applied Mathematics, University of Washington, Seattle, Washington, September 1989.

[4] Yserentant, H. "On the Multi-Level Splitting of Finite Element Spaces," *Numerische Mathematik*, **49** (1986) 379–412.

[5] Zienkiewicz, O.C., D.W. Kelly, J. Gago, and I. Babuska. "Hierarchical finite element approaches, error estimates and adaptive refinement," in *The Mathematics of Finite Elements and Applications IV*, (J.R. Whiteman, ed.), Mafelap 1981, London, 1982.

Chapter 2

An Algebraic Framework for Hierarchical Basis Functions Multilevel Methods or the Search for 'Optimal' Preconditioners

OWE AXELSSON

University of Nijmegen, Netherlands

Dedicated to David M. Young a pioneer in iterative solution methods, still going strong, on the occasion of his sixty-fifth birthday.

Abstract

A purely algebraic framework for the solution of linear algebraic systems given on an hierarchical multilevel form is presented. It is shown that spectrally equivalent preconditioners can be constructed using certain polynomial approximations of Schur complements. For finite element matrices the computational complexity becomes proportional to the number of mesh points on the finest level.

1 Introduction

Many practically important mathematical models lead naturally to a sequence of linear equations of hierarchical form of successively increasing order when more and more details of information are added to the model on the previous level in the sequence. If one lets the added information result in corrections to the solution on the previous level the matrices take the form

$$(1.1) \qquad \mathbf{A}^{(k)} = \begin{bmatrix} \mathbf{A}_{11}^{(k)} & \mathbf{A}_{12}^{(k)} \\ \mathbf{A}_{21}^{(k)} & \mathbf{A}^{(k-1)} \end{bmatrix}, \quad k = 1, 2, \dots,$$

17

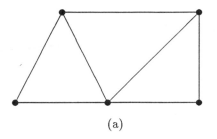 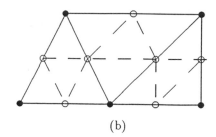

(a)   (b)

Figure 2.1 (a) First and (b) second level triangulations

where k is the level number and $\mathbf{A}^{(0)}$ is the matrix on the initial (coarse) level. We let n_k denote the order of $\mathbf{A}^{(k)}$, i.e., equivalently, the number of degrees of freedom on level k.

The most well known application where such a sequence of matrices occurs is for a nested sequence of finite element meshes used for the numerical solution of elliptic differential equations. We illustrate this for the case of piecewise linear finite element basis functions on a triangular mesh but other choices of basis functions and meshes are also possible. Consider then a polygonal domain with an initial triangulation conforming with the boundary of the domain as illustrated in Fig. 2.1(a), and with a stiffness matrix $\mathbf{A}^{(0)}$, corresponding to piecewise linear basis functions and a symmetric homogeneous bilinear form $a(u, v)$. Hence if $V_0 = \left\{ \varphi_i^{(0)}; \ i = 1, 2, \ldots, n_0 \right\}$ is the set of basis functions on the initial mesh Ω_0, then the entries of $A^{(0)}$ are defined by

$$\left(\mathbf{A}^{(0)} \right)_{i,j} = a \left(\varphi_j^{(0)}, \varphi_i^{(0)} \right).$$

Each triangle of the initial mesh is now divided into four congruent triangles by joining the midedge points of each triangle as illustrated in Fig. 2.1(b). The midedge points form the set of new mesh nodes $N_1^{(1)}$ and the total set of mesh nodes on the second level is $N_1 = N_1^{(1)} \cup N_0$, where N_0 is the set of node points (black points) of the first level. The new set of basis functions $V_1^{(1)} = \left\{ \varphi_i^{(1)}; \ i = 1, 2, \ldots, n_1 - n_0 \right\}$ corresponding to the new mesh nodes have support on the six triangles (or a fewer number of triangles for points on the boundary), which have the corresponding node as a vertex, i.e., such a basis function is linear on each of these triangles, continuous on the edges of the triangles, and zero outside the union of them. The total set of basis functions on the second level is $V_1 = V_1^{(1)} \cup V_0$. If we take the nodes in the order, first the new points and then the old, the corresponding stiffness matrix takes the form

$$\mathbf{A}^{(1)} = \begin{bmatrix} \mathbf{A}_{11}^{(1)} & \mathbf{A}_{12}^{(1)} \\ \mathbf{A}_{21}^{(1)} & \mathbf{A}^{(0)} \end{bmatrix}$$

where

$$\left(\mathbf{A}_{11}^{(1)}\right)_{ij} = a\left(\varphi_j^{(1)}, \varphi_i^{(1)}\right),$$

$$\left(\mathbf{A}_{12}^{(1)}\right)_{ij} = a\left(\varphi_j^{(0)}, \varphi_i^{(1)}\right),$$

$\mathbf{A}_{21}^{(1)} = \mathbf{A}_{12}^{(1)^T}$, and $\mathbf{A}^{(0)}$ is the standard nodal basis function matrix on the first level. For a general level k, we have the set of node points $N_k = N_k^{(1)} \cup N_{k-1}$, where $N_k^{(1)}$ is the new set of node points on this level and the set of basis functions $V_k = V_k^{(1)} \cup V_{k-1}$, where

$$V_k^{(1)} = \left\{\varphi_i^{(k)}; \ i = 1, 2, \ldots, n_k - n_{k-1}\right\}$$

is the new set of basis functions, corresponding to $N_k^{(1)}$. The stiffness matrix, corresponding to these hierarchical basis functions takes the form

$$\mathbf{A}^{(k)} = \begin{bmatrix} \mathbf{A}_{11}^{(k)} & \mathbf{A}_{12}^{(k)} \\ \mathbf{A}_{21}^{(k)} & \mathbf{A}^{(k-1)} \end{bmatrix}$$

where

$$\left(\mathbf{A}_{11}^{(k)}\right)_{i,j} = a\left(\varphi_j^{(k)}, \varphi_i^{(k)}\right),$$

$$\left(\mathbf{A}_{12}^{(k)}\right)_{i,j} = a\left(\varphi_j^{(k-1)}, \varphi_i^{(k)}\right),$$

$$\mathbf{A}_{21}^{(k)} = \mathbf{A}_{12}^{(k)^T}.$$

This mesh refinement can take place until we have found, for approximation purposes, a sufficiently fine mesh.

The initial triangulation can be irregular and it is actually not necessary to divide each triangle on every level in four congruent triangles. For instance, we can divide some (pairs of) triangles by median into two (four) triangles and some triangles do not have to be refined at all. However, we require that the corresponding finite element space is consistent, i.e., every linear combination, $\sum_i \alpha_i \varphi_i^{(k)}$ of basis functions on a level k is a continuous function. In Fig. 2.2, we have illustrated the three cases which can occur.

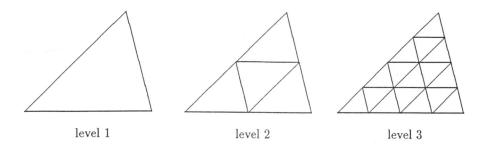

level 1   level 2   level 3

Figure 2.2 Three cases of refinements of triangles

The goal of the present paper is to construct a sequence of preconditioners $\mathbf{M}^{(k)}$, which are spectrally equivalent to $\mathbf{A}^{(k)}$, i.e., for which there exists positive numbers a, b (independent of the level number) such that

$$(1.2) \qquad a\left(\mathbf{A}^{(k)}\mathbf{u}, \mathbf{u}\right) \le \left(\mathbf{M}^{(k)}\mathbf{u}, \mathbf{u}\right) \le b\left(\mathbf{A}^{(k)}\mathbf{u}, \mathbf{u}\right)$$

for all $\mathbf{u} \in I\!\!R^{n_k}$, $k = 1, 2, \ldots$, where $(\cdot, \cdot)$ denotes the inner product in $I\!\!R^{n_k}$. On the initial level, we let $\mathbf{M}^{(0)} = \mathbf{A}^{(0)}$. Furthermore, we shall require that linear systems $\mathbf{M}^{(k)}\mathbf{x} = \mathbf{b}$ with the preconditioner $\mathbf{M}^{(k)}$ on level k can be solved with a computational complexity $O(n_k)$, i.e., proportional to the number of unknowns at that level. The preconditioner will be used in a basic iterative method such as

$$\mathbf{M}^{(k)}\boldsymbol{\delta}^{(s+1)} = \mathbf{b} - \mathbf{A}^{(k)}\mathbf{x}^{(s)}$$

$$(1.3) \qquad \mathbf{x}^{(s+1)} = \mathbf{x}^{(s)} + \frac{1}{\tau}\boldsymbol{\delta}^{(s+1)}, \quad s = 0, 1, \ldots,$$

where $\mathbf{x}^{(0)}$ is an initial vector. This method converges if $\tau > b/2$ and, as is well known, in general $\tau = (a + b)/2$ is the optimal choice. We can also use a sequence of variable parameters to generate Chebyshev polynomial approximations as in Richardson's iterative method, or we can use a conjugate gradient method; both of which result in faster convergence. For a survey of various iterative methods, see Young [19], for instance. For all of these methods, the number of iterations to converge to a given iteration error ϵ will be bounded by a number directly proportional to $\log(1/\epsilon)$, but independent of the level number, when (1.2) is satisfied. Therefore, the total computational complexity to solve $\mathbf{A}^{(k)}\mathbf{x} = \mathbf{b}$ is $O\left[n_k \log(1/\epsilon)\right]$, i.e., of optimal order.

The purpose of the present paper is to describe a method for which (1.2) is satisfied and for which systems with the preconditioner can be solved with an optimal order of computational complexity.

Perhaps the most well known previous class of methods of optimal order of computational complexity is the multigrid methods.

They are also based on multilevel meshes but not necessarily a nested sequence of meshes, and are usually applied to the standard nodal basis matrices or to difference matrices. Such methods use smoothing on each level (i.e., a Jacobi or Gauss-Seidel iterative method) to damp out the highly oscillatory components of the iteration error. After smoothing on a level, the residual is projected (restricted) to the next coarser level where a correction is to be computed. However, the same type of a procedure is applied also on this level, i.e., smoothing followed by a projection to the next coarser level and so on until the coarsest level has been reached, where the correction is usually computed by a direct solution method. This correction is then projected upwards (prolongated) to the next finer level until the finest level is reached, possibly using smoothing (postsmoothing) on each level. This constitutes a full cycle of the method (a V-cycle) which incidently can be viewed as a pre-conditioning step and therefore coupled with a global acceleration method, such as the conjugate gradient method (for a recent discussion of this, see Wachspress [18]. There is also the possibility to construct more involved cycles such as W-cycles. Multigrid methods were first considered by Fedorenko [11] and Bakhvalov [7] and have later been extended and further analyzed by Brandt [10], Nicolaides [16], Hackbusch [13], and others. The methods have optimal computational complexity, if the solution satisfies a certain elliptic regularity condition, $\|u\|_2 \leq C\|f\|_2$, where $\|\cdot\|_2$ is the norm in the second order Sobolev space (for a second order elliptic problem) and f is the source function. Unfortunately, such an elliptic regularity is not valid for problems with singularities such as with a re-entrant corner, for instance. In general, the number of iterations will depend on the level number if the elliptic regularity is not fulfilled.

There is another type of multilevel methods where the convergence occurs in a number of steps independent of the number of levels and the regularity of the solutions. These are based on the two-level hierarchical basis function finite element method where we only work on two levels, the fine and the coarser level. When we take the nodes in the same order as before, the matrix takes the same form as in (1.1), i.e.,

$$(1.4) \qquad \mathbf{A} = \begin{bmatrix} \mathbf{A}_{11} & \mathbf{A}_{12} \\ \mathbf{A}_{21} & \mathbf{A}_{22} \end{bmatrix}$$

where $\mathbf{A}_{11}$ corresponds to the added basis functions on the fine mesh and $\mathbf{A}_{22}$ to the standard nodal basis functions on the coarse mesh. Such methods have been considered by Bank and Dupont [8], Axelsson [1], Braess [9] and Axelsson and Gustafsson [3]. It has been shown that the block diagonal matrix, block-diag($\mathbf{A}_{11}, \mathbf{A}_{22}$) is spectrally equivalent to $\mathbf{A}$ (now independent on the mesh parameter h on the finest level) if the relation between the mesh size parameters on the fine and coarser meshes, h/H is bounded below by a constant, independent of h. It has also been shown in [1] and [3] that $\mathbf{A}_{22}$ is spectrally equivalent to the Schur complement, $\mathbf{A}_{22} - \mathbf{A}_{21}\mathbf{A}_{11}^{-1}\mathbf{A}_{12}$. However, the two-level method requires the solution of linear systems with $\mathbf{A}_{22}$ on the coarse mesh and $\mathbf{A}_{22}$ has still a spectral condition number $O(H^{-2})$ which becomes unbounded as $H \to 0$.

Now Yserentant [20] proposed to continue the use of hierarchical basis functions starting at the coarsest level and continuing to add new basis functions, typically in the edge midpoints, as has been illustrated above, until a final (fine) mesh is reached with a mesh parameter h. We then get a matrix structure as in (1.1), if we substitute all matrices with their explicit form on the previous coarse levels, where the number of levels is $O(\log h^{-1})$. Using an inverse estimate of the finite element functions in supremum norm, it was shown in [20] that the so derived hierarchical basis function finite element stiffness matrix has a condition number $O(\log h)^2$ for problems in two space dimensions. Using a conjugate gradient method to solve systems with it the number of iterations becomes proportional to $\log(h^{-1})$. For problems in three space dimensions, the condition number is $O(h^{-1})$. Hence, Yserentant's method is not of optimal order of convergence.

As will follow from the discussion in the present paper, we can use the method for hierarchical finite element basis functions recursively for the two-by-two block matrices (1.1) to construct spectrally equivalent preconditioners, thus leading to methods of optimal order of computational complexity.

Recursive use of the two-level version method was proposed in some of the earlier papers on two-level methods referred to above and a V-cycle version was first analyzed by Vassilevski [17] for the standard nodal basis stiffness matrices. He showed that the corresponding preconditioning matrix has a condition number $O(\log h^{-1})$. Kuznetsov [14] has analyzed a similar method, but for special difference type meshes.

More recently, the present author and Vassilevski [5,6], have analyzed ν-fold V-cycle methods for the standard nodal basis function stiffness matrices where the recursively occurring Schur complements are approximated by certain matrix poly-nomials, involving the preconditioner on the next coarser level. There it has been shown that with a proper choice of polynomials the preconditioner becomes spec-trally equivalent to the stiffness matrix, if the degree ν of the polynomial satisfies $\nu > (1 - \gamma^2)^{-1/2}$, where γ is the constant in the strengthened form of the C-B-S inequality.

Hence, if we use a mesh refinement as illustrated in Fig. 2.1, then the complexity of each level for a mesh in two space dimensions progresses in a geometric fashion with a factor 4, and the preconditioned iterative method is of optimal order of computational complexity if $\nu \leq 3$ (it is of optimal order, save a factor $\log h^{-1}$, if $\nu = 4$). Therefore, if $\gamma^2 < 8/9$, the method is of optimal order. If we choose piecewise linear basis functions for a problem in two space dimensions, then it was shown by Maitre and Musy [15] that $\gamma^2 < 3/4$ for any triangulation of a polygonal domain for the bilinear form $a(u, v) = \int \int_{\Omega} \nabla u \cdot \nabla v \, d\Omega$. Hence, it suffices to choose even $\nu = 2$ in this case. For problems in three space dimensions, we can choose ν up to 7, and still have a computational complexity per mesh point independent the level number, if the mesh elements are refined uniformly into eight congruent (or nearly congruent) pieces on each level.

The present paper will describe a general algebraic framework for deriving spectrally equivalent preconditioners for sequences of matrices of the hierarchical form (1.1), thereby identifying the basic algebraic properties which must be satisfied. This opens up a framework for constructing spectrally equivalent preconditioners also for matrices arising from other applications then from finite element approximations. However, the method will be illustrated for hierarchical basis function stiffness matrices, as arising from elliptic differential equations. Note that we can derive spectrally equivalent preconditioners for elliptic differential equations both in two and three space dimensions. This is especially important for problems in three space dimensions, where the hierarchical basis function method of Yserentant has a computational complexity of $O(h^{-1/2})$ (if a conjugate gradient iterative method is used) per mesh point as compared to $O(1)$ in our method. Furthermore, contrary to his method, our estimates are equally valid for problems with discontinuous coefficients assuming these are constant on the elements of the initial mesh. An additional advantage with the present method is that it permits an easy estimate of the spectral bounds, when the constant γ in the C-B-S inequality has been estimated. Therefore one can use a Richardson (Chebyshev) acceleration method instead of the conjugate gradient method, hence avoiding computations of the inner products required in the latter method.

The remainder of the paper is organized as follows. In Sec. 2, we present the basic assumptions of the hierarchical basis function method in an algebraic framework and illustrate these for finite element applications. In Sec. 3, we give the recursive definition of the preconditioner and describe its implementational details. The relative condition number is analyzed in Sec. 4, and we conclude with some remarks about extensions and variants of the method. The present work is based to a large extent on the previous papers [5] and [6].

2   The Algebraic Framework for Two-Level Hierarchical Basis Function Methods

In order to define and analyze the preconditioners, we must first specify some basic properties that the hierarchical sequence of matrices (1.1) shall satisfy.

Consider then a sequence of symmetric positive definite matrices on the two-level hierarchical form

$$(2.1) \qquad \mathbf{A}^{(k)} = \begin{bmatrix} \mathbf{A}^{(k)}_{11} & \mathbf{A}^{(k)}_{12} \\ \mathbf{A}^{(k)}_{21} & \mathbf{A}^{(k-1)} \end{bmatrix}, \qquad k = 1, 2, \ldots, \ell,$$

where k is the level number and $\mathbf{A}^{(0)}$ is the matrix on the initial level. Here $\mathbf{A}^{(k)}$ has order $n_k > n_{k-1}$, where n_{k-1} is the order of the block part $\mathbf{A}^{(k-1)}$.

The correspondingly partitioned vectors are denoted

$$\mathbf{v} = \begin{bmatrix} \mathbf{v}_1 \\ \mathbf{v}_2 \end{bmatrix}, \qquad \mathbf{v}_1 \in I\!\!R^{n_k - n_{k-1}}, \qquad \mathbf{v}_2 \in I\!\!R^{n_{k-1}}.$$

Let further

(2.2)
$$\mathbf{S}^{(k-1)} = \mathbf{A}^{(k-1)} - \mathbf{A}_{21}^{(k)} \mathbf{A}_{11}^{(k)^{-1}} \mathbf{A}_{12}^{(k)}$$

denote the Schur complement. Note that the Schur complement exists since $\mathbf{A}^{(k)}$ is positive definite.

In addition to the sequence $\mathbf{A}_{11}^{(k)}$, we shall assume that there exists a sequence $\mathbf{B}_{11}^{(k)}$ of the same order $(n_k - n_{k-1})$, where $\mathbf{B}_{11}^{(k)}$ approximates the inverse matrix $\mathbf{A}^{(k)^{-1}}$ and satisfies Assumptions 2 and 3 below. $\mathbf{B}_{11}^{(k)}$ will replace $\mathbf{A}^{(k)^{-1}}$ when we approximate the Schur complements. In practice, $\mathbf{B}_{11}^{(k)}$ will be a sparse matrix or a product of sparse matrices. Hence matrix-vector multiplications with $\mathbf{B}_{11}^{(k)}$ will not be costly.

We now present the basic assumptions required for the construction of preconditioners which are spectrally equivalent to $\mathbf{A}^{(k)}$. The first assumption is an algebraic formulation of the strengthened C-B-S inequality, used in [3], [8], and [9]. The second assumption means that the matrix sequence $\mathbf{B}_{11}^{(k)^{-1}}$ is spectrally equivalent to the sequence $\mathbf{A}_{11}^{(k)}$.

Basic Assumptions

1. There exists positive numbers γ_k and a constant γ smaller than 1, such that

$$\left| \mathbf{v}_1^T \mathbf{A}_{12}^{(k)} \mathbf{v}_2 \right| \le \gamma_k \left\{ \mathbf{v}_1^T \mathbf{A}_{11}^{(k)} \mathbf{v}_1 \right\}^{1/2} \left\{ \mathbf{v}_2^T \mathbf{A}^{(k-1)} \mathbf{v}_2 \right\}^{1/2}$$

$$\forall \mathbf{v}_1 \in I\!\!R^{n_k - n_{k-1}}, \ \mathbf{v}_2 \in I\!\!R^{n_{k-1}}, \ 0 < \gamma_k \le \gamma < 1.$$

2. There exists sparse matrices $\mathbf{B}_{11}^{(k)}$ such that

$$\mathbf{v}_1^T \mathbf{A}_{11}^{(k)} \mathbf{v}_1 \le \mathbf{v}_1^T \mathbf{B}_{11}^{(k)^{-1}} \mathbf{v}_1 \le (1 + b_k) \mathbf{v}_1^T \mathbf{A}_{11}^{(k)} \mathbf{v}_1$$

where $b \ge b_k > 0$, $\forall \mathbf{v}_1 \in I\!\!R^{n_k - n_{k-1}}$.

3. The number of nonzero entries in each row of $\mathbf{A}_{11}^{(k)}$ and $\mathbf{B}_{11}^{(k)}$ are bounded above by numbers independent of the level k.

Using Assumption 1, we can show that $\mathbf{A}^{(k-1)}$ is spectrally equivalent to the Schur complement $\mathbf{S}^{(k-1)}$.

Lemma 2.1 *Let $\mathbf{S}^{(k-1)}$ be defined by (2.2) and assume that the matrix blocks in $\mathbf{A}^{(k)}$ satisfy Assumption 1. Then*

$$1 - \gamma_k^2 \le \frac{\mathbf{v}_2^T \mathbf{S}^{(k-1)} \mathbf{v}_2}{\mathbf{v}_2^T \mathbf{A}^{(k-1)} \mathbf{v}_2} \le 1 \ \forall \mathbf{v}_2 \in I\!\!R^{n_{k-1}}.$$

Proof. The proof was presented in [1,3] and uses a block diagonal transformation of the two-by-two block matrix $\mathbf{A}^{(k)}$ to a form where the block diagonal part consists of the corresponding identity blocks and the off-diagonal blocks are $\tilde{\mathbf{A}}_{12}^{(k)} = \mathbf{A}_{11}^{(k)^{-1/2}} \mathbf{A}_{12}^{(k)} \mathbf{A}^{(k-1)^{-1/2}}$ and $\tilde{\mathbf{A}}_{21}^{(k)} = \tilde{\mathbf{A}}_{12}^{(k)^T}$. ∎

As we have already remarked in the introduction, the solution vector for linear systems with the hierarchical basis function matrices gives corrected values in the new components to the interpolated values of the solution at the previous level while the solution vector for systems with the standard basis function matrix gives the actual nodal values of the current approximations on that level.

In practical applications, it turns out that hierarchically structured matrices are less sparse than the standard matrices and, for reasons of efficiency, it can be advisable to use the standard basis function matrices in the actual implementation of the iterative algorithm. Therefore, we shall assume that there exists a transformation matrix taking one form of the matrix into the other. Hence, assume that there exists a transformation

$$\bar{\mathbf{J}}^{(k)} = \begin{bmatrix} \mathbf{I} & -\mathbf{J}_{12}^{(k)} \\ \mathbf{0} & \mathbf{I} \end{bmatrix}$$

taking $\mathbf{A}^{(k)}$ into the form

$$(2.3) \qquad \bar{\mathbf{A}}^{(k)} = \bar{\mathbf{J}}^{(k)^T} \mathbf{A}^{(k)} \bar{\mathbf{J}}^{(k)} = \begin{bmatrix} \bar{\mathbf{A}}_{11}^{(k)} & \bar{\mathbf{A}}_{12}^{(k)} \\ \bar{\mathbf{A}}_{21}^{(k)} & \bar{\mathbf{A}}_{22}^{(k)} \end{bmatrix}$$

or equivalently, since

$$\bar{\mathbf{J}}^{(k)^{-1}} = -\mathbf{J}^{(k)} = \begin{bmatrix} \mathbf{I} & -\mathbf{J}_{12}^{(k)} \\ \mathbf{0} & \mathbf{I} \end{bmatrix}$$

$$(2.4) \qquad \mathbf{A}^{(k)} = \mathbf{J}^{(k)^T} \bar{\mathbf{A}}^{(k)} \mathbf{J}^{(k)}.$$

Equation (2.4) implies

$$(2.5) \qquad \mathbf{A}_{11}^{(k)} = \bar{\mathbf{A}}_{11}^{(k)}, \quad \mathbf{A}_{12}^{(k)} = \bar{\mathbf{A}}_{12}^{(k)} + \bar{\mathbf{A}}_{11}^{(k)} \mathbf{J}_{12}^{(k)}$$

$$\mathbf{A}_{21}^{(k)} = \bar{\mathbf{A}}_{21}^{(k)} + \mathbf{J}_{12}^T \bar{\mathbf{A}}_{11}^{(k)}$$

and

$$\mathbf{A}^{(k-1)} = \bar{\mathbf{A}}_{22}^{(k)} + \bar{\mathbf{A}}_{21}^{(k)} \mathbf{J}_{12} + \mathbf{J}_{12}^T \bar{\mathbf{A}}_{12}^{(k)} + \mathbf{J}_{12}^T \bar{\mathbf{A}}_{11}^{(k)} \mathbf{J}_{12}.$$

The correspondingly transformed vector is denoted by $\bar{\mathbf{v}}$, i.e.,

$$\bar{\mathbf{v}} = \mathbf{J}^{(k)} \mathbf{v} = \begin{bmatrix} \mathbf{v}_1 + \mathbf{J}_{12} \mathbf{v}_2 \\ \mathbf{v}_2 \end{bmatrix}.$$

Therefore, the components of the solution of the standard nodal basis function matrix system at the added nodes at level k are equal to the corresponding components of $\mathbf{v}_1 + \mathbf{J}_{12}\mathbf{v}_2$, that is, the components of $\mathbf{v}_1$ are *corrections* to the interpolated values, $\mathbf{J}_{12}\mathbf{v}_2$ of the solution at the old nodes. The components of the solutions at the old nodes are the same for both systems. As pointed out by Vassilevski [17], there is a relation between the Schur complements to the hierarchical and standard matrices. They are in fact identical.

Lemma 2.2 *Let $\mathbf{A}^{(k)}$ and $\bar{\mathbf{A}}^{(k)}$ be defined by (2.1) and (2.3), respectively. Then the Schur complements*

$$\mathbf{S}^{(k-1)} = \mathbf{A}^{(k-1)} - \mathbf{A}_{21}^{(k-1)} \mathbf{A}_{11}^{(k)^{-1}} \mathbf{A}_{12}^{(k)}$$

and

$$\bar{\mathbf{S}}^{(k-1)} = \bar{\mathbf{A}}_{22}^{(k)} - \bar{\mathbf{A}}_{21}^{(k)} \mathbf{A}_{11}^{(k)^{-1}} \bar{\mathbf{A}}_{12}^{(k)}$$

are identical.

Proof. It follows from (2.5) that

$$\mathbf{S}^{(k-1)} = \bar{\mathbf{A}}_{22}^{(k)} + \bar{\mathbf{A}}_{21}^{(k)} \mathbf{J}_{12}^{(k)} + \mathbf{J}_{12}^{T} \bar{\mathbf{A}}_{12}^{(k)} + \mathbf{J}_{12}^{T} \bar{\mathbf{A}}_{11}^{(k)} \mathbf{J}_{12}$$

$$- \left(\bar{\mathbf{A}}_{21} + \mathbf{J}_{12}^{T} \bar{\mathbf{A}}_{11}^{(k)} \right) \bar{\mathbf{A}}_{11}^{(k)^{-1}} \left(\bar{\mathbf{A}}_{12}^{(k)} + \bar{\mathbf{A}}_{11}^{(k)} \mathbf{J}_{12}^{(k)} \right)$$

$$= \bar{\mathbf{A}}_{22}^{(k)} - \bar{\mathbf{A}}_{21}^{(k)} \mathbf{A}_{11}^{(k)^{-1}} \bar{\mathbf{A}}_{12}^{(k)} = \bar{\mathbf{S}}^{(k-1)}$$

∎

In addition to Assumption 3, we shall assume that

4. The number of nonzero entries in $\bar{\mathbf{A}}_{12}^{(k)}$, $\bar{\mathbf{A}}_{22}^{(k)}$, and $\mathbf{J}_{12}$ is bounded above by a number independent of the level number.

We now illustrate the above assumptions in the context of finite element methods for a self-adjoint elliptic operator in two space dimensions

$$Lu = - \sum_{r,s=1}^{2} \frac{\partial}{\partial x_r} \left(a_{rs} \frac{\partial u}{\partial x_s} \right), \quad \mathbf{x} \in \Omega \subset I\!\!R^2$$

where the matrix $[a_{rs}]$ of order 2 is uniformly positive definite on the polygonal domain Ω. For notational simplicity, we assume Dirichlet boundary conditions. We shall also assume that the coefficients a_{rs} are constant on each element of the initial triangulation.

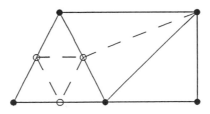

Figure 2.3

Then let $\mathbf{A}^{(k)}$ be the *hierarchical basis function matrix*, corresponding to the coersive bilinear form

$$a(u,v) = \int\int_\Omega \boldsymbol{\nabla} u^T \left[a_{rs}(x)\right] \boldsymbol{\nabla} v \, d\Omega.$$

Consider then a sequence of nested triangular meshes, consisting of uniformly refined triangles (Fig. 2.3). On each level, the basis functions are ordered in two sets as illustrated in Fig. 2.4 where $\{o\}$ nodes is the set of *new* node points,

$$V_1^{(k)} = \left\{\varphi_i^{(k)}; \mathbf{x}_i \in N_k \backslash N_{k-1}\right\}$$

is the corresponding set of new basis functions, $\{\cdot\}$ nodes is the set of *old* basis functions, and

$$V_2^{(k)} = \left\{\varphi_i^{(k-1)}; \ x_i \in N_{k-1}\right\}$$

is the corresponding set of old basis functions. Here N_k is the total set of node points on level k. The stiffness matrix is ordered and partitioned consistent with these two sets of node points. Note that the coarsest level matrix is the standard nodal basis function stiffness matrix on that level.

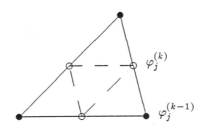

Figure 2.4

The corresponding two-level hierarchical basis function matrix $\mathbf{A}^{(k)}$ has therefore the following entries,

$$\left(\mathbf{A}_{11}^{(k)}\right)_{ij} = a\left(\varphi_j^{(k)},\ \varphi_i^{(k)}\right),$$

$$\left(\mathbf{A}_{12}^{(k)}\right)_{ij} = a\left(\varphi_j^{(k-1)},\ \varphi_i^{(k)}\right),$$

$$\left(\mathbf{A}_{21}^{(k)}\right)_{ij} = a\left(\varphi_j^{(k)},\ \varphi_i^{(k-1)}\right),$$

$$\left(\mathbf{A}_{22}^{(k)}\right)_{ij} = a\left(\varphi_j^{(k-1)},\ \varphi_i^{(k-1)}\right).$$

Note that $\mathbf{A}_{22}^{(k)}$ equals the hierarchical basis function matrix $\mathbf{A}^{(k-1)}$ on level $k-1$. The equivalent formulation of Assumption 1 takes the form:

$$|a(u,v)| \leq \gamma_k a(u,u)^{1/2} a(v,v)^{1/2}\ \forall u \in V_1^{(k)},\ v \in V^{(k-1)}.$$

$$V^{(k)} = V_1^{(k)} \cup V^{(k-1)}.$$

Note that $V^{(k)}$ can be represented by the hierarchical basis functions or by the corresponding standard basis functions, because they span the same function space. As has been shown in [2], γ_k can be computed locally for each triangle using just two levels of the hierarchical basis functions and $\gamma_k = \max\limits_{e \in \{\tau\}} \gamma_k^{(e)}$, where $\{\tau\}$ is the triangulation of the polygonal domain Ω.

Further, Maitre and Musy [15] have shown for the case of piecewise linear basis functions and $[a_{rs}] = a\begin{bmatrix} 1 & 0 \\ 0 & 1 \end{bmatrix}$ that:

$$\gamma^2 \leq \frac{3}{4} - \frac{1}{2}(3-d) \Big/ \left\{(4d-3)^{1/2} + 3\right\}$$

where $d = \max\limits_{e \in \{\tau\}} \sum\limits_{i=1}^{3} \cos^2 \theta_i$, and θ_i, $i = 1, 2, 3$, are the angles in the triangle e. Hence $\gamma^2 < 3/4$ for any triangulation of the domain.

In Bank and Dupont [8] and Axelsson and Gustafsson [3], it was shown that there exist matrices satisfying Assumptions 2 and 3. We can even choose $\mathbf{B}_{11}^{(k)}$ as a diagonal matrix, but for more accurate approximations the constant b in Assumption 2 gets smaller, of course. Another possible, but more expensive choice of $\mathbf{B}_{11}^{(k)}$ is a polynomial preconditioner of $\mathbf{A}_{11}^{(k)}{}^{-1}$. Assumptions 3 and 4 are clearly satisfied in the context of finite element methods.

3 Recursive Definition of Preconditioner

To construct a preconditioner to the hierarchical basis function matrix $\mathbf{A}^{(k)}$, we consider first the exact block matrix factorization of $\mathbf{A}^{(k)}$, namely,

$$\mathbf{A}^{(k)} = \begin{bmatrix} \mathbf{A}_{11}^{(k)} & 0 \\ \mathbf{A}_{21}^{(k)} & \mathbf{I} \end{bmatrix} \begin{bmatrix} \mathbf{I} & \mathbf{A}_{11}^{(k)^{-1}} \mathbf{A}_{12}^{(k)} \\ 0 & \mathbf{S}^{(k-1)} \end{bmatrix}$$

where

$$\mathbf{S}^{(k-1)} = \mathbf{A}^{(k-1)} - \mathbf{A}_{21}^{(k)} \mathbf{A}_{11}^{(k)^{-1}} \mathbf{A}_{12}^{(k)}.$$

We shall now recursively define the preconditioner $\mathbf{M}^{(k)}$ to $\mathbf{A}^{(k)}$. Let then

$$(3.1) \qquad \mathbf{M}^{(k)} = \begin{bmatrix} \mathbf{B}_{11}^{(k)^{-1}} & 0 \\ \mathbf{A}_{21}^{(k)} & \mathbf{I} \end{bmatrix} \begin{bmatrix} \mathbf{I} & \mathbf{B}_{11}^{(k)} \mathbf{A}_{12}^{(k)} \\ 0 & \tilde{\mathbf{A}}^{(k-1)} \end{bmatrix}, \qquad k = 1, 2, \ldots, \ell,$$

where $\mathbf{M}^{(0)} = \mathbf{A}^{(0)}$. Here $\tilde{\mathbf{A}}^{(k-1)}$ is an approximation of the Schur complement, defined by

$$(3.2a) \qquad \tilde{\mathbf{A}}^{(k-1)^{-1}} = \left[I - P_\nu \left(\mathbf{M}^{(k-1)^{-1}} \tilde{\mathbf{S}}^{(k-1)} \right) \right] \tilde{\mathbf{S}}^{(k-1)^{-1}} \quad \text{(version (i))}$$

$$(3.2b) \qquad \tilde{\mathbf{A}}^{(k-1)^{-1}} = \left[I - P_\nu \left(\mathbf{M}^{(k-1)^{-1}} \mathbf{A}^{(k-1)} \right) \right] \mathbf{A}^{(k-1)^{-1}} \quad \text{(version (ii))}$$

and $\tilde{\mathbf{S}}^{(k-1)}$ is the approximation of the Schur complement $\mathbf{S}^{(k-1)}$ where $\mathbf{A}_{11}^{(k-1)^{-1}}$ has been replaced by $\mathbf{B}_{11}^{(k)}$, i.e.,

$$(3.3) \qquad \tilde{\mathbf{S}}^{(k-1)} = \mathbf{A}^{(k-1)} - \mathbf{A}_{21}^{(k)} \mathbf{B}_{11}^{(k)} \mathbf{A}_{12}^{(k)}.$$

Further, P_ν is a polynomial of degree ν such that

$$(3.4) \qquad O \leq P_\nu(t) < 1, \ 0 < t \leq 1,$$

normalized such that $P_\nu(0) = 1$.

The form (3.1) and the normalization of P_ν permits the following efficient implementation of the preconditioner, i.e., the solution of a linear system $\mathbf{M}^{(k)} \mathbf{y} = \mathbf{v}$, where $\mathbf{y} = \begin{bmatrix} \mathbf{y}_1 \\ \mathbf{y}_2 \end{bmatrix}$, $\mathbf{v} = \begin{bmatrix} \mathbf{v}_1 \\ \mathbf{v}_2 \end{bmatrix}$ and $\mathbf{y}$, $\mathbf{v}$ are partitioned consistently with the partitioning of $\mathbf{M}^{(k)}$ (and $\mathbf{A}^{(k)}$).

Forward Substitution

Compute $\mathbf{z}_1 = \mathbf{B}_{11}^{(k)} \mathbf{v}_1$ and $\mathbf{z}_2 = \mathbf{v}_2 - \mathbf{A}_{21}^{(k)} \mathbf{z}_1$, i.e., make a matrix vector multiplication with $\mathbf{B}_{11}^{(k)}$ followed by a matrix vector multiplication with $\mathbf{A}_{21}^{(k)}$ and a vector addition.

Backward Substitution

Compute $\mathbf{y}_2 = \tilde{\mathbf{A}}^{(k-1)^{-1}} \mathbf{z}_2$ and $\mathbf{y}_1 = \mathbf{z}_1 - \mathbf{B}_{11}^{(k)} \mathbf{A}_{12}^{(k)} \mathbf{y}_2$. The last computation involves again two matrix vector multiplications and a vector addition.

The first step can be implemented efficiently in the following way. Let then

$$(3.5) \qquad Q_{\nu-1}(t) = \frac{1 - P_\nu(t)}{t} = q_0 + q_1 t + \cdots + q_{\nu-1} t^{\nu-1}.$$

Then in version (i),

$$(3.6) \qquad \mathbf{y}_2 = \tilde{\mathbf{A}}^{(k-1)^{-1}} \mathbf{z}_2 = Q_{\nu-1}\left(\mathbf{M}^{(k-1)^{-1}} \tilde{\mathbf{S}}^{(k-1)}\right) \mathbf{M}^{(k-1)^{-1}} \mathbf{z}_2$$

which can be computed by the following ν iteration steps:

Let $\mathbf{y}_2^{(0)} = 0$
For $r = 1$ to ν
    solve $\mathbf{M}^{(k-1)} \mathbf{y}_2^{(r)} = q_{\nu-r} \mathbf{z}_2 + \tilde{\mathbf{S}}^{(k-1)} \mathbf{y}_2^{(r-1)}$
Then $\mathbf{y}_2 = \mathbf{y}_2^{(\nu)}$

In version (ii), the Schur complement $\tilde{\mathbf{S}}^{(k-1)}$ above is replaced with $\mathbf{A}^{(k-1)}$. Note that it follows from (3.6) that the matrix $\tilde{\mathbf{A}}^{(k-1)}$ is symmetric (and also positive definite as we shall see) in both versions.

Solving a system with $\mathbf{M}^{(k)}$ involves among other things ν solutions of systems with $\mathbf{M}^{(k-1)}$, which in turn involves systems with $\mathbf{M}^{(k-2)}$ and so on. This means that the preconditioner $\mathbf{M}^{(\ell)}$, on the finest level, is only recursively defined and not explicitly given. Note that in the above implementation of the preconditioner, only matrix vector multiplications take place, except at the coarsest level where we solve the corresponding system (with $\mathbf{M}^{(0)} = \mathbf{A}^{(0)}$) with a direct solution method, for instance. The matrix vector multiplications can take place on element level, i.e., it is not required to assemble the matrices $\mathbf{A}_{ij}^{(k)}$.

The structure of the above implementation of the preconditioner resembles the steps taken in the classical multigrid method where we perform ν correction steps on every mesh level for every correction on the previous level in a ν-fold V-cycle method, where $\nu = 2$ corresponds to a W-cycle. The forward substitution step can be seen as a "presmoothing" step with $\mathbf{B}_{11}^{(k)}$ followed by a "restriction" by $\mathbf{A}_{21}^{(k)}$ of the vector $\mathbf{z}_1 \in V_1^{(k)}$ to vectors $\mathbf{v}_2$ in $V_2^{(k)} = V^{(k-1)}$. The backward substitution step begins with the computation of a correction (which involves ν calls to the preconditioner on the coarser level) followed by a "prolongation" with $\mathbf{A}_{12}^{(k)}$ from $\mathbf{y}_2 \in V^{(k-1)}$ to a vector in $V_1^{(k)}$, and a "postsmoothing" step with $\mathbf{B}_{11}^{(k)}$. Note, however, that we do not smooth on components which have just been corrected or have just already been corrected, which adds to the efficiency of the present method. However, more importantly, by the definition of $\tilde{\mathbf{A}}^{(k-1)}$ as in (3.2a) or (3.2b), we get a method which can be *analyzed completely algebraically* as we shall see in the next section and the conditions for optimal order of computational complexity are easily met in practice. This is in contrast to the classical multigrid method, where the regularity of the problem and the finite element mesh play a crucial role.

Computational Complexity

Consider now the computational complexity of the method. Assume that the degrees of freedom n_k increases in a geometric progression, $n_k = \sigma n_{k-1}$ for $\sigma \geq 2$. Then each iteration on the finest level ℓ involves arithmetic computations which are proportional to

$$C \sum_{k=1}^{\ell} \nu^{\ell-k} n_k = C \sum_{k=1}^{\ell} \left(\frac{\nu}{\sigma}\right)^{\ell-k} n_\ell \leq C \begin{cases} \frac{1}{1-\nu/\sigma} n_\ell & \text{if } \nu < \sigma \\ \ell n_\ell, & \text{if } \nu = \sigma \end{cases}$$

where C is a constant which depends on the relative sparsity of the matrices $\mathbf{A}_{i,j}^{(k)}$ and $\mathbf{B}_{11}^{(k)}$. In addition to this, we have a cost ν^ℓ times the cost to solve linear systems with matrix $\mathbf{A}^{(0)}$ (This level is visited ν^ℓ times). The order of $\mathbf{A}^{(0)}$ is n_ℓ/σ^ℓ. If we let $\ell = O(\log N)$ where $N = n_\ell$, then this order is fixed and the total computational complexity per iteration is $O(n_\ell)$ if $\nu < \sigma$.

Hence, the important question we must ask is if we can construct a spectrally equivalent preconditioner of the method. Assume that the degrees of freedom n_k increases in a geometric progression, $n_k = \sigma n_{k-1}$ for $\sigma \geq 2$. Then each iteration on the finest level ℓ involves arithmetic computations which are proportional to

$$C \sum_{k=1}^{\ell} \nu^{\ell-k} n_k = C \sum_{k=1}^{\ell} \left(\frac{\nu}{\sigma}\right)^{\ell-k} n_\ell \leq C \begin{cases} \frac{1}{1-\nu/\sigma} n_\ell & \text{if } \nu < \sigma \\ \ell n_\ell, & \text{if } \nu = \sigma \end{cases}$$

where C is a constant which depends on the relative sparsity of the matrices $\mathbf{A}_{i,j}^{(k)}$ and $\mathbf{B}_{11}^{(k)}$. In addition to this, we have a cost ν^ℓ times the cost to solve linear systems with matrix $\mathbf{A}^{(0)}$ (This level is visited ν^ℓ times). The order of $\mathbf{A}^{(0)}$ is n_ℓ/σ^ℓ. If we let $\ell = O(\log N)$ where $N = n_\ell$, then this order is fixed and the total computational complexity per iteration is $O(n_\ell)$ if $\nu < \sigma$.

Hence, the important question we must ask is if we can construct a spectrally equivalent preconditioner if $\nu < \sigma$. If the preconditioner is spectrally equivalent we know that there exist iterative methods, which converge to a relative iteration error ϵ in $O(\log \epsilon^{-1})$ steps.

The spectral equivalence property clearly must depend on the choice of the polynomial P_ν. A simple choice of a polynomial satisfying property (3.4) is

$$P_\nu(t) = (1-t)^\nu.$$

Here

$$Q_{\nu-1}(t) = \sum_{j=1}^{\nu} (-1)^{j-1} \binom{\nu}{j} t^{j-1}.$$

This and the following optimal choice where P_ν is defined as a shifted and normalized

Chebyshev polynomial were discussed in [5],

$$(3.7) \qquad P_\nu(t) = \left[T_\nu \left(\frac{1 + \alpha - 2t}{1 - \alpha} \right) + 1 \right] \Bigg/ \left[T_\nu \left(\frac{1 + \alpha}{1 - \alpha} \right) + 1 \right]$$

where

$$T_\nu(x) = \frac{1}{2} \left\{ \left(x + \sqrt{x^2 - 1} \right)^\nu + \left(x - \sqrt{x^2 - 1} \right)^\nu \right\}$$

is the Chebyshev polynomial of the first kind normalized on the interval (-1,1). Note that

$$P_\nu(\alpha) = 2 \Bigg/ \left[T_\nu \left(\frac{1 + \alpha}{1 - \alpha} \right) + 1 \right]$$

and

$$P_\nu(1) = \left[(-1)^\nu + 1 \right] \Bigg/ \left[T_\nu \left(\frac{1 + \alpha}{1 - \alpha} \right) + 1 \right].$$

P_ν in (3.7) gives the best polynomial approximation of the zero function on the interval $[\alpha,1]$ of all non-negative polynomials of degree ν normalized at the origin. The parameter α, which is positive, shall be specified in the next section. Two other choices were also discussed in [5], namely

$$Q_2(t) = 5 - 8t + 4t^2 \qquad\qquad (\nu = 3)$$

$$Q_4(t) = 11 - 45t + 85t^2 - 75t^3 + 25t^4 \qquad (\nu = 5).$$

Q_2 and Q_4 are defined from (3.5) and (3.7) with $\alpha = 1/\nu$ ($\nu = 3$ and 5, respectively) and satisfy the additional requirement that $P_\nu'(1) = -1$. Since the matrices $\mathbf{A}_{12}^{(k)}$ and $\mathbf{A}_{21}^{(k)}$ are less sparse [in fact, the number of nonzero entries in each row of $\mathbf{A}_{12}^{(k)}$ increases as $O(k)$] than the matrices $\bar{\mathbf{A}}_{12}^{(k)}$ and $\bar{\mathbf{A}}_{21}^{(k)}$ in the standard nodal basis function matrix, the efficient implementation of the iterative method uses the interpolation operator $J^{(k)}$ defined in (2.3) and (2.5). This means that we iterate with $\mathbf{J}^{(k)^T} \bar{\mathbf{A}}^{(k)} \mathbf{J}^{(k)}$ and $\mathbf{J}^{(k)^T} \bar{\mathbf{M}}^{(k)} \mathbf{J}^{(k)}$ where $\bar{\mathbf{M}}^{(k)} = \bar{\mathbf{J}}^{(k)^T} \mathbf{M}^{(k)} \bar{\mathbf{J}}^{(k)}$. This was recommended already by Yserentant [20] in connection with his method and has been further commented upon by Greenbaum, *et al.* [12].

Domain Decomposition

The iterative method (1.3) and the recursively defined preconditioner lend themselves to an efficient implementation on vector and parallel computers because they consist only of matrix vector operations and vector operations. The most efficient implementation on a parallel computer seems to be based on domain decomposition. Decompose, then, the domain into boxes or strips of about equal size. The number of such subdomains should equal the number of processors available or possibly one of the processors can act as a host processor for control of synchronization points and the computation of vector operations required for the iterations when adding the contributions from the various processors to the matrix vector operations parts on interior boundary points. Each processor (except the host) will then work on matrix vector multiplications for the interior points in its subdomain and contribute with its part to these products for the points on the interior boundaries.

4 The Relative Condition Number of $\mathbf{M}^{(\ell)}$ with Respect to $\mathbf{A}^{(\ell)}$

We shall show that the relative condition number of $\mathbf{M}^{(\ell)}$ with respect to $\mathbf{A}^{(\ell)}$ is bounded, independent of ℓ, i.e., that $\mathbf{M}^{(\ell)}$ is spectrally equivalent to $\mathbf{A}^{(\ell)}$. This will be done first for version (ii) and then for version (i). To this end note that it follows by (2.1) and (3.1) that for both versions

$$(4.1) \qquad \mathbf{M}^{(k)} - \mathbf{A}^{(k)} = \begin{bmatrix} \mathbf{B}_{11}^{(k)^{-1}} - \mathbf{A}_{11}^{(k)} & \mathbf{0} \\ \mathbf{0} & \tilde{\mathbf{A}}^{(k-1)} - \tilde{\mathbf{S}}^{(k-1)} \end{bmatrix}$$

where $\tilde{\mathbf{A}}^{(k-1)}$ and $\tilde{\mathbf{S}}^{(k-1)}$ are defined by (3.2) and (3.3), respectively. Here in version (ii)

$$
\begin{aligned}
(4.2) \quad \tilde{\mathbf{A}}^{(k-1)} - \tilde{\mathbf{S}}^{(k-1)} &= \mathbf{A}^{(k-1)} \left\{ \left[I - P_\nu \left(\mathbf{M}^{(k-1)^{-1}} \mathbf{A}^{(k-1)} \right) \right]^{-1} - I \right\} \\
&\quad + \mathbf{A}_{21}^{(k)} \mathbf{B}_{11}^{(k)} \mathbf{A}_{12}^{(k)} \\
&= \mathbf{A}^{(k-1)} P_\nu \left(\mathbf{M}^{(k-1)^{-1}} \mathbf{A}^{(k-1)} \right) \\
&\quad \times \left[I - P_\nu \left(\mathbf{M}^{(k-1)^{-1}} \mathbf{A}^{(k-1)} \right) \right]^{-1} + \mathbf{A}_{21}^{(k)} \mathbf{B}_{11}^{(k)} \mathbf{A}_{12}^{(k)}.
\end{aligned}
$$

It follows by (4.1) that

$$(4.3) \qquad \frac{\mathbf{v}^T \mathbf{M}^{(k)} \mathbf{v}}{\mathbf{v}^T \mathbf{A}^{(k)} \mathbf{v}} = 1 + \frac{\left[\mathbf{v}_1^T \left(\mathbf{B}_{11}^{(k)^{-1}} - \mathbf{A}_{11}^{(k)} \right) \mathbf{v}_1 + \mathbf{v}_2^T \left(\tilde{\mathbf{A}}^{(k-1)} - \tilde{\mathbf{S}}^{(k-1)} \right) \mathbf{v}_2 \right]}{\mathbf{v}^T \mathbf{A}^{(k)} \mathbf{v}}$$

and Assumption 2 and (3.4) show the lower bound

$$1 \leq \mathbf{v}^T \mathbf{M}^{(k)} \mathbf{v} / \mathbf{v}^T \mathbf{A}^{(k)} \mathbf{v} \ \forall \mathbf{v} \in I\!\!R^{n_k}.$$

To get an upper bound, we note first that

$$\mathbf{v}^T \mathbf{A}^{(k)} \mathbf{v} = \mathbf{v}_1^T \mathbf{A}_{11}^{(k)} \mathbf{v}_1 + \mathbf{v}_1^T \mathbf{A}_{12}^{(k)} \mathbf{v}_2 + \mathbf{v}_2^T \mathbf{A}_{21}^{(k)} \mathbf{v}_1 + \mathbf{v}_2^T \mathbf{A}^{(k-1)} \mathbf{v}_2$$

and by Assumption 1, and the arithmetic-geometric inequality, we have

(4.4) $\mathbf{v}^T \mathbf{A}^{(k)} \mathbf{v} \geq \left(1 - \zeta^{-1}\gamma_k\right) \mathbf{v}_1^T \mathbf{A}_{11}^{(k)} \mathbf{v}_1 + (1 - \zeta\gamma_k)\mathbf{v}_2^T \mathbf{A}^{(k-1)} \mathbf{v}_2$

$\forall \mathbf{v} = \begin{bmatrix} \mathbf{v}_1 \\ \mathbf{v}_2 \end{bmatrix} \in I\!\!R^{n_k}$, where $\gamma_k < \zeta < \gamma_k^{-1}$. The parameter ζ shall be chosen later to get a smallest upper bound in (4.3). Note now that Lemma 1 and Assumption 2 show that

$$\mathbf{v}_2^T \mathbf{A}_{21}^{(k)} \mathbf{B}_{11}^{(k)} \mathbf{A}_{12}^{(k)} \mathbf{v}_2 \leq \gamma_k^2 \mathbf{v}_2^T \mathbf{A}^{(k-1)} \mathbf{v}_2 \ \forall \, \mathbf{v}_2 \in I\!\!R^{n_{k-1}}.$$

This, together with (4.2) and (3.4) show that

(4.5) $\mathbf{v}_2^T \left(\tilde{\mathbf{A}}^{(k-1)} - \tilde{\mathbf{S}}^{(k-1)} \right) \mathbf{v}_2 / \mathbf{v}_2^T \mathbf{A}^{(k-1)} \mathbf{v}_2 \leq d_\nu^{(k)} + \gamma_k^2$

where

(4.6) $$d_\nu^{(k)} \equiv P_\nu(t_{k-1}^*) / \left(1 - P_\nu(t_{k-1}^*)\right)$$

$$P_\nu(t_{k-1}^*) = \sup P_\nu(t), \ \underline{t}_{k-1} \leq t \leq \bar{t}_{k-1}$$

and where $\underline{t}_{k-1}$ and $\bar{t}_{k-1}$ are the infimum and supremum of

$$\mathbf{v}_2^T \mathbf{A}^{(k-1)} \mathbf{v}_2 / \mathbf{v}_2^T \mathbf{M}^{(k-1)} \mathbf{v}_2,$$

respectively. Then (4.3) shows that

(4.7) $$\lambda_k \leq 1 + \max \left\{ \frac{b}{1 - \zeta^{-1}\gamma_k} \,, \ \frac{d_\nu^{(k-1)} + \gamma_k^2}{1 - \zeta\gamma_k} \right\}.$$

To minimize the upper bound of (4.7), we choose ζ such that

$$b/(1 - \zeta^{-1}\gamma_k) = d/(1 - \zeta\gamma_k)$$

where $d = d_\nu^{(k)} + \gamma^2$. This implies (assuming $b > 0$)

$$\zeta\gamma = \frac{1}{2}\left(1 - \frac{d}{b}\right) + \frac{1}{4}\left[\left(1 - \frac{d}{b}\right)^2 + \frac{d}{b}\gamma^2\right]^{1/2}$$

and

$$\frac{d}{1 - \zeta\gamma} = \frac{1}{1 - \gamma^2}\left\{\frac{1}{2}(b + d) + \left[\frac{1}{4}(b - d)^2 + db\gamma^2\right]^{1/2}\right\}.$$

Hence we have the recursion

(4.8) $$\lambda_k = f(\lambda_{k-1}), \quad k = 1, 2, \ldots, \qquad \lambda_0 = 1$$

where

$$f(\lambda_{k-1}) = 1 + \frac{1}{1-\lambda^2}\left\{\frac{1}{2}(b+d) + \left[\frac{1}{4}(b-d)^2 + db\,\gamma^2\right]^{1/2}\right\}$$

and $d = d_\nu^{(k)} + \gamma_k^2$. Here $d_\nu^{(k)}$ is defined by (4.6) where $P_\nu(t_{k-1}^*) = \sup P_\nu(t)$ and $\lambda_{k-1}^{-1} \le t \le 1$.

To analyze if (4.8) has a fixed-point, we write

$$f(\lambda_{k-1}) = 1 + \frac{b+d}{1-\lambda^2} - (bd)^{1/2}\frac{\delta}{\left[1 + (1 - \delta^2(1-\gamma^2))^{1/2}\right]} \le 1 + \frac{b+d}{1-\gamma^2}$$

where $\delta = 2(bd)^{1/2}/(b+d)$. Note that $0 \le \delta \le 1$.

Fixed-Point Analysis

Let $\left\{\widehat{\lambda}_k\right\}$ be a majorizing sequence of $\{\lambda_k\}$, where

$$\widehat{\lambda}_k = 1 + (b+d)/(1-\gamma^2), \quad d = \gamma^2 + P_\nu(\hat{t}_{k-1}^*)/(1 - P_\nu(\hat{t}_{k-1}^*))$$

and

$$P_\nu(\hat{t}_k^*) = \max P_\nu(t), \quad \widehat{\lambda}_{k-1}^{-1} = \hat{t}_{k-1} \le t \le 1.$$

Then

$$\widehat{\lambda}_k = \frac{1}{1-\gamma^2}\left[1 + b + \frac{P_\nu(\hat{t}_{k-1}^*)}{1 - P_\nu(\hat{t}_{k-1}^*)}\right] = \frac{1}{1-\gamma^2}\left[b + \frac{1}{1 - P_\nu(\hat{t}_{k-1}^*)}\right]$$

or

(4.9) $\qquad \hat{t}_k = (1-\gamma^2)\hat{t}_{k-1}Q_{\nu-1}(\hat{t}_{k-1})/(1 + b\hat{t}_{k-1}Q_{\nu-1}(\hat{t}_{k-1}))$

where we assumed that $\alpha \le \hat{t}_{k-1}$ in (3.7). It is readily seen that (4.9) has a fixed-point $t^* > 0$, if $(1-\gamma^2)Q_{\nu-1}(0) > 1$. Since

$$T_\nu\left(\frac{1+\alpha}{1-\alpha}\right) = \frac{1}{2}\left[\left(\frac{1+\alpha^{1/2}}{1-\alpha^{1/2}}\right)^\nu + \left(\frac{1-\alpha^{1/2}}{1+\alpha^{1/2}}\right)^\nu\right]$$

and as follows by Theorem 4.1 in [6], the above shows that

Theorem 4.1 (a) *The preconditioner* $\mathbf{M}^{(\ell)}$ *defined in (3.1) and (3.2b) where* P_ν *is defined in (3.7), is spectrally equivalent to* $\mathbf{A}^{(\ell)}$ *if* $\nu > (1-\gamma^2)^{-1/2}$.

(b) *For such a* ν, *there exists a positive root of the scalar equation*

$$1 - \gamma^2 = tb + \left[\frac{(1+t^{1/2})^\nu + (1-t^{1/2})^\nu}{2\sum_{s=1}^{\nu}(1+t^{1/2})^{\nu-s}(1-t^{1/2})^{s-1}}\right]^2$$

and the parameter α *in (3.7) is equal to or smaller than the smallest of such roots.*

(c) The relative condition number of $\mathbf{M}^{(\ell)}$ with respect to $\mathbf{A}^{(\ell)}$ is bounded by

$$\lambda \leq \frac{1}{1 - \gamma^2} \left\{ b + \left[\frac{(1 + \alpha^{1/2})^\nu + (1 - \alpha^{1/2})^\nu}{(1 + \alpha^{1/2})^\nu - (1 - \alpha^{1/2})^\nu} \right]^2 \right\}.$$

(d) For $\nu = 2$ we get

$$\alpha = \frac{(3 - 4\gamma^2)}{\left[1 + 2b + \left\{ (1 + 2b)^2 + (3 - 4\gamma^2) \right\}^{1/2} \right]}$$

and

$$\lambda \leq \frac{\left(b + \frac{1}{4\alpha} + \frac{1}{2} + \frac{1}{4}\alpha \right)}{(1 - \gamma^2)}.$$

The theorem shows that if $\gamma^2 < 8/9$ we have a spectrally equivalent preconditioner if $\nu = 3$ and if $\gamma^2 < 3/4$, it suffices to choose $\nu = 2$.

Consider now version (i). Here

$$\tilde{\mathbf{A}}^{(k-1)} - \tilde{\mathbf{S}}^{(k-1)} = \tilde{\mathbf{S}}^{(k-1)} P_\nu \left(\mathbf{M}^{(k-1)^{-1}} \tilde{\mathbf{S}}^{(k-1)} \right)$$

$$\times \left[I - P_\nu \left(\mathbf{M}^{(k-1)^{-1}} \tilde{\mathbf{S}}^{(k-1)} \right) \right]^{-1}$$

and a similar derivation as for version (ii) shows that the upper bound λ_k of $\mathbf{v}^T \mathbf{M}^{(k)} \mathbf{v} / \mathbf{v}^T \mathbf{A}^{(k)} \mathbf{v}$ satisfies

$$\lambda_k \leq 1 + \max \left\{ \frac{b}{1 - \zeta^{-1}\gamma_k} , \frac{d_\nu^{(k)}}{1 - \zeta\gamma_k} \right\}$$

where now

(4.10) $d_\nu^{(k)} \equiv P_\nu(t_{k-1}^*) / \left(1 - P_\nu(t_{k-1}^*) \right)$

$P_\nu(t_{k-1}^*) = \sup P_\nu(t), \ \underline{t}_{k-1} \leq t \leq 1$

where

$$\underline{t}_{k-1} = \inf_{\mathbf{v}_2} \frac{\mathbf{v}_2^T \tilde{\mathbf{S}}^{(k-1)} \mathbf{v}_2}{\mathbf{v}_2^T \mathbf{A}^{(k-1)} \mathbf{v}_2} \geq \inf_{\mathbf{v}_2} \frac{\mathbf{v}_2^T \tilde{\mathbf{S}}^{(k-1)} \mathbf{v}_2}{\mathbf{v}_2^T \mathbf{A}^{(k-1)} \mathbf{v}_2} \inf \frac{\mathbf{v}_2^T \mathbf{A}^{(k-1)} \mathbf{v}_2}{\mathbf{v}_2^T \mathbf{M}^{(k-1)} \mathbf{v}_2}$$

(4.11) $\geq (1 - \gamma_k^2)\lambda_{k-1}^{-1}.$

The recursion for the sequence λ_k now takes the form

$$\lambda_k = f(\lambda_{k-1}), \quad k = 1, 2, \ldots, \qquad \lambda_0 = 1,$$

where

$$f(\lambda_{k-1}) = 1 + \frac{1}{1-\gamma^2}\left\{(b+d) + \left[\frac{1}{4}(b-d)^2 + db\gamma^2\right]^{1/2}\right\}$$

$$\leq 1 + \frac{b+d}{1-\gamma^2}$$

and where now $d = d_\nu^{(k)}$ is defined in (4.10). In this case we get

(4.12) $$\widehat{\lambda}_k = 1 + \frac{1}{1-\gamma^2}\left[b + \frac{P_\nu(t_{k-1}^*)}{1-P_\nu(t_{k-1}^*)}\right]$$

and an analysis similar to the previous one shows that the recursion

$$\hat{t}_k = (1-\gamma^2)\hat{t}_{k-1}Q_{\nu-1}(\hat{t}_{k-1})/\left[1+\hat{t}_{k-1}(b-\gamma^2)Q_{\nu-1}(\hat{t}_{k-1})\right]$$

has a fixed point if $(1-\gamma^2)Q_{\nu-1}(0) > 1$ as before.

Theorem 4.2 (a) *The preconditioner $\mathbf{M}^{(\ell)}$ defined in (3.1) and (3.2a), where P_ν is defined in (3.7), is spectrally equivalent to $\mathbf{A}^{(\ell)}$ if $\nu > (1-\gamma^2)^{-1/2}$.*

(b) *For such a ν, there exists a positive root of the scalar equation*

$$1-\gamma^2 = t(b-\gamma^2) + \left[\frac{(1+t^{1/2})^\nu + (1-t^{1/2})^\nu}{2\sum_{s=1}^\nu (1+t^{1/2})^{\nu-s}(1-t^{1/2})^{s-1}}\right]^2$$

and the parameter α in (3.7) is equal to or smaller than the smallest of such roots.

(c) *The relative condition number of $\mathbf{M}^{(\ell)}$ with respect to $\mathbf{A}^{(\ell)}$ is bounded by*

$$\lambda \leq \frac{1}{1-\gamma^2}\left\{b-\gamma^2 + \left[\frac{(1+\alpha^{1/2})^\nu + (1-\alpha^{1/2})^\nu}{(1+\alpha^{1/2})^\nu - (1-\alpha^{1/2})^\nu}\right]^2\right\}.$$

(d) *For $\nu = 2$ we get*

$$\alpha = \frac{(3-4\gamma^2)}{\left[1+2(b-\gamma^2) + \left\{[1+2(b-\gamma^2)]^2 + (3-4\gamma^2)\right\}^{1/2}\right]}$$

and

$$\lambda \leq \left(b-\gamma^2 + \frac{1}{4\gamma} + \frac{1}{2} + \frac{1}{4}\alpha\right)\Big/(1-\gamma^2).$$

Theorem 4.2 does not actually always give the best upper bound. In particular, if $b = 0$ then Theorem 2 shows that

$$\lambda \le 1 + \frac{(1 - \alpha)^2}{4\alpha} \cdot \frac{1}{1 - \gamma^2}$$

for $\nu = 2$ while Theorem 4.1 in [5] shows that

(4.13) $$\lambda \le 1 + \frac{(1 - \alpha)^2}{4\alpha}.$$

As in [6] one can present a refined analysis where

(4.14) $$\lambda_k \le \frac{1}{1 + b} \left(1 + \frac{b}{1 - \gamma^2}\right) d_\nu^{(k)} + \frac{b}{1 - \gamma^2}$$

and

$$d_\nu^{(k)} = P_\nu(t_{k-1}^*) / \left(1 - P_\nu(t_{k-1}^*)\right).$$

If $b = 0$ this latter bound reduces to the best upper bound (4.13). If $b > 0$, (4.14) shows the bound

$$\lambda \le \frac{d_\nu^{(k)} + b}{1 - \gamma^2}$$

which is slightly better than (4.12).

Note that for problems in three space dimensions where the number of degrees of freedom progresses in a geometric fashion with a factor $\sigma = 8$, if we make uniform mesh refinements, we can choose the polynomial degree ν up to 7 and still get a method of optimal order of computational complexity.

As has been shown by Maître and Musy [15], for an arbitrary triangulation of a polygonal domain with constant coefficients and piecewise linear basis functions, we have $\gamma^2 < 3/4$. For a regular tetrahedron we have $\gamma^2 = 5/7$ for piecewise quadratic basis functions. Hence Theorems 4.1 and 4.2 show that it suffices to choose $\nu = 2$ in both cases.

5   Concluding Remarks

We have shown that for the hierarchical basis function matrices one can construct spectrally equivalent preconditioners recursively defined as shown in Sec. 3. For piecewise linear basis functions, it suffices to choose $\nu = 2$.

These results can be extended to higher order finite element approximations using the known spectral equivalence properties between the higher order finite element stiffness matrices and the piecewise linear one for meshes consisting of the same nodes. This has been shown in Chapter 7.4 of [2]. Therefore, the actual preconditioner for the higher order elements can be computed from the lower order finite element matrix and still be spectrally equivalent to the higher order stiffness matrix.

More generally, for problems with variable coefficients within elements the same idea can be applied. We then first compute the harmonic averaged matrix for piecewise linear basis functions as shown in Axelsson and Gustafsson [3] for instance, and use this to compute the preconditioner as shown in Sec. 3. This preconditioner is then spectrally equivalent to the higher order finite element matrix and in the iterative method (1.3) (on the global level ℓ) the residuals will be computed from the variable coefficient matrix (possible using element-by-element techniques, i.e., without actually assembling this matrix). In all cases, the resulting preconditioned iterative method will be of optimal order of computational complexity.

References

[1] Axelsson, O. "On multigrid methods of the two-level type," in *Multigrid Methods*, Proceedings, Köln-Porz, 1981 (W. Hackbusch and U. Trottenberg, eds.), LNM 960, Berlin: Springer-Verlag, 1982, 352–367.

[2] Axelsson, O. and V.A. Barker. *Finite Element Solution of Boundary Value Problems*. Orlando: Academic Press, 1984.

[3] Axelsson, O. and I. Gustafsson. "Preconditioning and two-level multigrid methods of arbitrary degree of approximation," *Math. Comp.* **40** (1983) 219–242.

[4] Axelsson, O. and I. Gustafsson. "An efficient finite element method for nonlinear diffusion problems," Report 84.06R, ISSN 0347-0946, Department of Computer Sciences, Chalmers University of Technology, Göteborg, Sweden, 1984.

[5] Axelsson, O. and P. Vassilevski. "Algebraic multilevel preconditioning methods, I," *Numer. Math.*, to appear.

[6] Axelsson, O. and P. Vassilevski. "Algebraic multilevel preconditioning methods, II," Report 1988-15, Institute for Scientific Computation, University of Wyoming, Laramie, Wyoming, 1988.

[7] Bakhvalov, N.S. "On the convergence of a relaxation method with natural constraints on the elliptic operator," *U.S.S.R. Computational Math. and Math. Phys.* **6** (1966) 101–135.

[8] Bank, R. and T. Dupont. "Analysis of a two-level scheme for solving finite element equations," Report CNA-159, Center for Numerical Analysis, The University of Texas at Austin, 1980.

[9] Braess, D. "The contraction number of a multigrid method for solving the Poisson equation," *Numer. Math.* **32** (1987) 387–404.

[10] Brandt, D. "Multi-level adaptive solutions to boundary-value problems," *Math. Comp.* **31** (1977) 333–390.

[11] Fedorenko, R.P. "A relaxation method for solving elliptic difference equations," *U.S.S.R. Comput. Math. and Math. Phys.* **1** (1962) 1092–1096.

[12] Greenbaum, D., C. Li, and H.Z. Chao. "Comparison of Linear System Solvers Applied to Diffusion-type Finite Element Equations," Ultracomputer Research Laboratory, Courant Institute of Mathematical Sciences, New York, 1987.

[13] Hackbusch, W. "Multigrid methods and applications," Springer Series in *Comp. Math.* **4**, Berlin: Springer-Verlag, 1985.

[14] Kuznetsóv, Y.A. "Multigrid domain decomposition methods for elliptic problems," in *Proceedings VIII International Conference on Computational Methods for Applied Science and Eng.*, Vol. 2, Versailles, 1987, 605–616.

[15] Maître, J.F. and F. Musy. "The contraction of a class of two-level methods; an exact evaluation for some finite element subspaces and model problems," in *Multigrid Methods*, Proceedings, Köln-Porz, 1981 (W. Hackbusch and N. Trottenberg, eds.) LNM 960, Berlin: Springer-Verlag, 1982, 535–544.

[16] Nicolaides, R.A. "On the ℓ-2 convergence of an algorithm for solving finite element equations," *Math. Comp.* **31** (1977) 892–906.

[17] Vassilevski, P. "Iterative methods for solving finite element equations based on multilevel splitting of the matrix," Preprint, Bulgarian Academy of Sciences, Sofia, Bulgaria, 1987.

[18] Wachspress, E.L. "Split-level iteration," *Comp. Maths. with Appls.* **10** (1984) 453–456.

[19] Young, D.M. *Iterative Solutions of Large Linear Systems*, New York: Academic Press, 1971.

[20] Yserentant, H. "On the multilevel splitting of finite element spaces," *Numer. Math.* **49** (1986) 379–412.

Chapter 3

ELLPACK and ITPACK as Research Tools for Solving Elliptic Problems

GARRETT BIRKHOFF
Harvard University

and

ROBERT E. LYNCH
Purdue University

Dedicated to David M. Young, Jr., on the occasion of his sixty-fifth birthday.

Abstract

A summary is given, in historical perspective, of some recent studies by the authors which utilized ELLPACK and ITPACK as research tools.

1 Background[1]

When David Young began his work on automating the numerical solution of (linear) elliptic problems, 40 years ago, one of us advised him to concentrate on Runge's 5-point difference approximation to the classical Dirichlet problem, i.e., solving $u_{xx} + u_{yy} = 0$ for Dirichlet boundary conditions on a square mesh. This topic seemed promising because Richard Southwell and his coworkers had successfully solved by hand a variety of Dirichlet problems using this approximation, and to automate their methods seemed more than adequate for a Ph.D. thesis.

At that time, the main practical bottleneck impeding the computation of accurate approximate solutions to Dirichlet problems was the time required to solve

---

[1]For fuller descriptions, see [14] and [19, pp. 17–38].

a large *set* of equations of the form $\sum_s a_{rs} U_s = b_r$. For any *ordering* of these equations, and of the components U_s of $\mathbf{U}$, this amounts to solving a single vector equation of the form $\mathbf{AU} = \mathbf{b}$, where $\mathbf{A}$ is a *large, sparse, nonsingular matrix*, typically having many other useful properties. For the Model Problem of the Poisson equation in a rectangle, and many others, these include being symmetric, positive definite, diagonally dominant, and having a positive inverse.

Young's brilliant and mathematically *rigorous* theoretical solution of this vector equation in terms of the *iterative* SOR method [22], applies on many domains to (self-adjoint) linear source problems of the form

$$(1.1) \qquad (A(x,y)u_x)_x + (C(x,y)u_y)_y + F(x,y)u = G(x,y)$$

with $AC > 0$ and Dirichlet boundary conditions, if one uses the standard 5-point difference approximation of [21, (6.36)] and [2, p. 57] on a rectangular mesh.

Young's results about solving such vector equations $\mathbf{AU} = \mathbf{b}$, together with further progress to 1962, were reviewed in two now classic monographs: [21] by Varga and [6] by Forsythe and Wasow (Part 3). Whereas Young's original theory utilized primarily the fact that the matrix $\mathbf{A} = (\|a_{rs}\|)$ has positive real eigenvalues, enjoys Property Y under "consistent orderings" of the equations and unknowns,[2] and is diagonally dominant if $AF \leq 0$, Varga's key Theorem 6.4 [21, p. 187] brings out many additional properties of $\mathbf{A}$, including that of being a *Stieltjes matrix*.[3]

The SOR method of Young and Frankel constituted the first of many new schemes for solving elliptic boundary value problems numerically. Already in 1971, Young's authoritative *Iterative Solution of Large Linear Systems* [23] analyzed 25 different methods with respect to 26 "fundamental matrix properties" (28 properties if *size* and *sparsity* are added to its list); cf. [23, pp. xxi–xxiv]. There as in [21], the emphasis was on iterative and semi-iterative techniques for solving $\mathbf{AU} = \mathbf{b}$, where $\mathbf{A}$ is typically a sparse Stieltjes matrix having Property Y. Also as in [21], the *matrix problem* of solving $\mathbf{AU} = \mathbf{b}$ (iteratively) was detached from the 'discretization problem' of *approximating* (1.1) by a suitable 'elliptic difference equation.'

Among the many methods analyzed in [23] and [7], there should be noted especially the CCSI (Cyclic Chebyshev Semi-Iterative) and SSOR SI (Symmetric SOR Semi-Iterative) methods. In particular, when properly 'accelerated' by Chebyshev Semi-Iterative (CSI) or conjugate gradient (CG) SSOR SI and SSOR CG can achieve an *asymptotic* order of convergence substantially faster than SOR for many coefficient-matrices $\mathbf{A}$;[4] see Sec. 10.

The matrix problem became reunited with the discretization problem about five years later, through the parallel development at Purdue University and The University of Texas at Austin of two compatible packages of Fortran programs,

---

[2] As in [2, Ch. 4, Sec. 14], we say that a matrix $\mathbf{A}$ has 'Property Y' when it is consistently ordered. For Property A, see [23, p. 43] or [2, p. 63]; only sets of equations having Property A can be consistently ordered.

[3] We recall that a matrix $\mathbf{A}$ is a "Stieltjes matrix" when it is symmetric, positive definite, and has a positive inverse.

[4] See [24] and [2, p. 163–4]. The condition on $\mathbf{A}$ is that the spectral radius of the product of its lower and upper triangular parts satisfies $\rho(\mathbf{LU}) \leq 1/4$.

ELLPACK and ITPACK, designed to "provide a tool for research in the evaluation and development of numerical methods for solving elliptic PDE's" [19, pp. 136, 163]. ELLPACK (cf. Sec. 2) is designed to treat boundary value problems for more general second-order linear elliptic DEs of the form

(1.2)
$$A(x,y)u_{xx} + 2B(x,y)u_{xy} + C(x,y)u_{yy}$$
$$+ D(x,y)u_x + E(x,y)u_y + F(x,y)u = G(x,y).$$

with $AC > B^2$, on general plane domains and for general linear boundary conditions. Unless $B = 0$, one cannot approximate (1.2) effectively with a 5-point stencil. For this reason, and because they are more accurate (cf. Sec. 8), many of the discretization modules of ELLPACK are based on 9-*point* stencils of unknowns. Among these, COLLOCATION[5] seems to be the most successful, but its equations cannot be solved by ITPACK [15, p. 303].

2   ELLPACK and ITPACK

For an authoritative description of the combined capabilities of ELLPACK and ITPACK, a package of iterative routines for solving large sparse linear systems, one should consult [15].[6] These packages make readily available, in debugged form, a large variety of algorithms, both for *discretizing* elliptic boundary value problems and for *solving* the resulting systems of linear algebraic equations.

Comparisons of the efficiency and accuracy of these algorithms are easiest to make and understand for special "model problems" defined by isotropic elliptic DE's with constant coefficients such as the Laplace, Poisson, and Helmholtz equations, in *rectangular domains*, and with Dirichlet-type boundary conditions.[7] For such problems, one can often derive local *orders of accuracy* such as $O(h)$, $O(h^2)$, $O(h^4)$, or $O(h^6)$, and asymptotic *solution costs* for solving $\mathbf{AU} = \mathbf{b}$, of the form $O(h^{-\gamma})$, for some "order of computational complexity" $\gamma > 0$, as the mesh length $h \downarrow 0$.

However, such model problems are atypical, because they can often be solved by special methods having low orders of computational complexity which are *only* applicable to elliptic DEs with constant coefficients on rectangular domains. For example, tensor product and FFT methods can be used to solve many of the resulting systems of equations, in a time proportional to the number N of unknowns or to $N \log N$. Likewise, multigrid methods are atypically easy to apply in rectangular domains. Therefore, although the study of such 'fast' methods is interesting and suggestive, we have concentrated our analysis on methods that are applicable

---

[5]See [15, pp. 146–9]; a more complete description is given in [8].

[6]For the applicability of ITPACK to the solution of *elliptic problems*, see [9] and [15, Ch. 7], which was written by Kincaid, Oppe, Respess, and Young. Earlier accounts of these two packages, also written by their principal developers, may be found in [19, pp. 135–86] and [3, pp. 3–22 and 53–63]. Other relevant packages are also discussed in [3, Part I], and in [5]. There (p. 235) only 5 POINT STAR is listed as usable in general plane domains.

[7]*Caution.* In [15, Ch. 9], the phrase "model problem" is used in a very different sense.

to elliptic DEs in *general domains* and for *general boundary conditions*. We have based this analysis on our 1984 book [2] which can be viewed both as a sequel to [6, Part 3] and as a second edition of [1]. In it, we tried to give a comprehensive survey of what was known about the numerical solution of elliptic problems as of that date, from an applied as well as from a theoretical standpoint. It concluded (in Chapter 9) with a description of "some capabilities of ELLPACK", following a chapter on "difference approximations", two on solving the resulting linear systems, and one on "finite element approximations." Referring to Table 6.1 on [15, p. 162],[8] we have concentrated our attention on the 5 POINT STAR discretization module, which typically has $O(h^2)$ accuracy [6], and the COLLOCATION module (with bicubic Hermite approximating functions), which typically gives nearly $O(h^4)$ accuracy [13], globally as well as locally (cf. Sec. 8).

For the *self-adjoint* linear source problem on *rectangles*, one version of 5 POINT STAR gives the "standard" 5-point approximation already alluded to. However, we felt unclear about the accuracy of its treatment of boundary conditions, especially since it gives a *non-symmetric* matrix of coefficients in most domains (cf. Sec. 6). We have therefore compared its effectiveness in the *unit disk* $x^2 + y^2 \leq 1$, for various values of $h = 1/n$ and for a few simple *boundary conditions*, with that of a new 5 POINT CUT modification of it which does give a symmetric coefficient matrix because it is based on the *network analogy* of [2, pp. 125, 292].

We have also compared the effectiveness of both approximations with those of COLLOCATION and a new HODIEG module developed by one of us, which will be described in the next paper. Both of these modules may be expected to have $O(h^4)$ accuracy in approximating *analytic* solutions if, as is usually done in analyzing the accuracy of discretization methods, *exact arithmetic* is assumed (cf. Sec. 8).

Most relevant for this conference, we compared the effectiveness of the *iterative* methods of ITPACK with standard *direct* methods for solving the resulting systems of linear equations, again as a function of the mesh length h—always restricting our attention to *square* meshes and the (elliptic) Laplace differential operator. These comparisons will be presented in Sec. 4 and Sec. 11 (cf. also Sec. 10).

3   Some Basic Questions

In preparation for this conference, we considered a large number of basic questions about the numerical solution of elliptic problems. We had become keenly aware of our inability to give clear and definitive answers to most of them while writing our recent book [2], but did not have adequate time to investigate them then. For example, [2, p. 57], we were surprised to find that the standard 5-point approximation does not give a *symmetric* matrix unless all mesh lines $x = x_i$ and $y = y_j$ intersect the boundary at mesh points (x_i, y_j). This makes it desirable to reexamine some of the implications of [21] and [23] and we will do this in Sec. 6.

---

[8]The "capacitance matrix" CMM modules, although listed in [15, Table 6.1], are not actually usable in ELLPACK. We have not yet tried P2CO TRIANGLES.

Utilizing **ELLPACK** and **ITPACK** where possible, and the discussions of them that have since appeared, especially [15], we will try to give at least partial answers to some of these basic questions in this report, namely:

A. For which elliptic problems are direct methods more efficient than iterative methods?

B. In general, *self-adjoint* linear elliptic problems are associated with *symmetric* operators, and so one would like their discretized form, $\mathbf{AU} = \mathbf{b}$, to involve a symmetric matrix $\mathbf{A}$. How practical is this?

C. For which elliptic problems does it pay to use 'Hodie', **COLLOCATION**, or other higher-order methods having more than $O(h^2)$ local accuracy in nonrectangular domains?

D. When using the **ELLPACK** two-dimensional *domain processor*, how should one choose the mesh lines?

E. For which elliptic problems are *finite element* methods preferable to *difference* methods?

F. How feasible is it to construct a useful analog of **ELLPACK** for solving three-dimensional (3D) problems?

G. How hard is it to treat *quasilinear* elliptic problems numerically?

H. The same question for *eigenproblems*.

I. How efficiently can *Frankel's method* [2, Ch. 4, Sec. 15] be adapted to *parallel* machines?

Any one of these questions would make a good Ph.D. thesis topic, and we hope that our partial answers to some of them will be found helpful by future research workers. Indeed, we think **ELLPACK** and **ITPACK** should be regarded as research "stimuli" as well as research "tools."

Finally, we mention two questions which will be discussed in Chapter 9 of these Proceedings.

J. How does the new **HODIEG** module of **ELLPACK** compare with previous discretization modules in efficiency and accuracy?

K. How should one discretize Neumann and mixed boundary conditions?

4   Direct vs. Iterative Methods

In their brilliant 1970 exposition of "The finite element method in solid mechanics" in [4, pp. 210–252], Felippa and Clough stated that "In the past years, direct solution methods have gradually replaced iterative techniques" for "solution of the stiffness equations" in applications of finite element methods to solid mechanics. Their Eqs. (7.2)–(7.3) seem to refer to the standard direct methods of solving equations of the form $\mathbf{A}\mathbf{U} = \mathbf{b}$: *Gauss band-elimination* in general, and *band-Cholesky* elimination when $\mathbf{A}$ is symmetric.

A brief theoretical analysis comparing the efficiency of band-elimination with SOR (on sequential computers) was given in [2, Ch. 4, Sec. 1]. In this section, we will use the preceding data, together with the data reported in [15, Ch. 11], to sharpen up the loose statement in [2, p. 98], that "in the range $300 < N < 3000$, the costs of good direct and good iterative methods usually differ by a factor of less than two" for "sparse matrices arising from linear source problems."

For *two-dimensional* problems as $h = 1/n \downarrow 0$, if the unknowns and equations are *naturally ordered* [2, p. 64], the time required to solve $\mathbf{A}\mathbf{U} = \mathbf{b}$ by band elimination is asymptotically

$$(4.1) \qquad\qquad t = C_1 N^2 + o(N^2) = C_1' n^4 + o(n^4).$$

For *iterative* schemes, if m is the number of iterations required to reduce the relative error $\epsilon = ||\mathbf{e}||/||u||$ to a prescribed allowable amount, it is asymptotically

$$(4.2) \qquad t = C_2 m N + o(mN), \qquad m = m(\epsilon) \sim m_0 + m_1 \log \epsilon.$$

For SOR with nearly "optimum ω", as h tends to zero for any fixed boundary value problem, $m(\epsilon)$ is asymptotically proportional to $\sqrt{\kappa(\mathbf{A})}$, where $\kappa(\mathbf{A})$ is the condition number of $\mathbf{A}$ [2, p. 139, (13.13)]. Hence it is proportional to $n = 1/h$, as in [2, p. 131], so that

$$(4.3) \qquad\qquad\qquad t = C_3 n^3 + o(n^4).$$

It follows that there will be a "crossover point" in $n \propto \sqrt{N}$, above which SOR is more efficient. Consequently, SOR should be more efficient for *sufficiently large* matrices. An early *quantitative* estimate of the 'crossover point' above which SOR becomes superior to band elimination for some simple finite element approximations was made in a 1971 paper by Fix and Larsen.[9]

Expanding on [23], Table 2-4.1 of Hageman-Young's *Applied Iterative Methods* [7, p. 23] lists schemes for accelerating five "basic" iterative methods: Jacobi, (first-order) Richardson, Gauss-Seidel, SOR, and SSOR. In the current decade there have been many further improvements in iterative techniques, with "preconditioned" conjugate gradient schemes [2, Ch. 5, Sec. 6], incomplete factorizations, variants of Lanczos's method, and so on. The new NSPCG system of The University of Texas

---

[9]G. J. Fix and K. Larsen, SIAM J. Numer. Anal. 8 (1971), 536–47.

at Austin's Center for Numerical Analysis[10] contains 19 preconditioners and 21 accelerators; various combinations of these are applicable in up to 7 different data storage modes, giving over 2000 combinations to choose from. Most good methods had nearly the same order of complexity (see Sec. 10).

Since [2] went to press, many more thoroughly documented quantitative results about the performance of ELLPACK "solution modules" have also been published in [15, Ch. 11], and the references listed there. These concern solving equations given by the 5 POINT STAR and the 7 POINT 3D methods, the COLLOCATION method, the finite element SPLINE GALERKIN method, and the HODIE HELMHOLTZ methods.

The comparisons in [15, Ch. 11] ignore the older and less efficient point-Jacobi and Gauss-Seidel methods. However, since the first step of JACOBI SI is a classical Jacobi step, one can set the number of iterations to be unity, and then call JACOBI SI repeatedly to get point-Jacobi iteration. Also, one can set the overrelaxation parameter $\omega = 1$ in SOR to get Gauss-Seidel and use nonadaptive procedures.

The experiments reported there (in Figs. 11.2 and 11.4) were for typical two-dimensional source-diffusion problems. They indicate that the "crossover point" above which good iterative methods (like SOR) become more efficient than Gauss band-elimination (say) is in the range of $N = 250$–500 unknowns, for the 5 POINT STAR approximation. Unfortunately, [15, p. 303]: "Normal iterative methods diverge for either way of writing the collocation equations" associated with the more accurate COLLOCATION discretization module of ELLPACK. This is because the 'finite element' ordering used by COLLOCATION gives rise to zeros on the main diagonal. The reordering produced by COLLORDER avoids this difficulty, but point iterative techniques applied to the resulting linear system fail to converge.

For three-dimensional problems, the crossover point comes for smaller N; see Sec. 11. On the other hand, several ingenious new direct methods were developed, as substitutes for Gauss and Cholesky band elimination by James Bunch, Alan George, Donald Rose, and others. These were briefly reviewed in [2, Ch. 4, Sec. 2–3],[11] and we have nothing to add to that discussion here.

5 Different Elliptic Problems

The ideas and numerical experiments with ELLPACK and ITPACK briefly reviewed above mostly concern second-order *source-diffusion* problems. In the self-adjoint case, these are naturally associated with *Stieltjes matrices* by the network analogy of Sec. 6. However, many different kinds of physical equilibrium problems also have elliptic boundary value problems as their most natural mathematical formulation. Two such classes of problems naturally come to mind; we digress to comment on them.

Solid mechanics. The choice of the method used to solve elliptic problems may well be guided by *physical* considerations. Thus, many problems in solid mechanics

---

[10]T. C. Oppe, W. D. Joubert, D. R. Kincaid, "NSPCG User's Guide, Version 1.0: A package for solving large sparse linear systems by various iterative methods", Report CNA-216.

[11]See also the articles by James Bunch and by Alan George and one of us in [20, pp. 197–220].

involve *fourth*-order DE's to which formulas (4.1)–(4.3) are inapplicable. Among these, loaded *clamped plate* problems defined by

$$(5.1) \qquad\qquad \nabla^4 u = f(x, y) \qquad \text{in} \quad \Omega,$$

with the boundary conditions

$$(5.2) \qquad\qquad u = \partial u / \partial n = 0 \qquad \text{on} \quad \Gamma,$$

are the simplest. For such problems, the condition number $\kappa(\mathbf{A})$ grows like n^4 and not n^2 (see [2, p. 132]), and so $m(\epsilon)$ is proportional to n^2 for SOR, and not to n. This fact, although not mentioned by Felippa and Clough, makes SOR and similar iterative methods much less attractive, and finite element methods more attractive. Because ELLPACK is currently restricted to second-order elliptic DEs,[12] we did not make any quantitative comparisons of solution times for elasticity problems.

Convection-diffusion problems. Formulas (4.1)–(4.3) are also inapplicable to convection-diffusion problems, such as the model problem discussed in [2, p. 111]. A more general (and more typical) class of problems is defined by the DE

$$(5.3) \qquad\qquad u_z = \epsilon \nabla^2 u + f(\mathbf{x}),$$

in a cylinder $\Omega = [0, \ell] \times \Omega_1$ where Ω_1 is bounded in the (x, y)-plane by a convex curve Γ_1. Here $\epsilon = \nu/U$ is the ratio of the diffusivity to the flow velocity, and it is plausible physically to assume inflow and outflow values

$$(5.4) \qquad u(x, y, 0) = g_0(x, y), \qquad u(x, y, \ell) = g_1(x, y),$$

and wall values

$$(5.5) \qquad u(x, y, z) = g_2(x, y, z) \qquad \text{for} \quad (x, y) \in \Gamma_1, \quad 0 < z < \ell,$$

if u signifies temperature, or

$$(5.6) \qquad \frac{\partial u}{\partial n}(x, y, z) = 0 \qquad \text{for} \quad (x, y) \in \Gamma_1, \quad 0 < z < \ell,$$

if $u = \rho$ denotes the concentration of some solute.

Finally, we note a connection between discretizations of such convection-diffusion problems and diagonally dominant *M-matrices*: nonsingular matrices $\mathbf{A}$ having positive diagonal entries $a_{j,j} > 0$, negative or zero off-diagonal entries, and strictly positive inverses. For example, observe the three difference approximations of [2, p. 71] to $U\rho_x = \nu \rho_{xx} + f(x)$: 'central', 'Allen's', and 'upwind'. In all cases, within the range of usefulness, $\mathbf{A}$ (and hence $-\mathbf{A}$) is *diagonally dominant*. Moreover, $-\mathbf{A}$ is an *essentially positive* matrix [21, p. 257], and correspondingly $\mathbf{A}$ is an *M-matrix* [21, p. 85].[13] We will rationalize this observation in Sec. 7.

---

[12] Atypically, the problem of a loaded simply supported plate is reducible to a system of two second-order DEs, and so can be treated using ELLPACK [2, p. 301].

[13] These two results are closely related, like Theorems 2.1 and 8.2 of [21].

6 Symmetry?

We now turn back to the second problem mentioned in Sec. 3: constructing discretizations $\mathbf{AU} = \mathbf{b}$ of linear self-adjoint boundary value problems which yield a symmetric $\mathbf{A}$. It is classic that linear source problems (1.1), associated with many of the most important boundary value problems arising in mathematical physics, have symmetric Green's functions $G(\mathbf{x}; \boldsymbol{\xi}) = G(\boldsymbol{\xi}; \mathbf{x})$. For given homogeneous linear boundary conditions, the discretizations $G_h(x_i, y_j; \xi_k, \eta_\ell) = G_h(\xi_k, \eta_\ell; x_i, y_j)$ should therefore approximate the corresponding matrices $\mathbf{A}^{-1}$; cf. [6, Sec. 23.6] and [2, p. 266].

Likewise, the *variational* considerations underlying most *finite element* methods typically minimize a quadratic functional

$$\mathbf{J}[u] = \langle \mathbf{L}[u] - G(\mathbf{x}), u \rangle = \langle u, \mathbf{L}[u] - G(\mathbf{x}) \rangle$$

associated with the inner product $\langle u, v \rangle = \int\int uv \, dx \, dy$. For example, for any *homogeneous* Dirichlet or Neumann boundary conditions, solutions of the self-adjoint DE (1.1) minimize the integral

$$(6.1) \qquad \mathbf{J}[u] = \int [A(x,y)u_x{}^2 + C(x,y)u_y{}^2 + F(x,y)u^2 - 2G(x,y)u] \, dx \, dy;$$

cf. [6, Sec. 20.5] and [2, Ch. 6, Sec. 1]. Any discretization of the quadratic functional $\mathbf{J}$ of will replace the corresponding boundary value problem by an algebraic problem of minimizing a quadratic function

$$(6.2) \qquad \sum a_{k\ell} U_k U_\ell - 2 \sum b_k U_k,$$

where the matrix $\mathbf{A} - (\|a_{k\ell}\|)$ is positive definite and symmetric. The minimum in U-space is located where $\mathbf{AU} = \mathbf{b}$.

From an aesthetic standpoint, one would surely like to have discretization preserve the natural symmetry of the exact problem. Yet very few of the discretizations now in ELLPACK yield symmetric matrices in general domains. The single exception is the 5 POINT STAR approximation to the Poisson equation, with Dirichlet boundary conditions, in polygons[14] whose sides are all vertical, horizontal, or have a 45° slope (cf. [6, p. 358, Fig. 25.1]), but not the general self-adjoint equation (1.1) for general boundaries or general boundary conditions. Moreover, as is cautioned in [2, p. 57], the standard 5 point approximation to (1.1) does not make $\mathbf{A}$ symmetric in general. How to make $\mathbf{A}$ symmetric is a second question that we have explored in various contexts. Our exploration would have been impossibly time-consuming without the help of carefully designed and debugged packages of programs like ELLPACK and ITPACK.

5 POINT CUT. One way to construct symmetric discretizations of linear source problems is to utilize a network analogy such as that proposed by MacNeal in 1953;

---

[14]Exceptions can be constructed, similar to those indicated by one of us on p. 203 of [18].

cf. [12], [6, Sec. 20.5], [21, p. 191], and [2, p. 125]. The basic idea is to replace *mesh segments* joining adjacent mesh points by conducting wires having suitable *conductances*. For example, in a square mesh, one can simply suppose all wires to have the same resistance per unit length, and to terminate where they meet the boundary; i.e., one can approximate $\nabla^2 u = 0$ with

$$(6.3) \quad \frac{U_{i,j} - U_{i-1,j}}{x_i - x_{i-1}} + \frac{U_{i,j} - U_{i+1,j}}{-x_i + x_{i+1}} + \frac{U_{i,j} - U_{i,j-1}}{y_j - y_{j-1}} + \frac{U_{i,j} - U_{i,j+1}}{-y_j + y_{j+1}} = 0.$$

This may be expected to give $O(h^2)$ accuracy (see Sec. 8).

We have made a very limited study of a new 5 POINT CUT module based on this analogy, primarily to compare its accuracy and efficiency with that of 5 POINT STAR, in the case of a disk overlaid with square meshes. As we expected, the 'solution time' required to solve the resulting vector equation $\mathbf{AU} = \mathbf{b}$ was cut nearly in half; but the resulting error was more than doubled in this example. Table 3.1 gives results[15] for the Laplace equation, with Dirichlet conditions on the unit circle chosen so that the exact solution is $e^x(\cos y + \sin y)/3.28$ whose maximum is unity. ELLPACK's 5 POINT STAR generates a nonsymmetric matrix whereas 5 POINT CUT yields a symmetric positive definite matrix. LINPACK BAND, a general band solver, was used with 5 POINT STAR and LINPACK SPD BAND, which takes advantage of the symmetry, was used with 5 POINT CUT; this accounts for the difference in the 'solve' time.

7   Extended Network Analogy

In this section, we will extend the network analogy of MacNeal and others from *source-diffusion* problems with leakage to *convection-diffusion* problems with sources and leakage. Conceptually, each index i, j, ... , can be imagined as a label for a tank or *cell* filled with fluid. Steady (forced) *convection* is one-way; hence it is natural to associate it with a special kind of *directed graph*, in which there cannot be both an arc from i to j and an arc from j to i. Such a graph may be called *unidirected*. The nonzero entries a_{ij} of any nonnegative matrix $\mathbf{A} = \|a_{ij}\|$ such that

$$(7.1) \qquad\qquad a_{ij}a_{ji} = 0 \qquad \text{for all } i, j$$

can be imagined as specifying the rates of mass-flow from cell i to cell j. Kirchhoff's conservation law,

$$(7.2) \qquad\qquad \sum_j a_{ij} = 0, \qquad \text{or} \qquad a_{ii} = -\sum_{j \neq i} a_{ij}$$

---

[15]These experimental results, as well as the others discussed in this paper, were obtained at Purdue with a Sequent Balance 21000 computer, using a single National Semiconductor 32032 processor with double precision arithmetic (accuracy about 1 part in 10^{15}). Times are in seconds and 'storage' is the approximate number of words required by a module.

5 POINT STAR						
$2/h$		Max		Times		Max
n	N	Band	Dis	Solve	Total	Error
4	9	3	0.33	0.01	0.34	1.03E–03
8	45	7	0.43	0.17	0.60	7.12E–04
16	193	15	0.73	2.05	2.78	2.00E–04
32	793	31	1.87	26.93	28.80	5.22E–05
5 POINT CUT						
4	9	3	0.28	0.02	0.30	2.50E–03
8	45	7	0.42	0.13	0.55	1.22E–03
16	193	15	0.62	1.40	2.02	8.72E–04
32	793	31	1.50	16.40	17.90	3.15E–04

Table 3.1

then yields an *essentially positive* matrix $\mathbf{Q}$ [21, Sec. 8.2].

Alternatively, it is suggestive to represent such a purely convective flow by a *skew-symmetric* matrix, with $\tilde{a}_{ij} + \tilde{a}_{ji} = 0$ for all i, j. The associated nonnegative matrix $\mathbf{A}$, in which all negative entries are replaced with 0, will then have a unidirected graph, but (7.2) will fail. (We recall the convenient vector lattice notation, in which $\mathbf{A} \vee \mathbf{B}$ (resp. $\mathbf{A} \wedge \mathbf{B}$) denotes the matrix $\mathbf{C}$ in which c_{ij} is the greater (resp. lesser) of a_{ij} and b_{ij}. In this notation, $\mathbf{A} = \tilde{\mathbf{A}} \vee \mathbf{0}$, so that $\mathbf{A} \wedge \mathbf{A}^T = \mathbf{D}$ is diagonal if (7.1) holds.)

A *cycle-free* unidirected graph naturally corresponds to *vortex-free* convection. This is the case that the transitive closure of the binary relation iRj (signifying that "there is an arc from i to j") is a *partial ordering*. Any strengthening of this partial ordering to a total ordering of the indices yields from $\mathbf{A}$ a strictly triangular, nonnegative matrix $\mathbf{T}$. Hence any matrix $\mathbf{M}$ which properly discretizes *vortex-free* convection with diffusion can be written in the form $\mathbf{M} = \mathbf{P}^{-1}(\mathbf{B} - \mathbf{T})\mathbf{P}$, where $\mathbf{B}$ is a Stieltjes matrix and $\mathbf{T}$ is nonnegative and *strictly upper triangular*.

Unlike convection, *diffusion* is two-way; hence it is naturally associated with a *symmetric* matrix (having a symmetric graph). If $\mathbf{B} = \|b_{ij}\|$ is such a diffusion matrix, we can imagine $\mathbf{C} = \mathbf{A} + \mathbf{B}$ as a *convection-diffusion* matrix, by interpreting u_i as the *concentration* of some chemical. Then the DE

(7.3) $$u'_i(t) = \sum c_{ij}(u_j - u_i)$$

is the natural expression for the rate-of-change of u_i. *Equilibrium*, with $\mathbf{u}'(t) = \mathbf{0}$ occurs precisely when

(7.4) $$-a_{ii}u_i = \sum_{j \neq i}(a_{ij} + b_{ij})u_j, \qquad -a_{ii} = \sum_{j \neq i} a_{ij}.$$

Leakage and *absorption* replace a_{ii} in (7.4) with some (real) number a'_{ii} of larger magnitude, while a *source* term adds $f_i(t)$ to the right side of (7.3). By making $f_i(t) = f_i$ time-independent, we get (finally) an equation of the form

(7.5) $$\mathbf{Mu} = \mathbf{f}, \qquad \mathbf{M} = \mathbf{A}' + \mathbf{B},$$

where $\mathbf{M}$ is *diagonally dominant* with positive diagonal and negative off-diagonal entries. If (the graph of) $\mathbf{M}$ is *irreducible*, and there is any leakage or absorption, it is an irreducibly diagonally dominant *M-matrix*; cf. [21, p. 23]

Conversely, let $\mathbf{M}$ be any diagonally dominant *M-matrix*. Setting also the diagonal entries of $\mathbf{M}$ equal to zero, we can form $\mathbf{M} \wedge \mathbf{M}^T = \mathbf{B}$, to get a nonnegative *diffusion* matrix. The off-diagonal entries of the difference $\mathbf{A} = \mathbf{M} - \mathbf{B}$ will then satisfy (7.1), and because $\mathbf{M}$ is diagonally dominant each diagonal entry $m_{ii} = a_{ii}$ will be at least the sum of the nonnegative off-diagonal entries in its *row*. By setting

(7.6) $$d_{ii} = m_{ii} - \sum_{j \neq i} a_{ij},$$

we get a diagonal *leakage* matrix, which will be nonzero if $\mathbf{M}$ is nonsingular.

As in [21, Ch. 8], the DE (7.3) can be viewed as a *semi-discretization* of a parabolic convection-diffusion equation with leakage, and such equations arise naturally in the multigroup diffusion equations of reactor physics. For the mathematical theory of *M*-matrices, see also A. Berman and R. J. Plemmons, *Nonnegative Matrices in the Mathematical Sciences*, Academic Press, 1979, Chs. 6–7. In particular, Theorem 2.3 in their Ch. 6 gives 50 necessary and sufficient conditions for a real $n \times n$ matrix with nonpositive off-diagonal entries to be an *M*-matrix.

8   Orders of Accuracy

A discretization method is said to have *order of accuracy* ν (in some norm in exact arithmetic) when the global *discretization error* $\mathbf{e} = \mathbf{U} - \mathbf{u}$ has (asymptotically) the magnitude $\|\mathbf{e}\| \sim Kh^\nu$, as $h \downarrow 0$. Let $\mathbf{AU} = \mathbf{b}$ signify the *family* of large, banded linear systems associated with a linear elliptic difference approximation on a two-dimensional domain in the natural ordering, so that (if Gauss band elimination is used) the solution time $t = O(n^4)$ as in (4.1). Then, if the discretization method has ν-th order accuracy, to reduce the magnitude of the error to ϵ or less requires a solution time of $O(\epsilon^{4/\nu})$. It follows that, for two discretization modules having different orders of accuracy, the one having higher order accuracy should always be preferable beyond some "crossover point", as $\epsilon \downarrow 0$.

For problems having analytic solutions in rectangular domains with Dirichlet boundary conditions, the ν just defined seems to be generally equal to the order of the local *truncation error*[16] defined as the smallest exponent of a nonzero term in the Taylor series of the difference $\mathbf{L}_h[u] - \mathbf{L}[u]$ between the difference operator and the differential operator when applied locally to an analytic function. Moreover, for the standard 5 point approximation to the Poisson equation on a square mesh, Bramble and Hubbard proved by ingenious methods 25 years ago [2, p. 86, ftnt. 41] that $O(h^2)$ accuracy would be achieved asymptotically in the interior in general plane domains (see [2, p. 86] and Math. Revs. 26, #7158). No equally general theoretical results have been proved for 9 point formulas having $O(h^4)$ local truncation errors or for COLLOCATION.

It is interesting to compare a *priori* ideas about orders of accuracy (in exact arithmetic) with *empirical* results. We have done this for several discretizations, making the usual assumption [2, p. 108], that the discretization error ϵ is related to the mesh length h by

$$(8.1) \qquad\qquad \epsilon = ||\mathbf{e}|| = C_4 h^q + o(h^q),$$

for some 'order of accuracy' q. (The stronger assumption $\mathbf{e}(\mathbf{x}) = h^q E(\mathbf{x}) + O(h^{q+1})$ provides the basis for Richardson extrapolation; cf. [2, pp. 76, 89]). As in [15, Ch. 10] we have made best straight line fits to experimental data relating $\log \epsilon$ to $\log h$—whose slope ν may be interpreted as the empirical "order of accuracy" of a given discretization scheme.

Among elliptic partial differential operators, the *Laplace operator* of potential theory is by far the most important. Moreover, its 9-point difference approximation on a square mesh,

$$(8.2) \qquad\qquad \mathbf{N}_h[U] = \frac{1}{6h^2} \left[-20\,U_0 + 4\sum_{j=1}^{4} U_j + \sum_{j=5}^{8} U_j \right],$$

has been very thoroughly studied. Thus, the difference equation $\mathbf{N}_h[U] = 0$ is well-known to approximate the Laplace DE $\nabla^2 u = 0$ with $O(h^6)$ local truncation error [2, Ch. 3, Sec. 10].

Dirichlet boundary conditions. We first considered Dirichlet boundary conditions, and tested the *global* orders of accuracy for the max norm of the 5 POINT STAR, COLLOCATION, and HODIEG methods for the (analytic) TRUE solution $e^x \sin y$ of the Laplace equation in four simple domains: the unit square $[0, 1]^2$, the (inscribed) disk with diameter 1, and the ellipses having horizontal and vertical axes 1,0.5 and 0.5,1 with centers (0.5,0.25) and (0.25,0.5), respectively. The empirical orders of accuracy were as listed in Table 3.2; cf. also Fig. 3.1. HODIEG was decidedly the most accurate; moreover, it required much less computing time than COLLOCATION.

---

[16]The entry "Truncation error (see Discretization error)" in the index of [6, p. 443] is easily misinterpreted. Contrast with [2, p. 61].

	Square	Disk	2:1 Ellipse	1:2 Ellipse
5 POINT STAR	1.96	2.12	2.77	2.59
COLLOCATION	3.85	3.65	3.75	3.78
HODIEG	5.99	4.49	4.32	4.60

Table 3.2 Empirical orders of accuracy, Dirichlet boundary conditions.

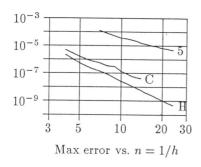
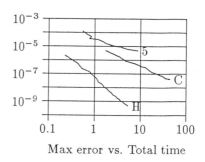

Max error vs. $n = 1/h$ Max error vs. Total time

Figure 3.1 Dirichlet conditions, 1:2 ellipse.

	Square	Triangle	Disk	1:2 Ellipse
5 POINT STAR	1.96	0.99	1.99	2.73
COLLOCATION	3.96	3.90	3.14	3.33
HODIEG	3.54	2.79	2.59	2.97

Table 3.3 Empirical orders of accuracy, mixed conditions $u + u_n = g$.

In the case of the square, the agreement with expected orders of accuracy was satisfactory. However, for the other domains, there were substantial deviations from the expected $O(h^2)$ (for 5 POINT STAR) and $O(h^4)$ (for COLLOCATION and HODIEG) accuracies. Recent empirical studies [17] and [13] by Rice and Mo Mu, of discretization errors for the Poisson DE and Dirichlet boundary conditions with the solution $3e^x(x - x^2)e^y(y - y^2)$ in 20 domains gave similar results. For 5 POINT STAR, about 72% of the observed convergence rates lay in the interval [1.5,2.5], while for COLLOCATION about 43% lay in the interval [3.5,4.5].

We also tried out a new 9 POINT CUT discretization based on the network analogy, in which (to obtain the discretization (8.2) on a square mesh), the resistance per unit length of all diagonal wires was assumed to be $2\sqrt{2}$ times that of horizontal and vertical wires. In spite of its local $O(h^6)$ accuracy in the interior, its error was about the same as for 5 POINT CUT in a unit disk.

Mixed boundary conditions. The advantage of HODIEG was much less decisive for the mixed boundary conditions $u_n + u = g$ and solution $e^x \sin x$. The empirical orders of accuracy were as listed in Table 3.3; cf. also Fig. 3.2. In all cases, the 5 POINT STAR discretization module was much less efficient than the COLLOCATION and HODIEG approximations.

It is surprising that on the isosceles right triangle, 5 POINT STAR is $O(h)$; on the

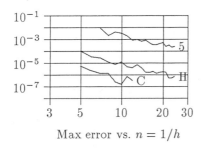

Max error vs. $n = 1/h$

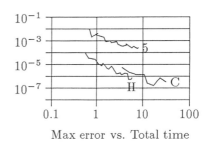

Max error vs. Total time

Figure 3.2 Mixed conditions, $u + u_n = g$, 1:2 ellipse.

square and disk it is nearly $O(h^2)$. As in the case of Dirichlet conditions, HODIEG is more accurate on the square and on domains with curved boundaries.

The presence of *irregular stencils* lowers the local orders of accuracy of HODIEG and COLLOCATION along curved boundaries, and where the boundary conditions involve normal derivatives. This complication is avoided in many "model problems" defined by Dirichlet boundary conditions on rectangular domains with side parallel to the axes [2, pp. 5, 23, 74, 130], but even in this case the order of accuracy depends on the smooth variation of the boundary values, as is shown by the following example due to Hadamard.

Hadamard's example. Because the order of accuracy of the discretization (8.2) depends on the smoothness of u, we thought it interesting to apply it to Hadamard's example, using ELLPACK's discretization module COLLOCATION. In polar coordinates, Hadamard's example concerns

$$(8.3) \qquad\qquad \nabla^2 u = 0 \qquad \text{for} \quad r < 1,$$

with the boundary condition $u(1, \theta)$ equal to 1 if $0 < \theta < \pi$, to -1 if $-\pi < \theta < 0$, and to 0 for $\theta = 0$ or $\theta = \pi$. As we expected, the maximum error did not decrease with h at all. The same was true in the rectangle $[-1, 1] \times [0, 1]$, with $u = [\arctan(y/x)]/\pi$ on the boundary, and $1/2$ at $(0,0)$. In contrast, the discrete ℓ_1 and ℓ_2 norms of the error appeared to be about $O(h^{1.5})$ and $O(h^{0.8})$, respectively.

The symmetry of the problem in the lines $y = \pm x$, suggested discretizing it on short *slits* emanating from the two points of discontinuity. Thus, in the second example, we set $U(0^+, 0) = 1/4$ and $U(0^-, 0) = 3/4$. We found that this change reduced the maximum error from about $1/70$ to about $1/140$. Likewise, in Hadamard's example of the disk, we achieved a similar reduction in the maximum error, by setting $U(\pm 1, 0^+) = 1/2$ and $U(\pm 1, 0^-) = -1/2$, all in rectangular coordinates.

Alternatively, for any classical Dirichlet problem where the boundary values have a jump of $\Delta u = C$ at a point P, one can take as unknown the function $v = u - C\theta/\pi$, where θ is the counterclockwise angle made with the horizontal by the tangent to Γ at P.

9   Choice of Mesh

For a given elliptic problem, the choice of mesh lines $x = x_i$ and $y = y_j$ is user-specified in the current ELLPACK DOMAIN PROCESSOR, with a square mesh of side h as an option; cf. [16] and [15, pp. 146–49]. It leaves untouched the interesting problems of *optimizing* (a) the choice of coordinate axes if Ω is to be overlaid with the *square* mesh $x = ih$, $y = jh$, and (b) the choice of mesh lines $x = x_i$ and $y = y_j$ for a given *variable* rectangular mesh. We have made some preliminary experiments using HODIEG and 5 POINT STAR concerned with these two problems of *mesh optimization*.

Square diamond. The effect of changing axes is nicely illustrated by taking for Ω the *square diamond* (alias "tilted square") with vertices at $(\pm 1, 0)$ and $(0, \pm 1)$,

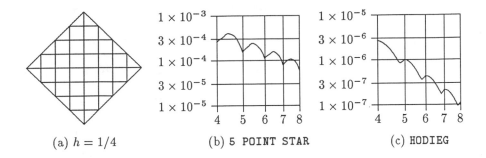

(a) $h = 1/4$ (b) 5 POINT STAR   (c) HODIEG

Figure 3.3 (a) Square diamond; (b) and (c) max error vs. $1/h$.

overlaid with a *square mesh* with mesh length h and grid points $(x_i, y_j) = (ih, jh)$. The *interior* mesh points are those for which $(|i| + |j|)h < 1$; the configuration of the *boundary* points depends on the choice of the parameter h; four cases may be distinguished.

A. When $h = 1/n$, there are $2n^2 - 2n + 1$ "interior" mesh points, and $4n$ "boundary" mesh points located at the mesh points $(x_{\pm i}, y_{\pm(n-i)})$. In addition, the diagonals issuing from interior points $(x_{\pm k}, y_{\pm(n-k-1)})$ next to the boundary meet the boundary at the $4n$ *midpoints* of the segments joining adjacent boundary mesh points, as indicated in Fig. 3.3(a).

B. When $h = 2/(2n+1)$, the grid lines intersect the boundary segment $x + y = 1$ at the *midpoints* $(ih + h/2, 1 - ih - h/2)$ of grid segments crossing the boundary. There are then, *two* kinds of interior points adjacent to the boundary: those with $|i| + |j| = n$ and those with $|i| + |j| = n - 1$. Those of the first kind are separated from the boundary by an isosceles right *triangle*, while those of the second kind are separated from it by a *pentagon* with one oblique side.

C. Now let h *increase* continuously from $1/n$ to $2/(2n - 1)$. For $h = (1/n) + \epsilon$ with small ϵ the boundary points will occur in clumps of three, just *inside* the mesh points (x_i, y_j) with $|i| + |j| = n$, separated by single boundary points at the midpoints of the diagonal sides of the *boundary* triangles.

D. Similar questions arise when h is allowed to *decrease* continuously from $1/n$ to $2/(2n + 1)$. Figure 3.3(b) shows the behavior of the error in the 5 POINT STAR solution $e^x \sin x$ of the Dirichlet Problem on the square diamond as h changes from $1/4$ to $1/8$. Stencils having some arms very much smaller than others did not seem to produce substantially greater error than those with equal arms ($h = 1/4, 1/5, 1/6, 1/7, 1/8$). In similar experiments by Rice[17] using COLLOCATION, very large mesh 'aspect ratios' did not seriously affect its error.

---

[17]J. R. Rice, "Is the aspect ratio significant for finite element problems?," Purdue University, Department of Computer Science, Report CSD-TR-535, Sept. 1985.

10   Computational Complexity

Basically, the 'computational complexity' of a method for solving an elliptic problem refers to the growth in the *cost* \$ of computing a solution as a function \$ $= f(\epsilon)$ of the *relative error* ϵ in the computed solution. This is traditionally approximated for suitable $K > 0$ and r as in [2, Ch. 4, Sec. 4], by an *asymptotic* formula of the form

$$(10.1) \qquad\qquad \$ \doteq K\epsilon^{-r}.$$

The exponent r in (10.1) is called the *order* of the computational complexity of the method. Some idea of the *complexity* of the concept of 'computational complexity' is given by the discussion of the preceding sections. The remarks made below are intended to amplify these discussions.

For example, if the time required to solve the resulting system of (linear) algebraic equations is $t \sim CN^\alpha$, then since $N = O(h^{-d})$ in d dimensions, to make $||\mathbf{e}|| < \epsilon$ requires

$$(10.2) \qquad\qquad t \sim C(K')^\alpha h^{-d\alpha} \sim K^* \epsilon^{-d\alpha/\nu}$$

This suggests that the *order of complexity* r in (10.1) is $d\alpha/\nu$.

In [2, Ch. 4], our dimensional analysis was based on the conventions of Eisenstat and Schultz presented in [20, pp. 271–83]. There only self-adjoint DEs of the form (1.1) in rectangular domains were treated, but their conventions are applicable generally. These conventions (summarized in [2, Ch. 4, Sec. 4]) suggest that the 'cost', \$, and relative error ϵ are related by (10.1), the dimension of the domain, and the order of accuracy of the discretization. In particular, the execution time t, which is roughly proportional to the cost of solution on a given computer, should be related to the mesh length h by a formula of the form $\log t \sim \log C - \gamma \log h$. Numerical *experiments* with variable h should therefore correspondingly be fitted well by straight lines in log-log plots of t against the error $\epsilon \sim Kh^\nu$. This is typically observed in practice (cf. [15, Ch. 11]), and was also the case in our experiments.

We computed approximate numerical solutions of the classical Dirichlet problem having $u = e^x \sin y$ as solution, in the unit disk. Table 3.4 lists coefficients of least square fits to experimental data. The ELLPACK module BAND GE was used to solve the linear systems which were generated by 5 POINT STAR, COLLOCATION, and the new module HODIEG. We do not know why the order of growth of the solve time seems to be significantly less than the expected $O(h^4)$.

Actually, the "cost" of solving a problem is a highly ambiguous concept, because of the many variables (including the computer) which influence it (cf. [15, pp. 418–23]) and the many steps which contribute to it. The programming time can be costly; for example it took more than one day to modify 5 POINT STAR to obtain a debugged 5 POINT CUT. As another example, because of 'fill-in' during the elimination process, elimination requires much more *storage* than SOR for approximations to elliptic problems. Moreover in any case, the storage required

	5 POINT STAR		COLLOCATION		HODIEG	
	c	ν	c	ν	c	ν
$\epsilon \approx c\,h^\nu$	1.2E-2	2.12	9.7E-4	3.65	1.4E-3	4.49
$\epsilon \approx c\,t^\nu$	1.9E-4	-0.80	3.2E-5	-1.46	3.3E-7	-2.28
$t_D \approx c\,h^\nu$	5.4E-2	-1.16	9.6E-2	-1.41	8.6E-2	-1.26
$t_S \approx c\,h^\nu$	7.8E-4	-3.28	6.4E-2	-2.64	2.1E-4	-3.42
$t \approx c\,h^\nu$	7.5E-3	-2.54	9.7E-2	-2.50	2.7E-2	-1.95

Table 3.4

increases indefinitely as $h \downarrow 0$, like h^{-2} or h^{-3} in two- and in three-dimensional problems.

If we neglect such variations in *storage* requirement, it seems reasonable to set $\$ \doteq Ct$, where t is the *time* required to compute an approximate solution and C is a constant depending on the computer used. In turn, t can be taken as the sum $t = t_D + t_S$ of the *discretization* time t_D and the *solution* time t_S. Since t_D is ordinarily proportional to the number N of unknowns, whereas t_S is proportional to a higher power of N, this suggests estimating $\$$ by a formula of the form

$$(10.3) \qquad \$ \doteq C_D N + C_S N^\gamma, \qquad \text{where} \qquad \gamma > 1,$$

and $N \propto h^{-d}$ in d-dimensional problems. Since $\gamma > 1$, there should be a 'crossover point' N_C above which the $t_S = C_S N^\gamma$ term becomes dominant. Thus, for their experiments, Eisenstat and Schultz derive the approximate formula [20, p. 277]

$$(10.4) \qquad t \doteq K(1600n^2 + 9n^4) = K(1600N + 9N^2),$$

so that 'crossover' occurs around $N = 177$, when $n = 13.3$ and $h = 0.075$. (One should really add a program-writing, debugging, and set-up cost C_0 to the right side of (10.3).)

Finally, it should be realized that ν may depend on the *norm* in which the total error $\epsilon = \|e\|$ is measured. In such cases, one should write ν_∞, ν_1, and ν_2 to distinguish errors in the ℓ_∞ (max), ℓ_1, and ℓ_2 norms, respectively. Moreover, there are two different sources of error: roundoff errors and discretization errors. Although the first are usually negligible, the roundoff errors will ultimately dominate as $h \downarrow 0$ in any fixed computer arithmetic. Furthermore, the dominant discretization errors may stem from approximate *boundary conditions* or from *singularities*. These can result in changing one or all of the ν_i.

11   3D Problems

Generally speaking, existing methods and machines seem to be adequate for solving most really important *two*-dimensional problems. It is primarily to solve *three*-dimensional problems in general domains that the supercomputers and parallel computers of tomorrow will be needed. In three dimensions, the asymptotic formulas for solution times corresponding to (4.1) and (4.2) are

$$(11.1) \qquad t = C_5 n^7 + o(n^7) \qquad \text{for band elimination,}$$

and

$$(11.2) \qquad t = C_6 n^4 + o(n^4) \qquad \text{for SOR, nearly optimal } \omega.$$

These formulas strongly suggest that iterative methods will be more efficient than direct methods for solving large three-dimensional problems; our experiments with ELLPACK and ITPACK support this conclusion.

Iterative techniques. We used least squares to obtain log-log fits to experimental data from ELLPACK's 7 POINT 3D discretization and its band Cholesky solver LINPACK SPD BAND, and ITPACK's routines for the Dirichlet problem on the unit cube with TRUE solution

$$u = e^{\sqrt{2}\,x} \cos y \sin z + e^{\sqrt{2}\,y} \cos z \sin x + e^{\sqrt{2}\,z} \cos x \sin y.$$

For the maximum error ϵ and the discretization time, t_D, we obtained

$$(11.3) \qquad \epsilon \approx 0.053\, h^{1.8} \qquad \text{and} \qquad t_D \approx 0.0061\, h^{-2.8},$$

whose exponents are close to the expected 2 and -3, respectively. The least square coefficients for error versus solve time t_S and total time $t = t_D + t_S$, and t versus h are listed in Table 3.5.

For the specified problem, the number of unknowns is $O(h^{-3})$, so, neglecting the time required for adaptive parameter selection, one expects solution time to be $O(h^{-4})$ for SOR and $O(h^{-3.5})$ for SSOR SI and CG.

In these experiments, there was little difference among the 'crossover points' of the iterative methods. Already with $h = 1/6$ and $N = 125$ interior unknowns, the total time for each of the iterative methods was less than for the direct solver; i.e., the 'crossover point' was between $n = 5$ and 6, or between $N = 64$ and 125. The graphs in Fig. 3.4 show the 'performance' of LINPACK SPD BAND, marked with 'bullets' ($\bullet$), and SOR, marked with 'diamonds' ($\diamond$); the dots show the extrapolation, using the least squares fits.

For $h = 1/10$, LINPACK SPD BAND used about 180,000 words of storage, whereas the storage for the iterative methods varied from 4,100 (SOR) to 8,900 (JACOBI SI). In the same order as listed in the table, the number of iterations required to reduce the error to 10^{-6} were: 56, 56, 32, 11, 14, 28, and 15, respectively.

	$\epsilon \approx A\,t^a$		$t_S \approx B\,h^b$		$t \approx C\,h^c$	
	A	a	B	b	C	c
LINPACK SPD BAND	2.6E–3	–0.28	3.3E–6	–7.3	8.8E–6	–6.9
JACOBI SI	3.7E–3	–0.50	1.9E–3	–4.1	4.3E–3	–3.8
JACOBI CG	3.3E–3	–0.50	1.3E–3	–4.1	3.9E–3	–3.7
SOR	3.2E–3	–0.50	1.2E–3	–4.1	3.7E–3	–3.7
SSOR SI	3.6E–3	–0.54	2.7E–3	–3.8	6.4E–3	–3.5
SSOR CG	3.6E–3	–0.53	2.5E–3	–3.8	5.8E–3	–3.5
REDUCED SYSTEM SI	3.5E–3	–0.59	9.9E–4	–4.1	9.5E–3	–3.2
REDUCED SYSTEM CG	3.3E–3	–0.61	1.1E–3	–3.9	1.1E–2	–3.0

Table 3.5 Least squares fits.

Acknowledgement

The authors thank the National Science Foundation for partial support under grant CCR-8704826. This paper was supplied by the authors in LaTeX and reformatted by Lisa Laguna.

References

[1] Birkhoff, G. *The Numerical Solution of Elliptic Equations*, Regional Conference Series in Applied Mathematics, Vol. 1, SIAM Publications, Philadelphia, 1972.

[2] Birkhoff, G., and R. E. Lynch. *Numerical Solution of Elliptic Problems*, SIAM Publications, Philadelphia, 1984.

[3] Birkhoff, G., and A. Schoenstadt (eds.). *Elliptic Problem Solvers* II, Academic Press, New York, 1984.

[4] Birkhoff, G., and R. S. Varga (eds.). *Numerical Solution of Field Problems in Continuum Physics*, SIAM-AMS Proceedings II, Amer. Math. Soc., Providence, RI, 1970.

[5] Cowell, W., (ed.). *Sources and Development of Mathematical Software*, Prentice-Hall, Englewood Cliffs, New Jersey, 1984.

[6] Forsythe, G. E., and W. R. Wasow. *Finite Difference Methods for Partial Differential Equations*, Wiley, New York, 1960.

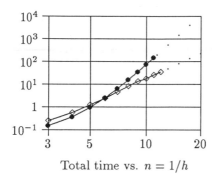 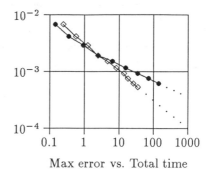

Total time vs. $n = 1/h$ Max error vs. Total time

Figure 3.4

[7] Hageman, L. A., and D. M. Young. *Applied Iterative Methods*, Academic Press, New York, 1981.

[8] Houstis, E. N., W. F. Mitchell, and J. R. Rice. "Collocation software for second-order elliptic partial differential equations; Algorithm 637 GENCOL; Algorithm 638 INTCOL and HERMCOL," *ACM Trans. on Math. Software* **11** (1985), 379–418.

[9] Kincaid, D. R., J. R. Respess, D. M. Young, and R. G. Grimes, "ITPACK 2C: A Fortran package for solving large sparse linear systems by adaptive accelerated iterative methods," *ACM Trans. on Math. Software* **8** (1981) 302–322.

[10] Lynch, R. E., and J. R. Rice. "High accuracy finite difference approximation to solutions of elliptic partial differential equations," *Proc. Nat. Acad. Sci.* **75**, (1978) 2541–2544.

[11] Lynch, R. E. "HODIEG: A new ELLPACK discretization module," Purdue University, Department of Computer Science, Report CSD-TR-871, March, 1989.

[12] MacNeal, R. H. "An asymmetrical finite difference network," *Quart. of Appl. Math.* **11** (1954) 295–310.

[13] Mu, Mo, and J. R. Rice. "An experimental performance analysis for the rate of convergence of COLLOCATION on general domains," Purdue University, Department of Computer Science, Report CSD-TR-738, Jan. 1988.

[14] Nash, S. G., (ed.). *A History of Scientific and Numeric Computation*, Addison-Wesley, Reading, Massachusetts, 1989.

[15] Rice, J. R., and R. F. Boisvert. *Solving Elliptic Problems Using ELLPACK*, Springer-Verlag, New York, 1984.

[16] Rice, J. R. "Numerical computation with general two-dimensional domains," and "Algorithm 625: A two-dimensional domain processor," *ACM Trans. on Math. Software* **10** (1984) 443–62.

[17] Rice, J. R. and Mo Mu. "An experimental performance analysis for the rate of convergence of 5 POINT STAR on general domains," Purdue University, Department of Computer Science, Report CSD-TR-747, Jan. 1988.

[18] Schoenberg, I., (ed.). *Approximations with Special Emphasis on Spline Functions*, Academic Press, New York, 1969.

[19] Schultz, M. H., (ed.). *Elliptic Problem Solvers*, Academic Press, New York, 1981.

[20] Traub, J. F., (ed.). *Complexity of Sequential and Parallel Algorithms*, Academic Press, New York, 1973.

[21] Varga, R. S. *Matrix Iterative Analysis*, Prentice-Hall, Englewood Cliffs, New Jersey, 1962.

[22] Young, D. M. "Iterative methods for solving partial difference equations of elliptic type," *Trans. Amer. Math. Soc.* **76** (1954) 92–111.

[23] Young, D. M. *Iterative Solution of Large Linear Systems*, Academic Press, New York, 1971.

[24] Young, D. M. "On the accelerated SSOR method for solving large linear systems," *Advances in Math.* **23** (1977) 215–71.

Chapter 4

Preconditioned Iterative Methods for Indefinite Symmetric Toeplitz Systems

PAUL CONCUS
Lawrence Berkeley Laboratory
   and University of California, Berkeley

and

PAUL E. SAYLOR
University of Illinois at Urbana-Champaign

Dedicated to David M. Young, Jr., on the occasion of his sixty-fifth birthday.

Abstract

Stable fast direct methods for solving symmetric positive-definite Toeplitz systems of linear equations have been known for a number of years. Recently, a conjugate-gradient method with circulant preconditioning has been proposed as an effective means for solving these equations. For the (non-singular) indefinite case, the only stable algorithms that appear to be known are the general $O(n^3)$ direct methods, such as **LU** decomposition, which do not exploit the Toeplitz structure. We depict here some initial numerical results on the feasibility of circulant preconditioned iterative methods for the indefinite symmetric case.

1   Introduction

The use of iterative methods for the solution of linear systems of equations $\mathbf{Ax} = \mathbf{b}$ for which $\mathbf{A}$ is a Toeplitz matrix has been stimulated by the recent work of G. Strang [14]. It was proposed in [14] that for the symmetric positive definite case the use of a circulant matrix as preconditioner could be particularly effective for the conjugate gradient method. Since circulant systems can be solved rapidly with the Fast Fourier Transform (FFT) and since for a significant class of matrices the spectrum of the preconditioned matrix turns out to have strong clustering with only a few isolated extremal eigenvalues, the conjugate gradient iteration can converge with striking efficiency. Subsequent related work can be found in [7], [8], [9], [10], and [15].

We report here on our initial findings for extending the use of circulant preconditioners to the case in which $\mathbf{A}$ is a symmetric Toeplitz matrix that is indefinite (both some positive and some negative eigenvalues). For this case the need for efficient iterative methods is more pressing than for the positive definite case, because the specialized rapid direct methods for Toeplitz matrices can be unstable when the matrix is indefinite.

2   Toeplitz and Circulant Matrices

Let $\mathbf{A}$ denote a real symmetric $n \times n$ Toeplitz matrix. A Toeplitz matrix is constant along its diagonals and thus, in the symmetric case, is determined by the n elements of the first row, $a_0, a_1, \ldots, a_{n-1}$,

$$(2.1) \qquad \mathbf{A} = \begin{bmatrix} a_0 & a_1 & \cdots & a_{n-2} & a_{n-1} \\ a_1 & a_0 & a_1 & \cdots & a_{n-2} \\ a_2 & a_1 & a_0 & \ddots & \vdots \\ \vdots & \ddots & \ddots & \ddots & a_1 \\ a_{n-1} & \cdots & a_2 & a_1 & a_0 \end{bmatrix} .$$

Such matrices arise in applications such as time series analysis, Padé approximation, and differential and integral equations.

A circulant matrix, which is a special case of a Toeplitz matrix, is determined by only the first $\frac{n}{2} + 1$ elements of the first row if n is even. Each successive row contains the elements of the row above shifted one to the right, with the last element wrapped around to become the first, i.e.,

$$\begin{bmatrix} c_0 & c_1 & \cdots & c_2 & c_1 \\ c_1 & c_0 & c_1 & \cdots & c_2 \\ c_2 & c_1 & c_0 & \ddots & \vdots \\ \vdots & \ddots & \ddots & \ddots & c_1 \\ c_1 & \cdots & c_2 & c_1 & c_0 \end{bmatrix} .$$

Since the eigenvectors of a circulant matrix are given by successive powers of the nth roots of unity, systems with circulant coefficient matrix are amenable to solution by the FFT in $O(n \log n)$ operations.

To obtain a preconditioner $\mathbf{S}$ for symmetric positive definite $\mathbf{A}$, Strang proposed keeping the central diagonals of $\mathbf{A}$ and replacing the outer diagonals by reflected values from the central ones to complete the circulant. For the matrix $\mathbf{A}$ in (1) the circulant preconditioner $\mathbf{S}$ is

(2.2)
$$
\mathbf{S} = \begin{bmatrix}
a_0 & a_1 & \cdots & a_2 & a_1 \\
a_1 & a_0 & a_1 & \cdots & a_2 \\
a_2 & a_1 & a_0 & \ddots & \vdots \\
\vdots & \ddots & \ddots & \ddots & a_1 \\
a_1 & \cdots & a_2 & a_1 & a_0
\end{bmatrix}.
$$

For n even, element a_{n-j} is replaced with a_j for $j = 1, 2, \ldots, \frac{n}{2} - 1$. No arithmetic computation of new elements is required.

The preconditioner $\mathbf{S}$ was shown to be extremely effective for use with the conjugate gradient method. Because the FFT can be used to form the matrix-vector products with $\mathbf{A}$ in $O(n \log n)$ operations as well as to solve the preconditioning circulant systems in $O(n \log n)$ operations, each conjugate gradient step is very efficient. A remarkable feature is that the eigenvalues of $\mathbf{S}^{-1}\mathbf{A}$ cluster exceptionally favorably. Suppose that the elements $a_0, a_1, \ldots, a_{n-1}$ are part of a sequence $\{a_k\}_{k=0}^{\infty}$, which defines a limiting symmetric positive definite Toeplitz matrix of infinite order. It was proved in [9] by R. Chan and G. Strang that for $n \to \infty$, if the underlying (real-valued) generating function $f(\theta) = \sum_{-\infty}^{\infty} a_{|k|} e^{ik\theta}$ is positive and in the Wiener class $\sum_{-\infty}^{\infty} |a_k| < \infty$, then the $n \times n$ matrices $\mathbf{S}$ and $\mathbf{S}^{-1}$ are uniformly bounded and positive definite for all sufficiently large n, and the eigenvalues of $\mathbf{S}^{-1}\mathbf{A}$ cluster at unity. An interesting relationship between this clustering and approximation on the unit circle is discussed in [16].

3   Solution Methods

As part of a general study on iterative methods for solving symmetric indefinite systems of linear equations, we describe here our initial investigations for Toeplitz coefficient matrices. We are interested in methods that work on the original system of equations and avoid transforming to the associated positive-definite normal-equation system.

Numerical methods for solving symmetric Toeplitz systems of equations can be categorized as follows.

A. *General direct methods.* General direct methods such as $\mathbf{LU}$ and $\mathbf{QR}$ decomposition require $O(n^3)$ operations and can be used for the positive-definite case or, with pivoting, for the indefinite case. They do not take specific advantage of the Toeplitz structure.

B. *Specialized direct methods.* For the positive definite case specialized direct methods have been devised that take advantage of the matrix structure to solve a system in fewer operations than required for general direct methods. The stability of these methods has been studied by Bunch in [6]. Both the "classical" $O(n^2)$ specialized methods of Levinson [11] and Trench [17] and the more recent "fast" $O(n \log^2 n)$ specialized methods (e.g., [1], [4], [5]) are generally unstable if the matrix is not positive definite.

C. *Iterative methods.* For the positive definite case the circulant preconditioned conjugate gradient method proposed by Strang requiring $O(n \log n)$ operations per iteration can be very effective. For the indefinite case, which is the area of our study, iterative methods can be of comparatively greater interest because of instability of the specialized direct methods. Three iterative-method possibilities for consideration are (a) the conjugate residual method on the original system of equations; (b) circulant preconditioning with acceleration, such as adaptive Chebyshev [12] or a conjugate gradient type method [3]; and (c) stabilization of the specialized direct methods with iterative techniques.

We are concerned here with the iterative method possibility (b) of circulant preconditioning for symmetric indefinite Toeplitz matrices and report on some preliminary numerical experiments.

4   Test Matrix Preconditioners

The circulant preconditioners we investigate for our test matrices are the Strang preconditioner $\mathbf{S}$ in (2) and the circulant matrix that best approximates $\mathbf{A}$ in the Frobenius norm, which was suggested by T. Chan [10].

The optimal Frobenius norm circulant preconditioner is obtained from $\mathbf{A}$ by replacing both a_i and a_{n+1-i} by a weighted average. Let $\mathbf{C}$ denote the preconditioner. Then there holds

$$(4.1) \qquad \mathbf{C} = \begin{bmatrix} a_0 & \bar{a}_1 & \cdots & \bar{a}_2 & \bar{a}_1 \\ \bar{a}_1 & a_0 & \bar{a}_1 & \cdots & \bar{a}_2 \\ \bar{a}_2 & \bar{a}_1 & a_0 & \ddots & \vdots \\ \vdots & \ddots & \ddots & \ddots & \bar{a}_1 \\ \bar{a}_1 & \cdots & \bar{a}_2 & \bar{a}_1 & a_0 \end{bmatrix},$$

where $\bar{a}_i = \frac{n-i}{n} a_i + \frac{i}{n} a_{n-i}$, $i = 1, 2, \ldots, \frac{n}{2}$ (for n even). The pairing of the off-diagonal elements of $\mathbf{A}$ has the same pattern for $\mathbf{C}$ as for $\mathbf{S}$. For $\mathbf{S}$ the lower-index element simply replaces the larger, whereas for $\mathbf{C}$ a weighted average of the two replaces both.

Recently, R. Chan has proved that the preconditioner $\mathbf{C}$, and others having the same pairing/replacement pattern, enjoy the same asymptotic clustering properties as does $\mathbf{S}$ for the same class of matrices—symmetric positive definite Toeplitz matrices with positive generating function in the Wiener class [7]. He showed also that the $\mathbf{S}$ preconditioning is the best circulant approximation to $\mathbf{A}$ in the 1 and ∞ norms [8]. For non-square matrices, other circulant preconditionings based on Cybenko's $\mathbf{QR}$ factorization are considered by Olkin in [13].

Although the asymptotic spectra are the same for the **C** and **S** preconditionings, for finite n there are differences that can have practical effects on the relative convergence rates for the preconditioned conjugate gradient method. The **S** preconditioning appears to cluster eigenvalues more sharply than does the **C** preconditioning, particularly when a_k decreases rapidly away from the main diagonal, although the **C** preconditioning appears to result in smaller condition number [10]. The sharper clustering for **S** results in enhanced convergence for the preconditioned conjugate gradient method for some problems.

5 Test Matrices

The test matrices considered here for the symmetric indefinite case are taken from the positive definite examples used by T. Chan and by Strang, which are then shifted by constant diagonal matrices to make some of the eigenvalues negative. The matrices before shifting are the following.

(i) $a_k = 1/(k+1)$, $k = 0, 1, \ldots, n$. Since the decay away from the diagonal is only arithmetic, the limiting underlying generating function does not belong to the Wiener class and hence would not satisfy the hypotheses for the asymptotic clustering and other results of [7], [9].

(ii) $a_k = \left(\frac{1}{2}\right)^k$, $k = 0, 1, \ldots, n$. This is a matrix studied by Kac, Murdock, and Szegö. Decay away from the diagonal is geometric, and the hypotheses of [7], [9] on the limiting underlying generating function are satisfied. In fact, the **S** preconditioner is unusually effective for this matrix, which is the inverse of a tridiagonal matrix (not Toeplitz). As pointed out in [15], the preconditioned matrix $\mathbf{S}^{-1}\mathbf{A}$ has only five distinct eigenvalues (for n even). These are $\frac{1}{1+\frac{1}{2}}$, $\frac{1}{1+\left(\frac{1}{2}\right)^{n/2}}$, 1, $\frac{1}{1-\left(\frac{1}{2}\right)^{n/2}}$, $\frac{1}{1-\frac{1}{2}}$. The smallest and largest eigenvalues are simple and independent of n. The eigenvalue 1 is double, and the other two are repeated $\left(\frac{n}{2} - 2\right)$ times each. Thus the clustering toward 1 is exponential. These striking properties, well suited for the conjugate gradient method, appear to have encouraged much of the work in [9], [14], and [15].

(iii) $a_k = \frac{\cos k}{k+1}$, $k = 0, 1, \ldots, n$. This is a variation of case (1). The $\cos k$ factor introduces some negative elements in the matrix and alters the smooth arithmetic decay away from the diagonal. As for (1), its limiting underlying generating function does not belong to the Wiener class.

(iv) $a_0 = 2 + \frac{1}{n^2}$, $a_1 = -1$, $a_2 = a_3 = \cdots = a_{n-1} = 0$. This tridiagonal matrix is a discretization of the one-dimensional Helmholtz operator $-u_{xx} + u$ with Dirichlet boundary conditions on a uniform mesh on $[0, 1]$. The limiting matrix A satisfies the hypotheses of [7], [9] so that the results there apply. The preconditioner **S** in this case replaces a_{n-1} with -1 and leaves the other a_i, $i = 0, \ldots, n-2$ unchanged. It corresponds to the Helmholtz operator with periodic, instead of Dirichlet, boundary conditions. Since $\mathbf{S}$ and $\mathbf{A}$ differ only by a rank-two matrix, $\mathbf{S}^{-1}\mathbf{A}$ has an $(n-2)$-fold eigenvalue of 1. The $\mathbf{C}$ preconditioner for this case replaces a_1 and a_{n-1} by their weighted average $\frac{n-1}{n}a_1$, and there holds

$$\mathbf{C} = \frac{n-1}{n}\mathbf{S} + \frac{1}{n}\mathbf{I}.$$

In general, $\mathbf{C}$ and $\mathbf{A}$ differ by a matrix of full rank.

Note that the discrete Laplace operator $-u_{xx}$, which would result simply in 2 on the diagonal instead of $2 + \frac{1}{n^2}$ as in (iv), would not provide a suitable test matrix, as the $\mathbf{S}$ preconditioner, which would correspond to a Laplace operator with periodic boundary conditions, is singular [9]. The generating function of the matrix, in this case, does not satisfy the positivity condition.

For the numerical experiments the above matrices are shifted by a constant diagonal matrix $-\alpha\mathbf{I}$, with values of α such that $\mathbf{A} - \alpha\mathbf{I}$ is indefinite, but nonsingular. As only the diagonal elements of $\mathbf{A}$ are changed by the shift, the preconditioners for $\mathbf{A} - \alpha\mathbf{I}$ are $\mathbf{S} - \alpha\mathbf{I}$ and $\mathbf{C} - \alpha\mathbf{I}$, where $\mathbf{S}$ and $\mathbf{C}$ are the preconditioners for $\mathbf{A}$. Note that if an unshifted preconditioned matrix has an eigenvalue 1 corresponding to eigenvector $\boldsymbol{\phi}$, e.g.,

$$\mathbf{A}\boldsymbol{\phi} = \mathbf{S}\boldsymbol{\phi},$$

then

$$(\mathbf{A} - \alpha\mathbf{I})\boldsymbol{\phi} = (\mathbf{S} - \alpha\mathbf{I})\boldsymbol{\phi},$$

so that the shifted preconditioned matrix also has eigenvalue 1 with eigenvector $\boldsymbol{\phi}$. The shifting preserves eigenvalues unity of the preconditioned matrix. There holds, more generally, that shifting preserves eigenvalues of the difference between $\mathbf{A}$ and the preconditioner, e.g., if μ and $\boldsymbol{\psi}$ are an eigenvalue-eigenvector pair such that $(\mathbf{A} - \mathbf{S})\boldsymbol{\psi} = \mu\boldsymbol{\psi}$, then $[(\mathbf{A} - \alpha\mathbf{I}) - (\mathbf{S} - \alpha\mathbf{I})]\boldsymbol{\psi} = \mu\boldsymbol{\psi}$. This implies that shifting preserves any clustering of the spectrum of $\mathbf{A} - \mathbf{S}$, for example as in Theorem 2 of [8] (dependence of $\mathbf{A}$ and $\mathbf{S}$ on the order n is denoted explicitly here): *Let f be a positive function in the Wiener class, then for all $\epsilon > 0$, there exist M and $N > 0$ such that for all $n > N$, at most M eigenvalues of $\mathbf{A}(n) - \mathbf{S}(n)$ have absolute values exceeding ϵ.* If also $(\mathbf{S} - \alpha\mathbf{I})$ and $(\mathbf{S} - \alpha\mathbf{I})^{-1}$ are uniformly bounded for all $n > N$, then the shifting would preserve the result of [8], [9] that the eigenvalues of $\mathbf{S}^{-1}\mathbf{A}$ cluster at unity.

6   Computed Spectra

The spectra for $n = 16$, as computed using a public-domain version of MATLAB, are given in Figs. 4.1–4.5. For Figs. 4.1–4.4 the value $\alpha = \frac{\nu_5(\mathbf{A}) + \nu_6(\mathbf{A})}{2}$ is chosen, where $\nu_i(\mathbf{A})$ denotes the ith eigenvalue of $\mathbf{A}$, $0 < \nu_1 \leq \nu_2 \leq \cdots \leq \nu_n$. For Fig. 4.5, $\alpha = \frac{\nu_4(\mathbf{A}) + \nu_5(\mathbf{A})}{2}$ is chosen. For all the test problems, $\mathbf{A}$ has distinct eigenvalues, so that $\nu_4 \neq \nu_5 \neq \nu_6$ and $\mathbf{A} - \alpha\mathbf{I}$ is nonsingular. For these problems $\mathbf{S} - \alpha\mathbf{I}$ and $\mathbf{C} - \alpha\mathbf{I}$ are nonsingular also. Our interest is to observe whether the shifted preconditioners yield clustering of the eigenvalues, and to compare the clustering properties. For the standard adaptive Chebyshev method [12], the eigenvalues of the preconditioned matrix must lie in a half-plane. If they do not, then an adaptive Chebyshev method using two disjoint regions—one in each half-plane—to enclose the eigenvalues might be employed, but such techniques are not as highly developed. If the eigenvalues do not lie in a half-plane, then a conjugate-gradient-like acceleration might be employed, but these methods are not robust. They also are not robust if the symmetric part is indefinite, even if the eigenvalues are in a half plane; thus this property is of interest also.

In each figure, the top three rows depict the eigenvalues of $\mathbf{A}$ and of the preconditioned matrices $\mathbf{S}^{-1}\mathbf{A}$ and $\mathbf{C}^{-1}\mathbf{A}$, with plotting symbols of plusses, circles, and triangles, respectively. The fourth row depicts the spectrum of the shifted matrix $\mathbf{A} - \alpha\mathbf{I}$. Plotted below are the imaginary vs. real part of the eigenvalues λ of $(\mathbf{S} - \alpha\mathbf{I})^{-1}(\mathbf{A} - \alpha\mathbf{I})$ and $(\mathbf{C} - \alpha\mathbf{I})^{-1}(\mathbf{A} - \alpha\mathbf{I})$, using the same plotting symbols as for the unshifted preconditioned matrices.

In Figs. 4.1–4.3, the spectra of the unshifted matrices are consistent with those given in [10] for the case $n = 15$. (We have chosen n to be even so that the property for case (ii) of only 5 distinct eigenvalues holds.) As observed in [10], for these cases the $\mathbf{C}$ preconditioning results in a spectrum that lies strictly within that for the $\mathbf{S}$ preconditioning. However, for examples such as (ii) (Fig. 4.2), the eigenvalue clustering is smeared out. For example (iv) (Fig. 4.4) the spectrum for the $\mathbf{C}$ preconditioning does not lie strictly within that for the $\mathbf{S}$ preconditioning, but the condition number for $\mathbf{C}$ is still smaller. For this case the $\mathbf{C}$ preconditioning smears out the clustering considerably. For the shifted matrices, behavior of the two preconditionings can differ more appreciably.

In Fig. 4.1 (the arithmetic decay case), one sees that for the shifted matrices the $\mathbf{S}$ preconditioning spectrum lies within that for the $\mathbf{C}$ preconditioning, in contrast to the case for the unshifted matrices. The imaginary parts for the $\mathbf{S}$ preconditioning are much smaller than for the $\mathbf{C}$ preconditioning, and the real parts are all positive. The complex conjugate pair of eigenvalues for the $\mathbf{C}$ preconditioning to the left of the imaginary axis would pose difficulties for, say, adaptive Chebyshev acceleration that would not be present for the $\mathbf{S}$ preconditioning.

In Fig. 4.2 (the geometric decay case), the strong clustering at 1 for the $\mathbf{S}$ preconditioning for the unshifted matrix can be seen. Even though the order of the matrix is only 16, the separation between the double eigenvalue at 1 and the sevenfold ones at $(1 + (\frac{1}{2})^8)^{-1}$ and $(1 - (\frac{1}{2})^8)^{-1}$ can be distinguished as only a slight thickening of the circle designating the eigenvalue at 1. For the shifted matrices, both the $\mathbf{C}$ and $\mathbf{S}$ preconditionings yield eigenvalues entirely in the right half plane. The $\mathbf{S}$ preconditioning yields much stronger clustering. Some eigenvalues have departed only slightly from the real axis, whereas for the $\mathbf{C}$ preconditioning significant imaginary parts have appeared. Again, the $\mathbf{C}$ preconditioning would appear less favorable for acceleration than the $\mathbf{S}$ preconditioning.

In Fig. 4.3 (the altered arithmetic decay case) the $\mathbf{S}$ preconditioning does not cluster the eigenvalues as well as the $\mathbf{C}$ preconditioning does for the unshifted

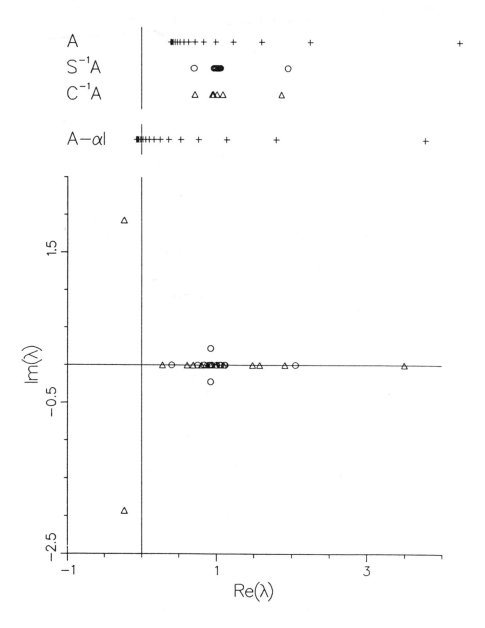

Figure 4.1 Spectra for $a_k = \frac{1}{k+1}$ and $\alpha = \frac{1}{2}\left(\nu_5(\mathbf{A}) + \nu_6(\mathbf{A})\right)$.

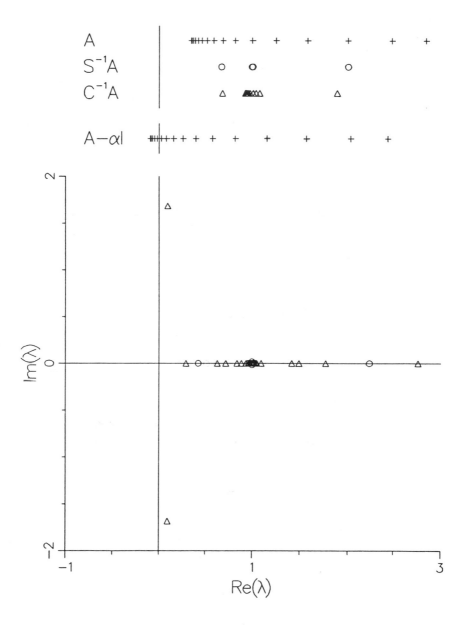

Figure 4.2 Spectra for $a_k = \left(\frac{1}{2}\right)^k$ and $\alpha = \frac{1}{2}\left(\nu_5(\mathbf{A}) + \nu_6(\mathbf{A})\right)$.

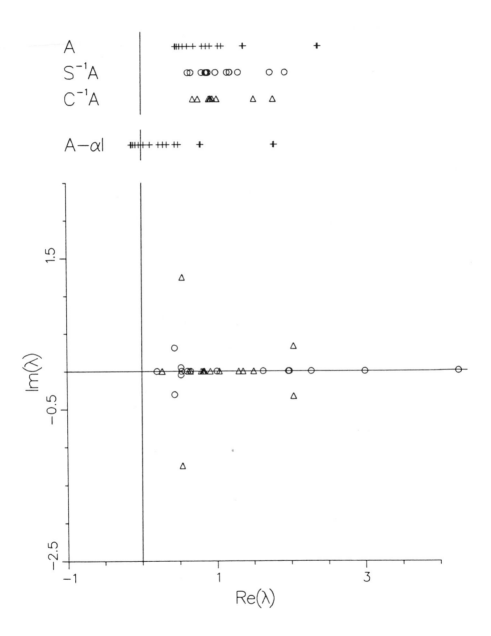

Figure 4.3 Spectra for $a_k = \frac{\cos k}{k+1}$ and $\alpha = \frac{1}{2}\left(\nu_5(\mathbf{A}) + \nu_6(\mathbf{A})\right)$.

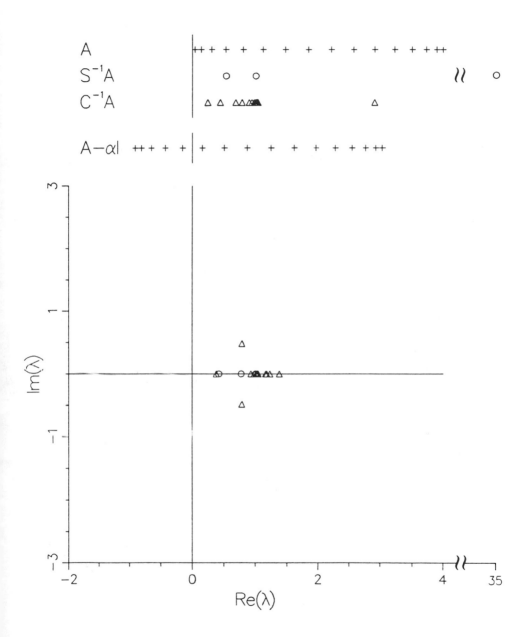

Figure 4.4 Spectra for $\mathbf{A}(1,\cdot) = (2 \ -1 \ 0 \cdots 0)$ and $\alpha = \frac{1}{2}\left(\nu_5(\mathbf{A}) + \nu_6(\mathbf{A})\right)$.

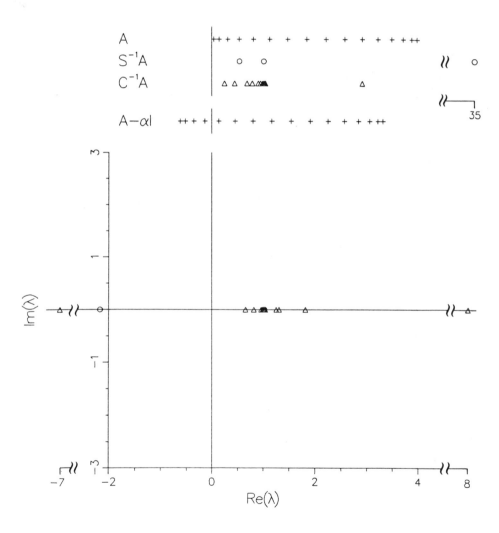

Figure 4.5 Spectra for $\mathbf{A}(1,\cdot) = (2 \ -1 \ 0 \cdots 0)$ and $\alpha = \frac{1}{2}(\nu_4(\mathbf{A}) + \nu_5(\mathbf{A}))$.

matrices. For the shifted matrices the real parts of the **C** preconditioning eigenvalues are interior to the interval of the real parts of the **S** preconditioning eigenvalues, but the imaginary parts are larger. An adaptive acceleration method that estimates an ellipse enclosing the eigenvalues might deal with the **S** preconditioning matrix more easily, although with other shifts for this problem we observed that neither preconditioning suggested itself as being substantially better than the other.

In Fig. 4.4 (the Helmholtz equation case) the special spectral properties for the **S** preconditioning can be observed. For both the shifted and unshifted case this preconditioning differs from the original matrix by only the rank two matrix whose elements are equal to -1 in the $(1, n)$ and $(n, 1)$ positions and zero elsewhere. Thus, the preconditioned matrix has an $(n - 2)$-fold eigenvalue of unity. For this case the **C** preconditioned spectrum for the unshifted matrix does not lie interior to the **S** preconditioned spectrum, as for the other cases, but overlaps it. (But it does have smaller condition number.) For the shifted matrices, the **C** preconditioned matrix has a complex-conjugate pair.

Fig. 4.5 illustrates the different behavior for the Helmholtz-equation test matrix if four rather than five eigenvalues of **A** are shifted to be negative. In the experiments for this matrix, we observed that there was some correlation between the behavior of the spectrum and whether an odd or an even number of eigenvalues of **A** had been shifted to be negative. Fig. 4.5 is representative of even shift cases (except for a shift of $\alpha = \frac{\nu_8(\mathbf{A})+\nu_9(\mathbf{A})}{2}$, for which matrices become singular). The eigenvalues of the preconditioned shifted matrix are not all in the right-half plane, as for the odd-shift case of Fig. 4.4, and for the **S** preconditioning they are all real. The property that the interior eigenvalues of the circulant preconditioners are double (the two extremal ones are simple) may have some bearing on the odd vs. even shift behavior that was observed.

For the examples, circulant preconditioning appears to be of benefit for conditioning the spectra of the shifted matrices. The **S** preconditioning seems to be of greater benefit than the **C** preconditioning in yielding compact spectra, particularly for the examples satisfying the conditions of [7], [9] that the limiting underlying generating function be positive and in the Wiener class. For the conjugate gradient type accelerations that rely on the symmetric part of the preconditioned matrix being positive definite, we note that this was generally not the case in the experiments—symmetric parts were indefinite. This suggests that an adaptive Chebyshev acceleration would be preferable, particularly when the spectrum of the preconditioned matrix lies in a half plane, which is satisfied often but not always in the experiments; the standard adaptive algorithm packages [2] could then be used. If the preconditioned spectrum does not lie in a half plane, then the preconditioning might be effective in conjunction with Chebyshev acceleration employing two disjoint spectra-enclosing regions or with the conjugate gradient method applied to the normal equations of the preconditioned system.

As a final note, we remark that the unshifted preconditioners, which are positive definite, were generally not nearly as effective in conditioning the spectra of the shifted matrices as were the shifted preconditioners. We believe that the numerical results are sufficiently encouraging to warrant further consideration of circulant preconditioning for the iterative solution of symmetric indefinite Toeplitz systems of linear equations.

Acknowledgements

This work was supported in part by the Applied Mathematical Sciences subprogram of the Office of Energy Research of the U.S. Department of Energy under contract DE-AC03-76SF000098 and by National Science Foundation grant DMS87-03226. The word processing for this paper was carried out originally in TEX by V. Heatlie and by the authors, and it was reformatted in LATEX by Lisa Laguna.

References

[1] Ammar, G.S. and W.B. Gragg. "Superfast solution of real positive definite Toeplitz systems," *SIAM J. Matrix Anal. Appl.* **9** (1988) 61–76.

[2] Ashby, S.F. CHEBYCODE: *A* FORTRAN *Implementation of Manteuffel's adaptive Chebyshev algorithm*, Report UIUCDCS-R-85-1203, Dept. Comp. Sci., University of Illinois, Urbana-Champaign, IL, 1985.

[3] Ashby, S.F., T.A. Manteuffel, and P.E. Saylor. *A Taxonomy of Conjugate Gradient Methods*, Report UIUCDCS-R-88-1414, Dept. Comp. Sci., University of Illinois, Urbana-Champaign, IL, 1988.

[4] Bitmead, R.R. and B.D.O. Anderson, "Asymptotically fast solution of Toeplitz and related systems of equations," *Linear Alg. Appl.* **34** (1980) 103–116.

[5] Brent, R.P, G.F. Gustavson, and D.Y.Y. Yun. "Fast solution of Toeplitz systems of equations and computation of Padé approximants," *J. Algorithms* **1** (1980) 259–295.

[6] Bunch, J.R. "Stability of methods for solving Toeplitz systems of equations," *SIAM J. Sci. Stat. Comput.* **6** (1985) 349–364.

[7] Chan, R.H. "The spectrum of a family of circulant preconditioned Toeplitz systems," *SIAM J. Num. Anal.*, to appear.

[8] ———, *Circulant Preconditioners for Hermitian Toeplitz Systems*, Tech. Report 406, Dept. Comp. Sci., Courant Inst., N.Y.U., 1988.

[9] Chan, R.H., and G. Strang. "Toeplitz equations by conjugate gradients with circulant preconditioner," *SIAM J. Sci. Stat. Comput.*, to appear.

[10] Chan, T.F. "An optimal circulant preconditioner for Toeplitz systems," *SIAM J. Sci. Stat. Comput.* **9** (1988) 766–771.

[11] Levinson, N. "The Wiener rms (root-mean-square) error criterion in filter design and prediction," *J. Math. Phys.* **25** (1947) 261–278.

[12] Manteuffel, T.A. "Adaptive procedure for estimating parameters for the non-symmetric Tchebyshev iteration," *Numer. Math.* **31** (1978) 183–208.

[13] Olkin, J.A. *Linear and Nonlinear Deconvolution Models*, Tech. Report 86-10, Dept. Math. Sci., Rice Univ., Houston, TX, 1986.

[14] Strang, G.A. "A Proposal for Toeplitz Matrix Calculations," *Studies in Appl. Math.* **74** (1986) 171–176.

[15] Strang, G., and Edelman, A. *The Toeplitz-circulant Eigenvalue Problem $Ax = \lambda Cx$*, Oakland Conf. on PDE's, Bragg, L. and Dettman, J., eds., Longman, 1987, 109–117.

[16] Trefethen, L.N. *Applications of Approximation Theory in Numerical Linear Algebra*, Preprint, Dept. Math., Mass. Inst. Tech., Boston, MA, 1988.

[17] Trench, W. "An algorithm for the inversion of finite Toeplitz matrices," *SIAM J. Appl. Math.* **12** (1964) 515–522.

Chapter 5

A Local Relaxation Scheme (Ad-Hoc SOR) Applied to Nine Point and Block Difference Equations

L.W. EHRLICH

John Hopkins Applied Physics Laboratory

Dedicated to David M. Young, Jr., on the occasion of his sixty-fifth birthday.

Abstract

A local relaxation scheme, applicable to five point difference equations, was introduced and referred to as Ad-Hoc SOR. Although developed for five point difference schemes, there has been some interest as to its applicability to nine point and block difference schemes. In this paper we review the method and present some ideas related to these questions.

1 History

Let me start with a brief history of the Ad-Hoc SOR method. The set of equations

$$(1.1) \qquad \nabla^2 \psi \;=\; -\Omega$$

$$(1.2) \qquad \nabla^2 \Omega \;=\; \phi$$

$$(1.3) \qquad \phi - W(\psi_y \phi_x - \psi_x \phi_y) \;=\; R(\psi_y \Omega_x - \psi_x \Omega_y)$$

was said to describe the flow of fluid of second grade. They were to be solved as a coupled system. The first two equations presented no problem, but the third one did. The system was to be solved only for a few values of the parameters, W and

R. Consequently, it was decided to use a method that was effective but possibly
not optimal. But what to use? Groping for something simple, yet effective (or at
least not too ineffective), the Ad-Hoc SOR method as devised and tried. It worked
surprisingly well. The results were reported at the 1979 SIAM National Meeting
in Denver, Colorado. Indeed, the last line of the abstract of that talk read, "An
Ad-Hoc SOR method...appears to have properties that my make it of more general
interest." It is in the light of a method that is effective, but may not be optimal
that we present the Ad-Hoc Sor technique.

The Ad-Hoc Sor method was devised to solve five point difference equations
iteratively. Essentially, the method is successive overrelaxation with a different
relaxation factor computed for each grid point.

2   The Method

The technique can be described as follows. At each point, we assume that the
difference equation coefficients for that point apply throughout at all points in a
square region and are thus constant. The resulting matrix is then separable and
one can find the eigenvalues of the corresponding Jacobi scheme. For the stencil

$$(2.1) \qquad \begin{pmatrix} & a_2 & \\ a_3 & -a_0 & a_1 \\ & a_4 & \end{pmatrix}$$

the Jacobi eigenvalues are

$$(2.2) \qquad \mu_{p,q} = \frac{2}{a_0} \left[\sqrt{a_1 a_3} \cos(\pi p/N + 1) + \sqrt{a_2 a_4} \cos(\pi q/M + 1) \right]$$

where N and M are the number of grid points in the x and y directions.

As is well known, if the matrix has Property A, one can find the eigenvalues of the
SOR matrix from the eigenvalues of the Jacobi matrix. The five point difference
equation has such property and thus the theory follows through. One can then
determine the optimal relaxation factor, ω, such that the spectral radius of the
SOR matrix is minimized with respect to ω. If the Jacobi eigenvalues, μ, are real,
then, for $\bar{\mu}$, the spectral radius,

$$(2.3) \qquad \omega = \frac{2}{1 + \sqrt{1 - \bar{\mu}^2}}$$

If the μ are purely imaginary, then

$$(2.4) \qquad \omega = \frac{2}{1 + \sqrt{1 + \bar{\mu}^2}}.$$

If the μ are complex, then the optimal ω is more difficult to find. In a paper
by Rigal [6], a method is presented which produces this optimal ω, but involves

formable computation. We can avoid this, as we shall see in a moment. Having relaxed the value at the point, we proceed to the next point and repeat the process. This is the Ad-Hoc method. If the problem is linear, the ω's need only be computed the first iteration and saved. If the problem is nonlinear, the ω's may need to be computed at each iteration.

In [3,4], the Ad-Hoc method is discussed in detail along with its variations, subject to boundary conditions and region shape. Also, comparison is made to other local relaxation techniques. One of the more successful applications of the Ad-Hoc method is in solving the Reynolds number equation in the Navier-Stokes equations for fluid flow. (See [4] for plots comparing spectral radii.)

Shortly after the first paper appeared describing the method, a paper by Botta and Veldman, [2], appeared. In this paper, the authors presented a local relaxation scheme which turned out to be the Ad-Hoc SOR method, but for one detail, namely the computation of the relaxation factor. They suggested using

$$(2.5) \qquad \omega = \frac{2}{1 + \sqrt{1 - \mu_r^2 + \mu_i^2(1 - \mu_r^{2/3})^{-1}}}$$

where μ_r is the real part and μ_i the imaginary part of μ. This equation is exact for μ real or purely imaginary and is an excellent approximation when μ is complex, being off by at most about 2% in the worst case. This, then, is the method as applied.

Unfortunately, there are no rigorous proofs that Ad-Hoc SOR will converge in general, in spite of the fact that the scheme appears to be quite robust. However, there is a proof that the technique is convergent and optimal if the matrix of the linear system is symmetric and positive definite. ([5] contains this proof.)

3   Nine Point Application: Cross Derivatives

As far as this author is aware, nine-point difference equations arise in two contexts. The first situation occurs when one uses finite difference to approximate an elliptic partial differential equation containing a cross derivative term. This cross term could appear in the original equation or be the result of a non-orthogonal mapping or transformation of the original region into a different shaped region, such as a rectangle. It is in this latter context that the author has had some experience with nine-point difference equations.

The equations used to map a region in the x,y plane into a region in the ϑ, η plane are (see e.g. [7])

$$(3.1) \qquad \alpha x_{\vartheta\vartheta} - 2\beta x_{\vartheta\eta} + \gamma x_{\eta\eta} = 0$$

$$(3.2) \qquad \alpha y_{\vartheta\vartheta} - 2\beta y_{\vartheta\eta} + \gamma y_{\eta\eta} = 0$$

where

$$(3.3a) \qquad \alpha = x_\eta^2 y_\eta^2$$

(3.3b) $\beta = x_\vartheta x_\eta + y_\vartheta y_\eta$

(3.3c) $\gamma = x_\vartheta^2 y_\vartheta^2$

The solution at a point in the (ϑ, η)–region contains the values of x and y which are mapped into that point from the physical plane. The equations are nonlinear and can themselves be solved using Ad-Hoc SOR. It should be noted that if the mapping is orthogonal, then β is zero and there are no cross derivative terms. Thus, β is a measure of the non-orthogonality of the mapping at each point. The Laplace Equation before and after the mapping is

(3.4) $u_{xx} + u_{yy} = 0$

(3.5) $\alpha u_{\vartheta\vartheta} - 2\beta u_{\vartheta\eta} + \gamma u_{\eta\eta} = 0.$

Differencing cross derivative terms produce nine point stencils such as

(3.6)
$$
\begin{pmatrix}
\beta/2 & \gamma & -\beta/2 \\
\alpha & -2(\alpha+\gamma) & \alpha \\
-\beta/2 & \gamma & \beta/2
\end{pmatrix}
$$

The resulting matrix is not separable in general.

The general approach using Ad-Hoc on these nine point stencils is to ignore the corner points in computing the relaxation factor, but use these points in the relaxation scheme. The effect of ignoring the corner points is generally small.

In [5], the authors also consider these nine point stencils. Again, assuming the given matrix is symmetric and positive definite, and assuming one uses the red-black ordering, they indicate that Fourier Analysis leads to the optimal ω as given in

(3.7) $\omega_{i,j} = \dfrac{2}{1 - \tau_{i,j} + \sqrt{(1 - \tau_{i,j})^2 - \mu_{i,j}^2}}$

where $\mu_{i,j}$ is an eigenvalue of the 5 point stencil and $\tau_{i,j}$ is an eigenvalue of the 4 corner points stencil. What is not clear is which eigenvalues of the corresponding matrices to use in the formula to obtain the optimal ω. In any event, the formula indicates the effect corner values have on the relaxation factor.

Table 5.1 contains numerical experiments using a stencil with varying corner point values. For this special example (constant), the Ad-Hoc ω was the same for each α, but in general this will not be true. Even so, Ad-Hoc was not a poor strategy to use. Further, the SOR factor may be difficult to determine.

The second way a nine point difference equation can arise is in high order approximations to the Laplace equation. Thus we have the stencil

(3.8)
$$
\begin{pmatrix}
1 & 4 & 1 \\
4 & 20 & 4 \\
1 & 4 & 1
\end{pmatrix}.
$$

$$\begin{pmatrix} -\alpha & 1 & \alpha \\ 1 & -4 & 1 \\ \alpha & 1 & \alpha \end{pmatrix}$$

8 × 8			
α	ω_0	$\lambda(\omega_0)$	$\lambda(\omega_H)$
.0	1.4903	.4903	.4903
.1	1.4880	.5100	.5123
.25	1.4746	.5236	.5372
.30	1.4646	.5435	.5626
.40	1.4330	.6218	.6367
.50	1.4533	.7276	.7303

31 × 31			
α	ω_0	$N(\omega_0)$	$N(\omega_H)$
.0	1.821	94	94
.1	1.820	95	97
.25	1.815	98	100
.30	1.813	98	104
.40	1.789	121	137
.50	1.774	315	355

$\lambda(\omega)$—spectral radius using ω
$N(\omega)$—number of iterations for convergence using ω

Table 5.1

A recent paper, [1], contains a study of SOR applied to this problem. The analysis in that paper indicates that both the relaxation factor and the spectral radius, ρ, are only affected slightly in going from 5 points to 9 points. Thus, for the 5-point SOR, $\omega = 2 - 2\pi h$, with $\rho = 1 - 2\pi h$, while for the 9-point SOR, $\omega = 2 - 2.116\pi h$, with $\rho = 1 - 1.79\pi h$, where h is the mesh size.

As an experiment we considered the stencil

(3.9)
$$\begin{pmatrix} \alpha & 1 & \alpha \\ 1 & -4(1+\alpha) & 1 \\ \alpha & 1 & \alpha \end{pmatrix}$$

with α varying [$\alpha = .25$ is equivalent to Eq. (3.8).] To use the Ad-Hoc method here, we again ignoring the corner points, but maintain their affect on the center point, i.e.

(3.10) $\mu_{p,q} = [\cos(\pi p/N + 1) + \cos(\pi q/M + 1)]/(2(1+\alpha)).$

Table 5.2 contains some numerical experiments using the optimal SOR and the Ad-Hoc SOR for this 9 point. The results are consistent with [1]. Indeed, the analysis of [1] can be used to analyze these cases. Again the Ad-Hoc SOR turns out to be a reasonable strategy for this nine point scheme, particularly if the optimum ω is difficult to obtain.

Shortly before the presentation of this paper, reference [8] came to the attention of the author. There, a generalization of the high order approximation is discussed, wherein the mesh size can differ in each direction. This leads to a stencil of the form

(3.11)
$$\begin{pmatrix} h^2 + k^2 & 2(5h^2 - k^2) & h^2 + k^2 \\ -2(h^2 - 5k^2) & -20(h^2 + k^2) & -2(h^2 - 5k^2) \\ h^2 + k^2 & 2(5h^2 - k^2) & h^2 + k^2 \end{pmatrix}$$

where h and k are the mesh size in each direction. Using the notation

(3.12)
$$\begin{pmatrix} b & c & b \\ a & -2(a + c + 2b) & a \\ b & c & b \end{pmatrix}$$

we consider the Jacobi eigenvalue

(3.13) $\mu = [|a| \cos(\pi h) + |c| \cos(\pi k)] / (a + c)$

However, a complication appears. If the two mesh sizes differ greatly, the signs of the off diagonal elements of the five point stencil can differ. In this case, to get a

$$\begin{pmatrix} \alpha & 1 & \alpha \\ 1 & -4(1+\alpha) & 1 \\ \alpha & 1 & \alpha \end{pmatrix}$$

8 × 8			
α	ω_0	$\lambda(\omega_0)$	$\lambda(\omega_H)$
.0	1.4903	.4903	.4903
.1	1.4850	.5135	.5186
.25	1.4709	.5270	.5449
.30	1.4664	.5285	.5506
.40	1.4581	.5293	.5586
.50	1.4511	.5279	.5633

31 × 31			
α	ω_0	$N(\omega_0)$	$N(\omega_H)$
.0	1.828 (1.822)	90 (.8220)	90 (.8220)
.1	1.826 (1.819)	85 (8325)	88 (.8353)
.25	1.816 (1.812)	88 (.8390)	91 (.8471)
.30	1.813 (1.810)	88 (.8392)	92 (.8495)
.50	1.803 (1.812)	90 (.7980)	97 (.8548)

Table 5.2

reasonable approximation to the relaxation factor one must take the absolute value of the contributions to the center point instead of the algebraic sum, i.e.

$$(3.14) \qquad \mu = [|a| \cos(\pi h) + |c| \cos(\pi k)] / (|a| + |c|).$$

Table 5.3 contains some numerical results. The values in parentheses were computed from formulas in [1] and are either ω, or if < 1, the spectral radius.

4 Block Iteration

Block iterative methods present a different problem. The expression for the Jacobi matrix eigenvalues using a block iteration is

$$(4.1) \qquad \bar{\mu} = \frac{2\sqrt{a_2 a_4} \cos(\pi/N + 1)}{a_0 - 2\sqrt{a_1 a_3} \cos(\pi/M + 1)}$$

where the points are block solved along the line with M number of grid points. The block method does enjoy Block Property A and thus SOR applies rigorously. It would seem that the Ad-Hoc method would be applicable, and indeed it is. However a danger lurks. Consider the stencil

$$(4.2) \qquad \begin{pmatrix} & 6 & \\ -2 & -4 & 3 \\ & -3 & \end{pmatrix}.$$

It appears innocently enough, but consider solving a 31×31 region with this difference equation using block Ad-Hoc SOR (or block SOR). The determining Jacobi eigenvalue is

$$(4.3) \qquad \bar{\mu}_L = \frac{2\sqrt{a_2 a_4} \cos(\pi/N + 1)}{a_0 - 2\sqrt{a_1 a_3} \cos(\pi/M + 1)} = -1.0332 + .84934i$$

which is a disaster since a necessary condition for SOR to converge is that the real part of the determining eigenvalue be less than 1. If this stencil occurred in the middle of a computation, the method would have to be aborted (and this has happened). But try point Ad-Hoc SOR (point SOR). The determining eigenvalue is

$$(4.4) \qquad \bar{\mu}_P = 2\left[\sqrt{a_1 a_3} \cos(\pi/M + 1) + \sqrt{a_2 a_4} \cos(\pi/N + 1)\right]/a_0 = 3.33i$$

and leads to no problem. Hence, the conclusion is that Ad-Hoc SOR is not recommended for general block methods.

The conclusion in general appears to be that Ad-Hoc is a relatively simple method to execute and is surprisingly efficient for its simplicity. For five point stencils, it may be optimal. For nine point stencils, it is a reasonably good estimate.

$$
\begin{pmatrix}
h^2 + k^2 & 2(5h^2 - k^2) & h^2 + k^2 \\
-2(h^2 - 5k^2) & -20(h^2 + k^2) & -2(h^2 - 5k^2) \\
h^2 + k^2 & 2(5h^2 - k^2) & h^2 + k^2
\end{pmatrix}
$$

h^{-1}	k^{-1}	ω_b	N_b	ω_H	N_H
32	32	1.802 (1.812)	68 (.8389)	1.821	74
32	22	1.777 (1.791)	60 (.8389)	1.802	65
32	12	1.752 (1.759)	53 (.8000)	1.795*	67
22	32	1.789 (1.784)	57 (.8046)	1.802	58
12	32	1.771 (1.745)	52 (.7594)	1.795[†]	61

*
$$
\begin{pmatrix}
.00792 & -.00412 & .00792 \\
.06749 & -.1584 & .06749 \\
.00792 & -.00412 & .00792
\end{pmatrix}
$$
[†]similar

Table 5.3

Acknowledgements

This work was supported by the Department of Navy, (SPAWAR) Contract N00039-89-C-5301. This paper was originally typed using Chiwriter by the author and retyped by Lisa Laguna.

References

[1] Adams, L.M., R.J. LeVeque, and D.M. Young, Jr. "Analysis of the SOR Iteration for the 9-point Laplacian," *SIAM Jr. Num. Anal.* **25** (1988) 1156–1180.

[2] Botta, E.F.F. and A.E.P. Veldman. "On Local Relaxation Methods and Their Application to Convection-Diffusion Equations," *Jr. of Comp. Physics* **48** (1982) 127–149.

[3] Ehrlich, L.W. "An Ad-Hoc SOR Method," *Jr. of Comp. Physics* **44** (1981), 31–45.

[4] Ehrlich, L.W. "The Ad-Hoc SOR Method: A Local Relaxation Scheme," *Elliptic Problem Solvers II*, Academic Press (1984), 257–269.

[5] Kuo, C.C.J., B.C. Levy, and B.R. Musicus. "A Local Relaxation Method for Solving Elliptic PDE's on Mesh-Connected Arrays," *SIAM Jr. Sci. Stat. Comput.* **8** (1987) 550–573.

[6] Rigal, A. "Convergence and Optimization of Successive Overrelaxation for Linear Systems of Equations with Complex Eigenvalues," *Jr. of Comp. Physics* **32** (1979) 10–23.

[7] Thompson, J.F., Zua Warsi, and C.W. Mastin. *Numerical Grid Generation–Foundations and Applications*, North-Holland (1985).

[8] van de Vooren, A.I., and A.C. Vliegenthart. "The 9-point Difference Formula for Laplace's Equations," *J. Engr. Math.* **1** (1967) 187–202.

Chapter 6

Block Iterative Methods for Cyclically Reduced Non-Self-Adjoint Elliptic Problems

HOWARD C. ELMAN
University of Maryland

and

GENE H. GOLUB
Stanford University

Dedicated to David M. Young, Jr., on the occasion of his sixty-fifth birthday. We are particularly pleased to dedicate this paper to Professor Young. His original work on iterative methods is a landmark of modern numerical analysis. Besides his great technical achievements, we all owe him a great debt of gratitude for his support of his younger colleagues and his gentlemanly manner.

Abstract

We study iterative methods for solving linear systems arising from two-cyclic discretizations of non-self-adjoint two-dimensional elliptic partial differential equations. The methods consist of applying one step of cyclic reduction, resulting in a "reduced system" of half the order of the original discrete problem, combined with a reordering and a block iterative technique for solving the reduced system. For constant coefficient problems, we present analytic bounds on the spectral radii of the iteration matrices in terms of cell Reynolds numbers that show the methods to be rapidly convergent. In addition, we describe numerical experiments that confirm and supplement the analysis. The paper summarizes results from [7], where further details can be found.

1   Introduction

We consider iterative methods for solving nonsymmetric linear systems

$$(1.1) \qquad\qquad \mathbf{Au} = \mathbf{f},$$

of the type that arise from finite difference discretizations of two-dimensional non-self-adjoint elliptic partial differential equations. For five-point finite difference discretizations, $\mathbf{A}$ has Property-A [15], i.e. its rows and columns can be ordered so that (1.1) has the form

$$(1.2) \qquad\qquad \begin{pmatrix} \mathbf{D} & \mathbf{C} \\ \mathbf{E} & \mathbf{F} \end{pmatrix} \begin{pmatrix} \mathbf{u}^{(r)} \\ \mathbf{u}^{(b)} \end{pmatrix} = \begin{pmatrix} \mathbf{f}^{(r)} \\ \mathbf{f}^{(b)} \end{pmatrix}$$

where $\mathbf{D}$ and $\mathbf{F}$ are diagonal matrices. The system (1.2) corresponds to a *red-black* ordering of the underlying grid. With one step of cyclic reduction, the "red" points $\mathbf{u}^{(r)}$ can be decoupled from the "black" points $\mathbf{u}^{(b)}$, producing a *reduced system*

$$(1.3) \qquad\qquad [\mathbf{F} - \mathbf{ED}^{-1}\mathbf{C}]\mathbf{u}^{(b)} = \mathbf{f}^{(b)} - \mathbf{ED}^{-1}\mathbf{f}^{(r)}.$$

We will show that the coefficient matrix

$$\mathbf{S} = \mathbf{F} - \mathbf{ED}^{-1}\mathbf{C}$$

is also sparse, so that (1.3) can be solved by some sparse iterative method. It has been observed empirically that preconditioned iterative methods are more effective for solving (1.3) than for solving (1.1) [5, 6].

In this paper, we outline a convergence analysis of some block iterative methods for solving (1.1) based on a *1-line* ordering of the reduced grid. The paper summarizes results from [7], where further details as well as comparisons with some methods for the full system (1.1) can be found. In the self-adjoint case, line methods of the type considered are known to be effective for (1.1), see e.g., [10, 14, 15]. They have also been applied successfully to non-self-adjoint problems [3, 4]. Our analysis applies to finite difference discretizations of constant coefficient elliptic problems with Dirichlet boundary conditions. We identify a wide variety of conditions under which the the coefficient matrix $\mathbf{S}$ is symmetrizable, and we use symmetrizability to derive bounds on the convergence in terms of cell Reynolds numbers. In addition, we present the results of numerical experiments on nonsymmetrizable and variable coefficient problems that supplement the analysis.

An outline of the paper is as follows. In Sec. 2, we describe the discrete constant coefficient two-dimensional convection-diffusion equation, and we present conditions under which the reduced matrix $\mathbf{S}$ is symmetrizable. In Sec. 3, we present bounds on the spectral radii of iteration matrices arising from a block Jacobi splitting of the reduced matrix where the underlying grid is ordered by diagonals. In Sec. 4, we present some numerical experiments that confirm the analysis of the symmetrizable case and demonstrate the effectiveness of the reduced system in other cases.

2 The Reduced System for the Convection-Diffusion Equation

Consider the constant coefficient convection-diffusion equation

$$(2.1) \qquad\qquad -\Delta u + \sigma u_x + \tau u_y = f$$

on the unit square $\Omega = (0,1) \times (0,1)$, with Dirichlet boundary conditions $u = g$ on $\partial\Omega$. We discretize (2.1) on a uniform $n \times n$ grid, using standard second order differences [14, 15]

$$\Delta u \approx \frac{u_{i+1,j} - 2u_{ij} + u_{i-1,j}}{h^2} + \frac{u_{i,j+1} - 2u_{ij} + u_{i,j-1}}{h^2}$$

for the Laplacian, where $h = 1/(n+1)$. We examine two choices of finite difference schemes for the first derivative terms:

centered differences: $u_x \approx \frac{u_{i+1,j}-u_{i-1,j}}{2h}$, $u_y \approx \frac{u_{i,j+1}-u_{i,j-1}}{2h}$,

upwind differences: $u_x \approx \frac{u_{ij}-u_{i-1,j}}{h}$, $u_y \approx \frac{u_{ij}-u_{i,j-1}}{h}$,

where the latter is applicable when $\sigma \geq 0$, $\tau \geq 0$.

Suppose the grid points are ordered using the rowwise natural ordering, i.e., the vector $\mathbf{u}$ is ordered lexicographically as $(u_{11}, u_{21}, \ldots, u_{nn})^T$. Then, for both discretizations of the first derivative terms, the coefficient matrix has the form

$$\mathbf{A} = \text{tri}\,[\,\mathbf{A}_{j,j-1},\, \mathbf{A}_{jj},\, \mathbf{A}_{j,j+1}\,]$$

where

$$\mathbf{A}_{j,j-1} = b\mathbf{I}, \qquad \mathbf{A}_{jj} = \text{tri}\,[c,\, a,\, d\,], \qquad \mathbf{A}_{j,j+1} = e\mathbf{I},$$

where "tri" indicates a tridiagonal matrix, $\mathbf{I}$ is the identity matrix, and all blocks are of order n. After scaling by h^2, the matrix entries are given by

$$a = 4, \quad b = -(1+\delta), \quad c = -(1+\gamma),$$
$$d = -(1-\gamma), \quad e = -(1-\delta),$$

for the centered difference scheme, where the quantities $\gamma = \sigma h/2$ and $\delta = \tau h/2$ are referred to as the cell Reynolds numbers; and

$$a = 4 + 2(\gamma + \delta), \quad b = -(1+2\delta), \quad c = -(1+2\gamma),$$
$$d = -1, \quad e = -1,$$

for the upwind scheme.

The left side of Fig. 6.1 shows the computational molecule for $\mathbf{A}$, and the center of the figure shows a portion of the graph of $\mathbf{A}$ relevant to the construction of the reduced matrix $\mathbf{S}$. For the reduction, the points numbered 3, 6, 8 and 11 (the "red

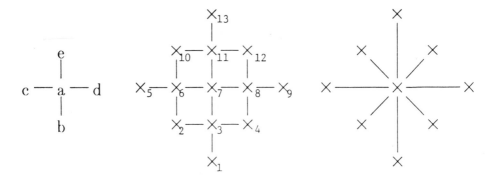

Figure 6.1 The computational molecule of the full system, and construction of the computational molecule of the reduced system.

points") are eliminated. The reduced matrix is a skewed nine-point operator whose computational molecule is shown on the right side of Fig. 6.1.

To see the entries of the reduced matrix, consider the submatrix of **A** consisting of the rows for points 3, 6, 7, 8 and 11, and the columns for all the points of the graph in the center of Fig. 6.1.

	Column Index →	3	6	8	11	7	1	2	4	5	9	10	12	13
Row Index ↓														
3		a				e	b	c	d					
6			a			d	b		c	e				
8				a		c		b	d		e			
11					a	b				c	d	e		
7		b	c	d	e	a								

Eliminating points 3, 6, 8 and 11 is equivalent to decoupling the first four rows of this matrix by Gaussian elimination. This modifies and produces fill-in in the last row. The computations performed for the elimination are shown in the following table. The new entries of the last row are obtained by summing the columns.

Col. →	7	1	2	4	5	9	10	12	13
Elim. :									
↓	a								
3	$-ba^{-1}e$	$-ba^{-1}b$	$-ba^{-1}c$	$-ba^{-1}d$					
6	$-ca^{-1}d$		$-ca^{-1}b$		$-ca^{-1}c$		$-ca^{-1}e$		
8	$-da^{-1}c$			$-da^{-1}b$		$-da^{-1}d$		$-da^{-1}e$	
11	$-ea^{-1}b$						$-ea^{-1}c$	$-ea^{-1}d$	$-ea^{-1}e$

Thus, the typical diagonal value in the reduced matrix is

$$(2.2) \qquad a - 2ba^{-1}e - 2ca^{-1}d,$$

which occurs at all interior grid points. After scaling by a, the computational molecule at an interior point for the reduced system is shown in Fig. 6.2. For grid points next to the boundary (see Fig. 6.3), some elimination steps are not required; for example, for a point next to the right boundary, it is not necessary to eliminate d. The diagonal values for mesh points next to the boundary are

$$a - 2ba^{-1}e - ca^{-1}d \text{ for pts. with one horizontal and two vertical neighbors}$$
$$(2.3) a - ba^{-1}e - 2cu^{-1}d \text{ for pts. with one vertical and two horizontal neighbors}$$
$$a - ba^{-1}e - ca^{-1}d \quad \text{for pts. with just two neighbors.}$$

Now, suppose the reduced grid is ordered by diagonal lines oriented in the NW–SE direction. An example of such an ordering derived from a 6×6 grid is shown in Fig. 6.3. The reduced matrix $\mathbf{S}$ then has block tridiagonal form

$$\begin{pmatrix} \mathbf{S}_{11} & \mathbf{S}_{12} & & & \\ \mathbf{S}_{21} & \mathbf{S}_{22} & \mathbf{S}_{23} & & \\ & \ddots & \ddots & \ddots & \\ & & & & \mathbf{S}_{l-1,l} \\ & & & \mathbf{S}_{l,l-1} & \mathbf{S}_{ll} \end{pmatrix},$$

where $l = n - 1$ is the number of diagonal lines. The diagonal blocks $\{\mathbf{S}_{jj}\}$ are tridiagonal,

$$\mathbf{S}_{jj} = \text{tri}[\quad -2ce, \quad *, \quad -2bd \quad],$$

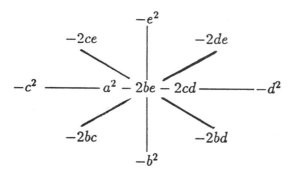

Figure 6.2 The computational molecule for the reduced system.

where "$*$" is defined as in (2.2) and (2.3) (scaled by a). The subdiagonal blocks $\{\mathbf{S}_{j,j-1}\}$ have nonzero structure

$$-\begin{pmatrix} d^2 & & & \\ 2de & d^2 & & \\ e^2 & 2de & \ddots & \\ & e^2 & \ddots & d^2 \\ & & \ddots & 2de \\ & & & e^2 \end{pmatrix}, \quad -\begin{pmatrix} 2de & d^2 & & \\ e^2 & 2de & d^2 & \\ & \ddots & \ddots & \ddots \\ & & & d^2 \\ & & e^2 & 2de \end{pmatrix}, \quad -\begin{pmatrix} e^2 & 2de & d^2 & \\ & e^2 & 2de & d^2 \\ & & \ddots & \ddots & \ddots \\ & & & e^2 & 2de & d^2 \end{pmatrix},$$

for $2 \leq j < l/2 + 1$, $j = l/2 + 1$ (l even), and $l/2 + 1 < j$, respectively. The corresponding superdiagonals $\{\mathbf{S}_{j-1,j}\}$ are

$$-\begin{pmatrix} c^2 & 2bc & b^2 & \\ & c^2 & 2bc & b^2 \\ & & \ddots & \ddots & \ddots \\ & & & c^2 & 2bc & b^2 \end{pmatrix}, \quad -\begin{pmatrix} 2bc & b^2 & & \\ c^2 & 2bc & b^2 & \\ & \ddots & \ddots & \ddots \\ & & & b^2 \\ & & c^2 & 2bc \end{pmatrix}, \quad -\begin{pmatrix} b^2 & & & \\ 2bc & b^2 & & \\ c^2 & 2bc & \ddots & \\ & c^2 & \ddots & b^2 \\ & & \ddots & 2bc \\ & & & c^2 \end{pmatrix}.$$

The following result gives circumstances under which $\mathbf{S}$ is symmetrizable.

Theorem 2.1 *The reduced matrix $\mathbf{S}$ can be symmetrized with a real diagonal similarity transformation if and only if the product $bcde$ is positive.*

This result says that there is a matrix $\mathbf{Q} = \mathrm{diag}(\mathbf{Q}_1, \mathbf{Q}_2, \ldots, \mathbf{Q}_l)$ where the submatrix $\mathbf{Q}_j = \mathrm{diag}(q_1^{(j)}, q_2^{(j)}, \ldots, q_{r_j}^{(j)})$ has the same order as $\mathbf{S}_{jj}$, such that $\mathbf{Q}^{-1}\mathbf{S}\,\mathbf{Q}$ is

$\times 7$	$\cdot$	$\times 3$	$\cdot$	$\times 1$	$\cdot$
$\cdot$	$\times 8$	$\cdot$	$\times 4$	$\cdot$	$\times 2$
$\times 13$	$\cdot$	$\times 9$	$\cdot$	$\times 5$	$\cdot$
$\cdot$	$\times 14$	$\cdot$	$\times 10$	$\cdot$	$\times 6$
$\times 17$	$\cdot$	$\times 15$	$\cdot$	$\times 11$	$\cdot$
$\cdot$	$\times 18$	$\cdot$	$\times 16$	$\cdot$	$\times 12$

Figure 6.3 The reduced grid derived from an 6×6 grid, with ordering by diagonals.

symmetric. A sketch of the proof is as follows. The individual blocks $\mathbf{S}_{jj}$ on the diagonal are symmetrized provided the entries of $\mathbf{Q}_j$ satisfy

$$q_i^{(j)} = \left(\frac{ce}{bd}\right)^{1/2} q_{i-1}^{(j)}.$$

This recurrence is well-defined if and only if $ce/(bd) = bcde/(bd)^2$ is positive. The off-diagonal part of $\mathbf{S}$ can be symmetrized if the first entries of each block, $\{q_1^{(j)}\}_{j=1}^l$, are chosen appropriately, see [7] for details. The result also follows from the analysis of [11].

Let $\mathbf{D} = \text{diag}(\mathbf{S}_{11}, \mathbf{S}_{22} \ldots, \mathbf{S}_{ll})$ denote the block diagonal of $\mathbf{S}$, and denote the block Jacobi splitting of $\mathbf{S}$ as $\mathbf{S} = \mathbf{D} - \mathbf{C}$. Let $\hat{\mathbf{S}}$ denote the symmetrized matrix $\mathbf{Q}^{-1}\mathbf{S}\mathbf{Q}$ (when it exists), and let $\hat{\mathbf{S}} = \hat{\mathbf{D}} - \hat{\mathbf{C}}$ denote the block Jacobi splitting. Here $\hat{\mathbf{D}} = \mathbf{Q}^{-1}\mathbf{D}\mathbf{Q}$ and $\hat{\mathbf{C}} = \mathbf{Q}^{-1}\mathbf{C}\mathbf{Q}$. The following corollary gives conditions for symmetrizablility of the reduced matrices arising from the two finite difference schemes under consideration. It applies for the upwind difference scheme, and it applies for the centered difference scheme if the absolute values of the cell Reynolds numbers are either both less than one or if they are both greater than one. See [7] for a proof. The computational molecule for the symmetrized reduced matrix is shown in Fig. 6.4.

Corollary 2.1 *If $\mathbf{A}$ is constructed using centered differences, then $\mathbf{S}$ is symmetrizable via a real diagonal matrix $\mathbf{Q}$ if and only if either $|\gamma| < 1$ and $|\delta| < 1$ both hold, or $|\gamma| > 1$ and $|\delta| > 1$ both hold. If $|\gamma| < 1$ and $|\delta| < 1$, then $\mathbf{S}$ is an irreducibly diagonally dominant M-matrix and $\mathbf{Q}$ can be chosen so that $\hat{\mathbf{S}}$ is an irreducibly diagonally dominant M-matrix. If $|\gamma| > 1$ and $|\delta| > 1$, then $\mathbf{Q}$ can be chosen so that $\hat{\mathbf{D}}$ is a diagonally dominant M-matrix. If $\mathbf{A}$ is constructed using upwind differences, then $\mathbf{S}$ is symmetrizable for all $\gamma \geq 0$ and $\delta \geq 0$, and $\mathbf{S}$ and (for appropriately chosen $\mathbf{Q}$) $\hat{\mathbf{S}}$ are irreducibly diagonally dominant M-matrices.*

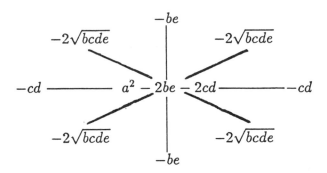

Figure 6.4 The computational molecule for the symmetrized reduced system.

3 Bounds for Solving the Convection-Diffusion Equation

In this section, bounds for the spectral radius of the iteration matrix $\mathbf{B} = \mathbf{D}^{-1}\mathbf{C}$ are presented based on the block Jacobi splitting of $\mathbf{S}$, in the case where $\mathbf{S}$ is symmetrizable. Note that

$$\mathbf{B} = \mathbf{Q}\hat{\mathbf{D}}^{-1}\hat{\mathbf{C}}\mathbf{Q}^{-1},$$

i.e., $\mathbf{B}$ is similar to $\hat{\mathbf{B}} = \hat{\mathbf{D}}^{-1}\hat{\mathbf{C}}$. Hence, we can restrict our attention to $\hat{\mathbf{B}}$. The analysis is essentially based on the result

(3.1)
$$\rho(\hat{\mathbf{D}}^{-1}\hat{\mathbf{C}}) \leq \|\hat{\mathbf{D}}^{-1}\|_2\|\hat{\mathbf{C}}\|_2 = \frac{\rho(\hat{\mathbf{C}})}{\lambda_{\min}(\hat{\mathbf{D}})},$$

where the equality follows from the symmetry of $\hat{\mathbf{D}}$ and $\hat{\mathbf{C}}$. $\hat{\mathbf{D}}$ is a block tridiagonal matrix each of whose blocks is approximately equal to a constant coefficient tridiagonal matrix of the form $\hat{\mathbf{T}} = \mathrm{tri}\,[-\hat{b}, \hat{a}, -\hat{b}]$, where $\hat{a} = a^2 - 2be - 2cd$ and $\hat{b} = 2\sqrt{bcde}$. Using this fact, it is easily shown that when $be > 0$ and $cd > 0$,

(3.2)
$$\lambda_{\min}(\hat{\mathbf{D}}) \geq \hat{a} - 2\hat{b}\cos(\pi h).$$

The spectral radius of $\hat{\mathbf{C}}$ is bounded by Gerschgorin's theorem [14], see Fig. 6.4:

$$\rho(\hat{\mathbf{C}}) \leq 4\sqrt{bcde} + 2be + 2cd = 2(\sqrt{be} + \sqrt{cd}\,)^2.$$

When $be < 0$ and $cd < 0$, the inequality of (3.2) is not valid and (3.1) cannot be used directly. However, a careful perturbation analysis leads to a somewhat weaker result [7]. We summarize these results for the two finite difference schemes as follows:

Theorem 3.1 *For the centered difference scheme, if $|\gamma| < 1$ and $|\delta| < 1$, then the spectral radius of the block Jacobi iteration matrix for the reduced system is bounded by*

$$\frac{(\sqrt{1 - \gamma^2} + \sqrt{1 - \delta^2})^2}{8 - (\sqrt{1 - \gamma^2} + \sqrt{1 - \delta^2})^2 + 2\sqrt{(1 - \gamma^2)(1 - \delta^2)}\,(1 - \cos(\pi h))}.$$

If $|\gamma| > 1$, $|\delta| > 1$ and $\sqrt{(\gamma^2 - 1)(\delta^2 - 1)} \leq 4$, then the spectral radius is bounded by

$$\frac{\frac{1}{2}\mu(\gamma, \delta) + \gamma^2 - 1 + \delta^2 - 1}{8 + (\sqrt{\gamma^2 - 1} - \sqrt{\delta^2 - 1})^2 + 2\sqrt{(\gamma^2 - 1)(\delta^2 - 1)}\,(1 - \cos(\pi h))},$$

where

$$\mu(\gamma, \delta) \equiv \max\left\{4\sqrt{(\gamma^2 - 1)(\delta^2 - 1)},\ 2\sqrt{(\gamma^2 - 1)(\delta^2 - 1)} + \gamma^2 - 1,\right.$$
$$\left. 2\sqrt{(\gamma^2 - 1)(\delta^2 - 1)} + \delta^2 - 1,\ \gamma^2 - 1 + \delta^2 - 1\right\}.$$

For the upwind difference scheme, the spectral radius is bounded by

$$\frac{(\sqrt{1 + 2\gamma} + \sqrt{1 + 2\delta})^2}{2(2 + \gamma + \delta)^2 - (\sqrt{1 + 2\gamma} + \sqrt{1 + 2\delta})^2 + 2\sqrt{(1 + 2\gamma)(1 + 2\delta)}\,(1 - \cos(\pi h))}.$$

We will show in Sec. 5 that these bounds agree with the results of numerical computations when $be > 0$ and $cd > 0$, but they are pessimistic when $be < 0$ and $cd < 0$. We now present a Fourier analysis of a variant of the symmetrized reduced operator using the methodology of [2]. Consider the discrete nine-point operator of Fig. 6.4. This operator is based on the version of $\hat{S}$ of Fig. 6.4, except that it is defined on a rectilinear grid with periodic boundary conditions. The horizontal lines of the rectilinear grid correspond to the lines oriented in the NW–SE direction of the skewed grid. (In the figure, the orientation of the skewed grid is indicated in parentheses.) We refer to this operator as the rectilinear periodic reduced operator.

The Fourier analysis is defined as follows, see [2] for a more detailed description. Suppose the rectilinear grid is contained in a square domain with n interior points in each direction and periodic boundary conditions. Let $\hat{S}_P$ denote the operator defined by the computational molecule of Fig. 6.5. That is, if v is a mesh function with value v_{jk} at the (j, k) mesh point, $1 \leq j, k \leq n$, then

$$(\hat{S}_P v)_{jk} \equiv (a^2 - 2be - 2cd)\,v_{jk} - 2\sqrt{bcde}\,v_{j-1,k} - 2\sqrt{bcde}\,v_{j+1,k}$$
$$- be\,v_{j-1,k+1} - 2\sqrt{bcde}\,v_{j,k+1} - cd\,v_{j+1,k+1}$$
$$- cd\,v_{j-1,k-1} - 2\sqrt{bcde}\,v_{j,k-1} - be\,v_{j+1,k-1}.$$

The analogue of the line Jacobi splitting considered above is

$$\hat{S}_P = \hat{D}_P - \hat{C}_P,$$

where $\hat{D}_P$ corresponds to the horizontal connections of Fig. 6.5 (indices $(j - 1, k)$, (j, k) and $(j, k + 1)$), and $\hat{C}_P$ corresponds to the other connections. Let $\mathbf{v} = \mathbf{v}^{(s,t)}$

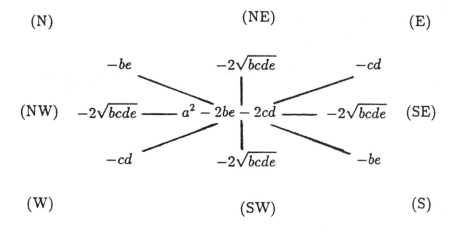

Figure 6.5 The computational molecule for the rectilinear periodic reduced operator.

have values $v_{jk}^{(s,t)} = e^{ij\theta_s} e^{ik\phi_t}$, where $\theta_s = 2\pi sh$, $\phi_t = 2\pi th$ $1 \leq s,t \leq n$, and $h = 1/(n+1)$. It is straightforward to show by direct substitution that

$$(\hat{\mathbf{S}}_P\mathbf{v})_{jk} = \lambda v_{jk}, \qquad (\hat{\mathbf{D}}_P\mathbf{v})_{jk} = \psi v_{jk}, \qquad (\hat{\mathbf{C}}_P\mathbf{v})_{jk} = \mu \mathbf{v}_{jk},$$

where

$$\psi = \psi_{st} = a^2 - 2be - 2cd - 4\sqrt{bcde} \cos\theta_s,$$
$$\mu = \mu_{st} = 4\sqrt{bcde} \cos\phi_t + 2cd\cos(\theta_s + \phi_t) + 2be\cos(\theta_s - \phi_t),$$

and $\lambda = \psi - \mu$. The quantities μ and ψ are the eigenvalues of $\hat{\mathbf{D}}_P$ and $\hat{\mathbf{C}}_P$, respectively, corresponding to the (shared) eigenvector $\mathbf{v}$. The analogous eigenvalue of $\hat{\mathbf{D}}_P^{-1}\hat{\mathbf{C}}_P$ is

(3.3) $$\frac{4\sqrt{bcde} \cos\phi_t + 2cd\cos(\theta_s + \phi_t) + 2be\cos(\theta_s - \phi_t)}{a^2 - 2be - 2cd - 4\sqrt{bcde} \cos\theta_s}.$$

The maximal value of this expression over all θ_s, ϕ_t, $1 \leq s,t \leq n$, is a heuristic bound for the maximal eigenvalue for the analogous Dirichlet operator. For simplicity we examine (3.3) for all θ_s, $\phi_t \in [0, 2\pi]$. This ignores some $O(h^2)$ effects that are significant only when $\gamma = \delta = 0$. The resulting bounds are summarized as follows, see [7] for a proof.

Theorem 3.2 *Fourier analysis of the rectilinear periodic reduced operator yields the following asymptotic bounds on the Jacobi iteration matrix.*

For centered differences with $|\gamma| < 1$, $|\delta| < 1$:

$$\frac{(\sqrt{1-\gamma^2} + \sqrt{1-\delta^2})^2}{8 - (\sqrt{1-\gamma^2} + \sqrt{1-\delta^2})^2}.$$

For centered differences with $|\gamma| > 1$, $|\delta| > 1$:

$$\frac{(\sqrt{\gamma^2-1} + \sqrt{\delta^2-1})^2}{8 + (\sqrt{\gamma^2-1} + \sqrt{\delta^2-1})^2}.$$

For upwind differences:

$$\frac{(\sqrt{1+2\gamma} + \sqrt{1+2\delta})^2}{2(2 + \gamma + \delta)^2 - (\sqrt{1+2\gamma} + \sqrt{1+2\delta})^2}.$$

The first and third of these bounds agree with the analogous asymptotic results from Theorem 3.1. The second bound does not depend on any restrictions on γ and δ.

   Finally, note that in the symmetrizable cases, the eigenvalues of the block Jacobi matrices are real. Since $\mathbf{S}$ has block Property-A, Young's SOR analysis [14,15] applies, i.e., the optimal SOR iteration parameter can be obtained from the spectral radius of the block Jacobi matrix. Symmetrization is needed only for the analysis; the actual computations can be performed with the nonsymmetric matrices.

4   Numerical Experiments

In this section, we present the results of numerical experiments that confirm and supplement the analysis of Sec. 3, for centered difference discretizations. In all cases, we present our results for the block Gauss-Seidel splitting $\mathbf{S} = [\mathbf{D} - \mathbf{L}] - \mathbf{U}$, where $\mathbf{L}$ and $\mathbf{U}$ are the the lower and upper triangles of $\mathbf{C}$, respectively. (Since $\mathbf{S}$ has block Property-A, $\rho((\mathbf{D}-\mathbf{L})^{-1}\mathbf{U}) = [\rho(\mathbf{D}^{-1}\mathbf{C})]^2$.) In particular, we compare the bounds of Theorem 3.2 with computed values for several different mesh sizes, and we also examine the effectiveness of the block Gauss-Seidel method in cases where the analysis is not applicable (e.g., for the centered difference discretization where $|\gamma| > 1$ and $|\delta| < 1$). In addition, we present numerical results for several variable coefficient problems. All experiments were performed on a VAX-8600 in double precision Fortran. The reduced matrices were computed using PCGPAK [12]. The spectral radii $\rho((\mathbf{D} - \mathbf{L})^{-1}\mathbf{U})$ were determined using the QZ algorithm in EISPACK [8, 9]. See [7] for numerical experiments for problems with other boundary conditions and for comparisons with block Jacobi methods applied directly to the full system (1.1).

   Table 6.1 shows computed values of the spectral radii of the block Gauss-Seidel iteration matrices derived from the centered difference discretization of (2.1), for

---

*We suspect that when $\gamma = 1$, the Gauss-Seidel matrix has nonlinear elementary divisors and that this is why this computed spectral radius exceeds the asymptotic bound.

γ	$h = 1/8$	$h = 1/16$	$h = 1/32$	Asymptotic Bound
.2	.50	.79	.89	.92
.4	.40	.62	.69	.72
.6	.26	.40	.45	.46
.8	.13	.19	.21	.22
1.0	.01	.02	.05*	.02
1.2	.03	.03	.04	–
1.4	.04	.05	.06	–
1.6	.08	.08	.08	–
1.8	.10	.10	.11	–
2.0	.10	.10	.15	–

Table 6.1 Spectral radii and bounds for the Gauss-Seidel iteration matrices, centered differences, $\delta = 0$.

$\tau = 0$ (so that $\delta = 0$). For $\gamma \leq 1$, the table also shows the asymptotic bound for these quantities derived by squaring the first expression of Theorem 3.1, with $h = 0$. The value for $\gamma = 1$ is the limit as $\gamma \to 1$. Numerical experiments with $\gamma = 0$ and varying δ produced the same spectral radii; since the bounds of Theorem 3.1 are symmetric with respect to γ and δ, Table 6.1 also applies for the case $\gamma = 0$ and δ taking on the values in the first column. The results for $\gamma < 1$ show that the limiting values of the spectral radii tend to the the bounding value as $h \to 0$ (for γ fixed). The analytic bounds for the values of h in the table are closer to the asymptotic bounds than to the computed spectral radii. The results also show that the method is highly effective for $\gamma > 1$, where **S** is not symmetrizable and our analysis does not apply. In this case, some computed eigenvalues of each Gauss-Seidel matrix are complex.

Table 6.2 shows the computed spectral radii of the block Gauss-Seidel iteration matrices for the centered difference discretization and $\gamma = \delta$. This table contains both the rigorous bounds from Theorem 3.1 and the bounds of Theorem 3.2 derived using Fourier analysis. The two bounds agree when $\gamma < 1$ and $\delta < 1$. The rigorous bounds are pessimistic when $\gamma > 1$ and $\delta > 1$, and the Fourier results agree with experimental results. Note that in both Tables 6.1 and 6.2, for moderate values of γ, the more highly nonsymmetric problems are easier to solve than the nearly symmetric problems.

Finally, we describe some results for problems with variable coefficients, of the form

$$-\Delta u + f(x, y)u_x + g(x, y)u_y$$

on $\Omega = (0, 1) \times (0, 1)$, $u = g$ on $\partial\Omega$. We consider four problems, which are taken from [1, 13]:

$$1. \quad -\Delta u + \sigma x^2 u_x + \sigma x^2 u_y = 0,$$

γ	$h = 1/8$	$h = 1/16$	$h = 1/32$	Asymptotic Bound	Fourier
.2	.46	.73	.82	.85	.85
.4	.30	.46	.51	.52	.52
.6	.13	.19	.21	.22	.22
.8	.03	.04	.05	.05	.05
1.0	0	0	0	0	0
1.2	.02	.03	.03	.05	.03
1.4	.07	.10	.10	.23	.11
1.6	.14	.18	.19	.61	.19
1.8	.21	.26	.27	1.25	.28
2.0	.27	.33	.35	2.25	.36

Table 6.2 Spectral radii and bounds for the Gauss-Seidel iteration matrices, centered differences, $\gamma = \delta$.

$$2. \quad -\Delta u + \tfrac{1}{2}\sigma(1 + x^2)u_x + 100u_y = 0,$$
$$3. \quad -\Delta u + \sigma x^2 u_x = 0,$$
$$4. \quad -\Delta u + \sigma(1 - 2x)u_x + \sigma(1 - 2y)u_y = 0,$$

with $u = 0$ on $\partial\Omega$. As in [1, 13], we discretized each problem using centered differences and mesh size $h = 1/20$. The spectral radii of the Gauss-Seidel iteration matrices are shown in Table 6.3. For reference, the table also reports $\gamma = \sigma h/2$. For Problems 1 and 3, γ represents the maximum cell Reynolds number on the mesh, the minimum being 0. For Problem 2, it represents the maximum cell Reynolds number for the x-coordinate; the minimum is $\gamma/2$, and the y-coordinate has constant value $\delta = 5$. For Problem 4, the coefficients change sign in Ω; the cell Reynolds numbers in each coordinate vary between $-\gamma$ and γ.

Although it is difficult to make definitive statements about these results, they appear to be consistent with the analysis of the constant coefficient case. In particular, for moderate σ (the three smaller values), the spectral radii are bounded well below one for all four problems. For Problems 1, 3, and 4, performance improves as $|\gamma|$ increases from 0; and for Problem 2 (where $\delta = 5$) it is very good for moderate γ. Performance declines when γ gets very large; in these cases finer meshes will improve both performance of the iterative solver and accuracy of the discrete solution. We remark that for Problem 2 with $\sigma \leq 10$, the Gauss-Seidel matrices have complex eigenvalues close in modulus to their spectral radii. In many of the other cases (e.g., all instances of Problem 1), some computed eigenvalues contain small imaginary parts, of order at most 10^{-2}.

σ	$\gamma = \frac{\sigma h}{2}$	Problem 1	Problem 2	Problem 3	Problem 4
1	.03	.91	.23	.91	.90
10	.26	.91	.23	.92	.80
100	2.6	.78	.40	.83	.18
1000	26	.96	.94	.89	.95
10000	263	.998	.999	.994	.999

Table 6.3 Spectral radii for the Gauss-Seidel iteration matrices of four problems with variable coefficients, centered differences, $h = 1/20$.

Acknowledgements

The work of Howard Elman was supported by the National Science Foundation under grant DMS-8607478, and by the U. S. Army Research Office under grant DAAL-0389-K-0016. The work of Gene Golub was supported by the National Science Foundation under grant DCR-8412314, the Simon Guggenheim Memorial Foundation, and the University of Maryland's Institute for Advanced Computer Studies, whose support is gratefully acknowledged. This paper was originally typed by the authors in TEX and reformatted by Lisa Laguna in LaTeX.

References

[1] Botta, E.F.F., and A. E. P. Veldman. "On local relaxation methods and their application to convection-diffusion equations," *J. Comput. Phys.* **48** (1981) 127–149.

[2] Chan, T. and H. C. Elman. "Fourier Analysis of Iterative Methods for Elliptic Problems," *SIAM Review* **31** (1989) 20–49.

[3] Chin, R.C.Y. and T. A. Manteuffel. "An analysis of block successive overrelaxation for a class of matrices with complex spectra," *SIAM J. Numer. Anal.* **25** (1988) 564–585.

[4] Chin, R.C.Y., T.A. Manteuffel, and J. de Pillis. "ADI as a preconditioning for solving the convection-diffusion equation," *SIAM J. Sci. Statj. Comput.* **5** (1984) 281–299.

[5] Eisenstat, S.C., H.C. Elman and M.H. Schultz. "Block-preconditioned conjugategradientlike methods for numerical reservoir simulation," *SPE Reservoir Engineering*, Feb. 1988, 307-312.

[6] Elman, H.C. *Iterative Methods for Large, Sparse, Nonsymmetric Systems of Linear Equations*, Ph.D. Thesis, Department of Computer Science, Yale University, 1982.

[7] *Iterative Methods for Cyclically Reduced Non-Self-Adjoint Linear Systems*, Report UMIACS-TR-88-87, University of Maryland, College Park, MD, Nov. 1988. To appear in *Math. Comp.*

[8] Garbow, R.S., J.M. Boyle, J.J. Dongarra, and C.B. Moler. *Matrix Eigensystem Routines:* EISPACK *Guide Extension*, Springer-Verlag, New York, 1972.

[9] Golub, G.H., and C.F. van Loan. *Matrix Computations*, The Johns Hopkins University Press, Baltimore, MD, 1983. (2nd Edition, 1989.)

[10] Parter, S.V. "On estimating the 'rates of convergence' of iterative methods for elliptic difference equations," *Trans. Amer. Math. Soc.* **114** (1965) 320-354.

[11] Parter, S.V., and J.W.T. Youngs. "The symmetrization of matrices by diagonal matrices," *J. Math. Anal. Appl.* **4** (1962) 102–110.

[12] PCGPAK *User's Guide*, Version 1.04, Scientific Computing Associates, New Haven, CT, 1987.

[13] Thompson, M.C., J.H. Ferziger, and G.H. Golub. "Block SOR applied to the cyclically-reduced equations as an efficient solution technique for convection-diffusion equations," in *Computational Techniques and Applications: CTAC-87*, J. Noye, C.A.J. Fletcher and G. de Vahl Davis, eds., North Holland, 1988.

[14] Varga, R.S. *Matrix Iterative Analysis*, Prentice-Hall, New Jersey, 1962.

[15] Young, D.M. *Iterative Solution of Large Linear Systems*, Academic Press, New York, 1971.

Chapter 7

Toward an Effective Two-Parameter SOR Method

GENE H. GOLUB
Stanford University

and

JOHN E. DE PILLIS
University of California, Riverside

This paper is dedicated to David Young on the occasion of his sixty-fifth birthday. Professor Young has made profound and imaginative contributions to the field. But it is a pleasure to acknowledge his character, too—the embodiment of a gentleman and a scholar.

Abstract

Let $\mathbf{A}$ be a symmetric $n \times n$ matrix where $\mathbf{B} = \mathbf{I} - \mathbf{A}$ is the corresponding Jacobi matrix: We assume $\mathbf{B}$ has been re-ordered using the red/black ordering so that

$$\mathbf{B} = \begin{bmatrix} \mathbf{0} & \mathbf{M_{p \times q}} \\ \mathbf{M_{q \times p}^T} & \mathbf{0} \end{bmatrix}.$$

Using the Singular Value Decomposition of $\mathbf{M}$, we are able to derive an equation for computing the eigenvalues of the two-parameter SOR iteration matrix L_{ω_1,ω_2}. In addition, we give a method for computing the spectral norm of L_{ω_1,ω_2}^k for any integer k. A numerical example shows that the spectral radius and the spectral norm are minimized for different values of ω_1 and ω_2.

1 Background

The theory for solving linear systems by successive overrelaxation (SOR) developed by David Young [4] is certainly one of the most elegant of modern numerical analysis. Stationary iterative methods, including SOR, solve the $n \times n$ linear system

$$\mathbf{Ax} = \mathbf{f}$$

by first splitting $\mathbf{A}$ into two terms

(1.1) $\mathbf{A} = \mathbf{A}_0 - \mathbf{A}_1$

where $\mathbf{A}_0^{-1}$ is, in some sense, easy to compute. Then, by choosing any starting vector $\mathbf{x}_0$, the splitting (1.1) is used to generate the vector sequence $\{\mathbf{x}_k\}$ where

(1.2) $\mathbf{x}_k = \underbrace{\mathbf{A}_0^{-1}\mathbf{A}_1}_{\mathbf{B}} x_{k-1} + \mathbf{A}_0^{-1}\mathbf{f}, \qquad k = 1,2,3,\ldots.$

The sequence (1.2) defines the *iteration matrix* $\mathbf{B}$ by

(1.3) $\mathbf{B} = \mathbf{A}_0^{-1}\mathbf{A}_1 \qquad \text{whenever } \mathbf{A} = \mathbf{A}_0 - \mathbf{A}_1.$

Let $\sigma(\mathbf{B})$ denote the *spectrum* or *eigenvalues* of matrix $\mathbf{B}$. The largest modulus (absolute value) among the eigenvalues $\sigma(\mathbf{B})$ is called the *spectral radius* of $\mathbf{B}$ and is denoted $\rho(\mathbf{B})$. It is this quantity that determines whether $\{\mathbf{x}_k\}$ converges to solution $\mathbf{x}$ of $\mathbf{Ax} = \mathbf{f}$ for any $\mathbf{x}_0$. In fact,

$$\mathbf{x}_k \to \mathbf{x} \quad \text{iff} \quad \rho(\mathbf{B}) < 1.$$

Moreover, it can be shown that asymptotically, the number of iterations which produce each decimal place of accuracy is

(1.4) $\text{step\_count} = \dfrac{-1}{\log_{10}(\rho(\mathbf{B}))}.$

(cf. [3] Chp. 3) so that the smaller $\rho(\mathbf{B})$ implies faster convergence of (1.2).

Norms and Errors: For given linear system $\mathbf{Ax} = \mathbf{f}$ and iterative sequence (1.2), we define the *error vector* $\mathbf{e}_k = \mathbf{x}_k - \mathbf{x}$. Since solution vector $\mathbf{x}$ satisfies (1.2), i.e., we may write $\mathbf{x} = \mathbf{Bx} + \mathbf{A}_0^{-1}f$ which, when subtracted from (1.2), yields $\mathbf{e}_k = \mathbf{Be}_{k-1}$: This, in turn, implies (inductively)

(1.5) $\mathbf{e}_k = \mathbf{Be}_{k-1} = \mathbf{B}^2\mathbf{e}_{k-2} = \cdots = \mathbf{B}^k\mathbf{e}_0.$

Applying norms to (1.5) gives us the important error measure

(1.6a) $\|\mathbf{e}_k\| = \|\mathbf{B}^k\mathbf{e}_0\| \leq \|\mathbf{B}^k\| \cdot \|\mathbf{e}_0\|$

(1.6b) $\approx c_{k,p}\rho(\mathbf{B})^{k-p} \cdot \|\mathbf{e}_0\| \qquad \text{for large } k.$

The constant $c_{k,p}$ in (1.6b) is increasing in k: Integer p is the dimension of the largest Jordan block of $\mathbf{B}$ and there is no other $p \times p$ Jordan block. (cf. [3], Thm. 3.1.) The bound (1.6b) is consistent with the spectral theorem which says that $\lim_{k \to \infty} ||\mathbf{B}^k||^{1/k} = \rho(\mathbf{B})$ for any norm $|| \cdot ||$. The usual measure for speed of convergence, (1.4), derives from (1.6b). On the other hand, (1.6b), which depends on $||\mathbf{B}^k||$, is not restricted to large k.

From this we conclude that short-term convergence may be improved by minimizing $||\mathbf{B}||$ while long-term or asymptotic convergence is improved by minimizing the spectral radius $\rho(\mathbf{B})$.

In this paper, then, we characterize both the eigenvalues (hence, the spectral radius ρ) and the norms of competing iteration matrices which we shall denote $\mathbf{B}_J$ and L_ω.

For a comparison of two iteration matrices (1.3) resulting from two splittings (1.1), we turn to the SOR method as studied by Young. The symmetric systems $\mathbf{Ax} = \mathbf{f}$ will be given by matrices $\mathbf{A}$ with special form

$$\mathbf{Ax} = (\mathbf{I}_n - \mathbf{L} - \mathbf{U})\mathbf{x} = \mathbf{f}$$

where

$$(1.7) \qquad \mathbf{A}_{n \times n} = \begin{bmatrix} \mathbf{I}_p & -\mathbf{M} \\ -\mathbf{M}^T & \mathbf{I}_q \end{bmatrix} = \underbrace{\begin{bmatrix} \mathbf{I}_p & 0 \\ 0 & \mathbf{I}_q \end{bmatrix}}_{\mathbf{I}_n} - \underbrace{\begin{bmatrix} 0 & 0 \\ \mathbf{M}^T & 0 \end{bmatrix}}_{\mathbf{L}} - \underbrace{\begin{bmatrix} 0 & \mathbf{M} \\ 0 & 0 \end{bmatrix}}_{\mathbf{U}}$$

and $\mathbf{M}$ is a $p \times q$ matrix with $p \geq q$. Now two-by-two block matrices of form (1.7), whose *block* diagonals are themselves *scalar* diagonal matrices, are a special case of matrices with *Property A* having consistent ordering.

Given (1.7), Young compared the iteration matrices induced by two splittings $\mathbf{A} = \mathbf{A}_0 - \mathbf{A}_1$ (1.1): one with $\mathbf{A}_0 = \mathbf{I}_n$, and the other with $\mathbf{A}_0 = \frac{1}{\omega}\mathbf{I}_n - \mathbf{L}$. The result is the corresponding pair of iteration matrices (1.3) $\mathbf{B}_J$ and L_ω defined by

$$\mathbf{B}_J = \mathbf{I}_n^{-1}(\mathbf{L} + \mathbf{U}), \qquad \mathbf{A}_0 = \mathbf{I}_n$$

and

$$(1.8) \qquad L_\omega = \left(\frac{1}{\omega}\mathbf{I}_n - \mathbf{L}\right)^{-1}\left(\left(\frac{1}{\omega} - 1\right)\mathbf{I}_n - \mathbf{U}\right)$$

$$= (\mathbf{I}_n - \omega\mathbf{L})^{-1}((1 - \omega)\mathbf{I}_n - \omega\mathbf{U}), \qquad \mathbf{A}_0 = \left(\frac{1}{\omega}\right)\mathbf{I}_n - \mathbf{L}.$$

Writing (1.8) more explicitly, we obtain the so-called *Jacobi* matrix

$$(1.9) \qquad \mathbf{B}_J = \begin{bmatrix} 0 & \mathbf{M} \\ \mathbf{M}^T & 0 \end{bmatrix}$$

Also, since $\mathbf{L}^2 = \mathbf{0_n}$, it follows that $(\mathbf{I} - \omega\mathbf{L})^{-1} = (\mathbf{I} + \omega\mathbf{L})$ so that the ω-parameter *family* of iteration matrices L_ω of (1.8) has the form

$$(1.10) \qquad L_w = \begin{bmatrix} (1 - \omega)\mathbf{I}_p & \omega\mathbf{M} \\ \omega(1 - \omega)\mathbf{M}^T & (1 - \omega)\mathbf{I}_q + \omega^2\mathbf{M}^T\mathbf{M} \end{bmatrix}.$$

Young was then able to *explicitly* and in *closed form*, relate eigenvalues $\mathbf{B}_J$ (1.9) (the *Jacobi* iteration matrix) to those of L_ω (1.10) (the *SOR* iteration matrix.) In fact, each $\mu \in \sigma(\mathbf{B}_J)$ is linked to each $\lambda \in \sigma(L_\omega)$ by the equation

(1.11) $$(1 - \omega - \lambda)^2 = \lambda \omega^2 \mu^2.$$

(*This fundamental relation will be proved in* (2.14) *and also generalized to two parameters* ω_1, ω_2, *in* (3.3).)

2 Singular Value Decomposition and Orthogonal Similarities

Motivation: We first give a simple derivation of Young's theory for (1.11) which relates eigenvalues of μ of $\mathbf{B}_J$ and λ of L_ω. We will assume, henceforth, that our system is symmetric positive-definite and has form (1.7). We believe this derivation is quite useful in teaching since it quickly produces the basic elements of Young's theory.

An important tool is the Singular Value Decomposition SVD (cf. [1]) which will allow us to unitarily transform our two iteration matrices (1.8) into simpler forms which more easily surrender eigenvalues and 2-norms.

To begin the transformation process, let us use the SVD to decompose the corner block matrix $\mathbf{M}$ which appears in (1.7) and (1.10). We obtain

(2.1) $$\mathbf{M} = \mathbf{U}\mathbf{\Sigma}\mathbf{V}^T,$$

where $p \times p$ matrix $\mathbf{U}$ and $q \times q$ matrix $\mathbf{V}$ are orthogonal, i.e.,

$$\mathbf{U}^T\mathbf{U} = \mathbf{I}_p, \qquad \mathbf{V}^T\mathbf{V} = \mathbf{I}_q$$

and $\mathbf{\Sigma}$ is the $p \times q$ diagonal matrix (of singular values) defined by

(2.2)
$$\mathbf{\Sigma} = \left.\left[\begin{array}{cccccc} s_1 & 0 & \cdots & & \cdots & 0 \\ 0 & s_2 & 0 & & \cdots & 0 \\ \vdots & & \ddots & & & \vdots \\ \vdots & & & s_{q-1} & & 0 \\ 0 & \cdots & \cdots & & 0 & s_q \\ 0 & \cdots & \cdots & & \cdots & 0 \\ \vdots & & & & & \vdots \\ 0 & \cdots & \cdots & & \cdots & 0 \end{array}\right] \begin{array}{l} \left.\rule{0pt}{3.2em}\right\} q \times q \\[0.5em] \left.\rule{0pt}{2.4em}\right\} (p-q) \times q \end{array}\right..$$

From (2.1), we see that matrix $\mathbf{M}\mathbf{M}^T = \mathbf{U}\mathbf{\Sigma}\mathbf{\Sigma}^T\mathbf{U}^T$ has eigenvectors equal to the columns of orthogonal matrix $\mathbf{U}$. Similarly, $\mathbf{M}^T\mathbf{M} = \mathbf{V}\mathbf{\Sigma}^T\mathbf{\Sigma}\mathbf{V}^T$ has its eigenvectors equal to the columns of of orthogonal matrix $\mathbf{V}$.

The eigenvalues $\{s_i^2\}$ of $\mathbf{M}\mathbf{M}^T$ (and of $\mathbf{M}^T\mathbf{M}$) are the squares of the *singular values* of $\mathbf{M}$. That is,

$$s_i(\mathbf{M}) \stackrel{\text{def}}{=} \left[\lambda_i(\mathbf{M}^T\mathbf{M})\right]^{\frac{1}{2}} \geq 0, \qquad i = 1, 2, \ldots, q.$$

The number of nonzero singular values s_i of $\mathbf{M}$ equals the rank of $\mathbf{M}$. Re-ordering subscripts, if necessary, we may always assume that

$$s_1 \geq s_2 \geq \ldots \geq s_q \geq 0$$

Decomposition of $\mathbf{B}_J = \mathbf{L} + \mathbf{U}$: [1] By substituting the SVD decomposition (2.1) into the corner elements $\mathbf{M}$, $\mathbf{M}^T$ of (1.7), we obtain

$$(2.3) \qquad \mathbf{B}_J = \begin{bmatrix} \mathbf{0} & \mathbf{M} \\ \mathbf{M}^T & \mathbf{0} \end{bmatrix} = \begin{bmatrix} \mathbf{0} & \mathbf{U}\boldsymbol{\Sigma}\mathbf{V}^T \\ \mathbf{V}\boldsymbol{\Sigma}^T\mathbf{U}^T & \mathbf{0} \end{bmatrix}.$$

How do singular values s_i $i = 1, 2, \ldots, q$ (diagonal of $\boldsymbol{\Sigma}$) relate to the eigenvalues μ of $\mathbf{B}_J$? The following observation on the eigenvectors of $\mathbf{B}_J$ will unravel the riddle.

It is direct from (2.3) above, that for eigenvalues $\mu_i \neq 0$,

$$(2.4) \quad \mathbf{B}_J \begin{bmatrix} \mathbf{x}_i \\ \mathbf{y}_i \end{bmatrix} \begin{matrix} \}p \\ \}q \end{matrix} = \mu_i \begin{bmatrix} \mathbf{x}_i \\ \mathbf{y}_i \end{bmatrix} \quad \text{iff} \quad \mathbf{B}_J \begin{bmatrix} \mathbf{x}_i \\ -\mathbf{y}_i \end{bmatrix} = -\mu_i \begin{bmatrix} \mathbf{x}_i \\ -\mathbf{y}_i \end{bmatrix} \quad i = 1, 2, \ldots, t$$

so that t non-zero eigenvalues of Hermitian $\mathbf{B}_J$ come in $\pm\mu$ pairs. To account for zero eigenvalues (which need not come in $\pm$ pairs), we write

$$(2.5) \qquad \mathbf{B}_J \begin{bmatrix} \mathbf{z}_i \\ \mathbf{z}_i' \end{bmatrix} \begin{matrix} \}p \\ \}q \end{matrix} = 0 \cdot \begin{bmatrix} \mathbf{z}_i \\ \mathbf{z}_i' \end{bmatrix} = \begin{bmatrix} \mathbf{0} \\ \mathbf{0} \end{bmatrix} \quad i = 1, 2, \ldots, r.$$

We construct the $n \times n$ non-singular matrix $\mathbf{W}$ whose columns are the orthogonal eigenvectors of (2.4) and (2.5). (*Recall that $\mathbf{B}_J = \mathbf{B}_J^T$ so that there always exists an orthogonal basis of eigenvectors.*)

$$\mathbf{W} = \underbrace{\begin{bmatrix} \mathbf{X} & \mathbf{X} & \mathbf{Z} \\ \mathbf{Y} & -\mathbf{Y} & \mathbf{Z}^T \end{bmatrix}}_{t \quad\quad t \quad\quad r} \begin{matrix} \}p \text{ rows} \\ \}q \text{ rows} \end{matrix} \qquad n = p + q = 2t + r.$$

Note that the t columns of $p \times t$ matrix $\mathbf{X}$ and $q \times t$ matrix $\mathbf{Y}$ are the t respective eigenvectors of (2.4)—the r columns of $p \times r$ matrix $\mathbf{Z}$ and $q \times r$ matrix $\mathbf{Z}^T$ come from the r null vectors of (2.5).

Ordinarily, we would scale the columns of $\mathbf{W}$ to produce an orthogonal matrix. As a technical convenience, however, we assume that the columns of $\mathbf{W}$ are scaled so that

$$(2.6) \qquad \mathbf{W}^T\mathbf{W} = \mathbf{W}\mathbf{W}^T = 2\mathbf{I}_n.$$

Let the matrix $\mathbf{J}$ denote the $t \times t$ matrix whose diagonal elements are the t positive eigenvalues μ_i of (2.4). Then (2.4) and (2.5) can be combined to produce the single

---

[1] This section follows the exposition of Lanczos ([2]).

matrix equation

$$\mathbf{B}_J \underbrace{\begin{bmatrix} \mathbf{X} & \mathbf{X} & \mathbf{Z} \\ \mathbf{Y} & -\mathbf{Y} & \mathbf{Z}' \end{bmatrix}}_{\mathbf{W}} = \underbrace{\begin{bmatrix} \mathbf{X} & \mathbf{X} & \mathbf{Z} \\ \mathbf{Y} & -\mathbf{Y} & \mathbf{Z}' \end{bmatrix}}_{\mathbf{W}} \begin{bmatrix} \mathbf{J} & 0 & 0 \\ 0 & -\mathbf{J} & 0 \\ 0 & 0 & 0 \end{bmatrix}$$

which, when multiplied through on the right by $\mathbf{W}^T$, (*see* (2.6)) yields

(2.7) $$\mathbf{B}_J = \begin{bmatrix} 0 & \mathbf{XJY}^T \\ \mathbf{YJX}^T & 0 \end{bmatrix}.$$

Comparing the block entries of $\mathbf{B}_J$ in (2.3) and (2.7), we obtain the equalities

(2.8a) $$\mathbf{XJY}^T = \mathbf{M} = \mathbf{U\Sigma V}^T$$

which imply
(2.8b) $$\mathbf{M}^T\mathbf{M} = \mathbf{V\Sigma}^T\mathbf{\Sigma V}^T \text{ and } \mathbf{MM}^T = \mathbf{U\Sigma\Sigma}^T\mathbf{U}^T.$$

The following theorem relates eigenvalues of $\mathbf{B}_J$ to singular values of the component submatrices $\mathbf{M}$ and $\mathbf{M}^T$.

Theorem 2.1 *Given symmetric Jacobi iteration matrix $\mathbf{B}_J$ of (2.3) with corner submatrices $\mathbf{M}$ and $\mathbf{M}^T$. Then*

(2.9) $$\{\mu_i^2\} = \sigma(\mathbf{B}_J^2) = \sigma(\mathbf{B}_J)^2 = \sigma(\mathbf{MM}^T) = \{s_i^2\} \quad i = 1, 2, \ldots, q$$

where μ_i^2 are the eigenvalues of $\mathbf{B}_J^2$ and s_i^2 are the squares of the singular values of $\mathbf{M}$.

Proof: Note $\mathbf{B}_J^2$ is block diagonal with just the matrices $\mathbf{M}^T\mathbf{M}$ and $\mathbf{MM}^T$ on the diagonals. Thus, $\{\mu_i^2\}$, the eigenvalues of $\mathbf{B}_J^2$, are the eigenvalues $\mathbf{MM}^T$ and $\mathbf{M}^T\mathbf{M}$, which, from (2.8), are the eigenvalues of $\mathbf{\Sigma}^T\mathbf{\Sigma}$ and $\mathbf{\Sigma\Sigma}^T$. This ends the proof. ∎

Decomposition of L_ω : How do we relate eigenvalues $\mu_i \in \sigma(\mathbf{B}_J)$ to eigenvalues of L_ω? Observe what happens to L_ω when the SVD decomposition $\mathbf{M} = \mathbf{U\Sigma V}^T$ (2.1) is substituted for $\mathbf{M}$ in (1.10) and the orthogonal matrices $\mathbf{U}$ and $\mathbf{V}$ are "factored out":

(2.10) $$L_w = \underbrace{\begin{bmatrix} \mathbf{U} & 0 \\ 0 & \mathbf{V} \end{bmatrix}}_{\mathbf{Q}} \underbrace{\begin{bmatrix} (1-\omega)\mathbf{I}_p & \omega\mathbf{\Sigma} \\ \omega(1-\omega)\mathbf{\Sigma}^T & (1-\omega)\mathbf{I}_q + \omega^2\mathbf{\Sigma}^T\mathbf{\Sigma} \end{bmatrix}}_{\mathbf{\Gamma}_\omega} \underbrace{\begin{bmatrix} \mathbf{U}^T & 0 \\ 0 & \mathbf{V}^T \end{bmatrix}}_{\mathbf{Q}^T}.$$

Note that (2.10) reveals the unitarily equivalent matrix $\mathbf{\Gamma}_\omega$ with four block submatrices, each of which is a *diagonal* sub-matrix! This means that there is a permutation matrix [2] **P** which "pulls" the two corner diagonal matrices to the main diagonal, i.e., $\mathbf{P}\mathbf{\Gamma}_\omega\mathbf{P}^T$ has only 2×2 or 1×1 matrices along its main diagonal. When $\mathbf{\Gamma}_\omega$ of (2.10) is permuted into the block diagonal form, we obtain

$$(2.11a) \quad \mathbf{\Delta}(\omega) = \mathbf{P}\mathbf{\Gamma}_\omega\mathbf{P}^T = \begin{bmatrix} \mathbf{\Delta}_1(\omega) & 0 & \cdots & \cdots & & 0 \\ 0 & \ddots & & & & \vdots \\ \vdots & & \mathbf{\Delta}_q(\omega) & & & \vdots \\ \vdots & & & \ddots & & 0 \\ 0 & \cdots & \cdots & 0 & (1-\omega)\mathbf{I}_{p-q} \end{bmatrix}$$

where each 2×2 matrix $\mathbf{\Delta}_i(\omega)$ is given by

$$(2.11b) \quad \mathbf{\Delta}_i(\omega) = \begin{bmatrix} (1-\omega) & \omega s_i \\ \omega(1-\omega)s_i & (1-\omega)+\omega^2 s_i^2 \end{bmatrix}, \quad i = 1, 2, \ldots, q$$

where s_i are the singular values of (2.2).

Remark: The diagonal entries $(1 - \omega)$ of (2.11a) are, equal in number to the "excess" rectangular dimension, $p - q$, of the corner matrices $\mathbf{M}$ and $\mathbf{M}^T$ of $\mathbf{B}_J$. At the same time, From (2.3), we see that there are (at least) $p - q$ null vectors of $\mathbf{B}_J$, i.e.,

(2.12)   *For the matrix $\mathbf{\Delta}(\omega)$ of (2.11a), each diagonal entry $(1-\omega)$ corresponds to a zero eigenvalue of $\mathbf{B}_J$ while each 2×2 matrix $\mathbf{\Delta}_i(\omega)$ (2.11b) corresponds to a singular value s_i of $\mathbf{M}$ (2.1).*

We have seen that each member of the ω-family of SOR iteration matrices L_ω is unitarily equivalent to a matrix $\mathbf{\Delta}(\omega)$ having only 2×2 or 1×1 matrices on the diagonal. That is, from (2.10) and (2.11a),

$$(2.13) \qquad L_\omega = \mathbf{Q}\mathbf{P}^T\mathbf{\Delta}(w)\mathbf{P}\mathbf{Q}^T \qquad \text{for unitary } (\mathbf{P}\mathbf{Q}^T).$$

Unitary equivalence (2.13) implies that both the eigenvalues and the 2-norms agree for both (ω-families of) matrices L_ω and $\mathbf{\Delta}(\omega)$. We now summarize these equivalences, in the following theorem:

Theorem 2.2 *Given matrices $\mathbf{B}_J$ of (1.9) and L_ω of (3.1) Then the eigenvalues $\mu_i \in \sigma(\mathbf{B}_J)$ and $\lambda_i \in \sigma(L_\omega)$ are linked by the functional relation*

$$(2.14) \qquad (1 - \omega - \lambda_i)^2 - \omega^2\mu_i^2\lambda_i = 0.$$

Moreover, eigenvalues and 2-norms of matrices L_ω and $\mathbf{\Delta}(\omega)$ of (3.2a)–(3.2b) are related as follows:

---

[2]See the Appendix for construction of the permutation matrix **P**.

(2.15a) $\sigma(L_\omega) = \sigma(\Delta(\omega))$

(2.15b) $\rho(L_\omega) = \rho(\Delta(\omega)) = \max\limits_{1 \le i \le r} \|\rho(\Delta_i(\omega))\|$

(2.15c) $\|L_\omega^k\|_2 = \|\Delta^k(\omega)\|_2 = \max\limits_{1 \le i \le r} \|\Delta_i^k(\omega)\|_2$,   for all k.

Proof: Unitary equivalence of L_ω and $\Delta(\omega)$ in (2.13) assure that both matrices have identical eigenvalues λ so that

$$\det(\lambda \mathbf{I}_n - L_\omega) = 0 \quad \text{iff} \quad \det(\lambda \mathbf{I}_n - \Delta(\omega)) = 0$$

From the right-hand determinant above, we see, from (2.11a), (2.11b), that all λ are constrained by

$$\lambda = 1 - \omega, \quad \text{or} \quad \det(\lambda \mathbf{I}_n - \Delta_i(\omega)) = 0 \quad i = 1, 2, \ldots, q.$$

The determinant above reduces to the relation

$$\lambda = 1 - \omega, \quad \text{or} \quad (1 - \omega - \lambda)^2 - \omega^2 s_i^2 \lambda = 0 \quad i = 1, 2, \ldots, q$$

which, from (2.9), gives us

(2.16) $\lambda = 1 - \omega, \quad \text{or} \quad (1 - \omega - \lambda)^2 - \omega^2 \mu_i^2 \lambda = 0, \quad i = 1, 2, \ldots, q.$

Now the left-hand equation, $\lambda = 1 - \omega$ appears in (2.11a)–(2.11b) once for each occurrence of a zero eigenvalue for $\mathbf{B}_J$ (see (2.12).) But $\lambda = 1 - \omega$ is a special case of the *right*-hand side of (2.16), namely, when μ_i is set to zero. Therefore, (2.16) is described by the *single* relation (2.14), which is Young's identity.

Finally, noting that unitary equivalence always preserves eigenvalues and the *spectral norm*, i.e., the 2-norm $\| \cdot \|_2$, (see (2.13)), we establish (2.15a)–(2.15c). The proof is finished.   ∎

Remark: It was shown by Young that if $\rho(\mathbf{B}_J) < 1$ (*i.e.*, $\mathbf{A} = \mathbf{I} - \mathbf{B}_J$ *is positive definite*), then among all $\omega \in [0, 2]$, the spectral radius $\rho(L_\omega)$ of (2.15b) attains its minimum when

$$\omega = w_b = \frac{2}{1 + \sqrt{1 - s_1^2}}.$$

In this case, the minimum spectral radius (among all iteration matrices L_ω) is given in the beautiful closed form $\rho(L_\omega) = \omega_b - 1$.

Minimizing the norms in (2.15a)–(2.15c) is another matter. Unfortunately, we are unable to derive a closed form or analytical expression for $\min\limits_{\omega} \|L_\omega^k\|_2$.

3   Two-Parameter SOR

Definitions: For any two scalars ω_1, ω_2, define the two-parameter splitting for matrix $\mathbf{A} = \mathbf{I} - \mathbf{B}$ as follows:

$$\mathbf{A} = \mathbf{I}_n - \begin{bmatrix} \mathbf{0} & \mathbf{M} \\ \mathbf{M}^T & \mathbf{0} \end{bmatrix} = \underbrace{\begin{bmatrix} \frac{1}{\omega_1}\mathbf{I}_p & \mathbf{0} \\ -\mathbf{M}_{q\times p}^T & \frac{1}{\omega_2}\mathbf{I}_q \end{bmatrix}}_{\mathbf{A}_0} - \underbrace{\begin{bmatrix} \frac{(1-\omega_1)}{\omega_1}\mathbf{I}_p & \mathbf{M}_{p\times q} \\ \mathbf{0} & \frac{(1-\omega_2)}{\omega_2}\mathbf{I}_q \end{bmatrix}}_{\mathbf{A}_1}$$

where the "easy-to-invert" matrix $\mathbf{A}_0$ has the inverse

$$\mathbf{A}_0^{-1} = \begin{bmatrix} \omega_1\mathbf{I}_p & \mathbf{0} \\ \omega_1\omega_2\mathbf{M}_{q\times p}^T & \omega_2\mathbf{I}_q \end{bmatrix}$$

so that the two-parameter iteration matrix $\mathbf{A}_0^{-1}\mathbf{A}_1 = L_{\omega_1,\omega_2}$ (1.3) has the form

$$(3.1) \quad L_{\omega_1,\omega_2} = \begin{bmatrix} (1-\omega_1)\mathbf{I}_p & \omega_1\mathbf{M}_{p\times q} \\ \omega_2(1-\omega_1)\mathbf{M}_{q\times p}^T & (1-\omega_2)\mathbf{I}_q + \omega_1\omega_2\mathbf{M}_{q\times p}^T\mathbf{M}_{p\times q} \end{bmatrix}.$$

As in (2.10), we set $\mathbf{M} = U\boldsymbol{\Sigma}V^T$ and factor out the orthogonal matrices $\mathbf{U}$, $\mathbf{V}$ to obtain

$$L_{\omega_1,\omega_2} = \underbrace{\begin{bmatrix} \mathbf{U} & \mathbf{0} \\ \mathbf{0} & \mathbf{V} \end{bmatrix}}_{\mathbf{Q}} \underbrace{\begin{bmatrix} (1-\omega_1)\mathbf{I}_p & \omega_1\boldsymbol{\Sigma} \\ \omega_2(1-\omega_1)\boldsymbol{\Sigma}^T & (1-\omega_2)\mathbf{I}_q + \omega_1\omega_2\boldsymbol{\Sigma}^T\boldsymbol{\Sigma} \end{bmatrix}}_{\boldsymbol{\Gamma}_{\omega_1,\omega_2}} \underbrace{\begin{bmatrix} \mathbf{U}^T & \mathbf{0} \\ \mathbf{0} & \mathbf{V}^T \end{bmatrix}}_{\mathbf{Q}^T}.$$

The same computations which produced block-diagonal orthogonal equivalent matrix $\boldsymbol{\Delta}(\omega)$ (2.11a)–(2.11b), now produce the analogous two-parameter orthogonal equivalent $\boldsymbol{\Delta}(\omega_1, \omega_2))$ defined by

$$\boldsymbol{\Delta}(\omega_1,\omega_2) = \mathbf{P}\boldsymbol{\Gamma}_{\omega_1,\omega_2}\mathbf{P}^T = \begin{bmatrix} \boldsymbol{\Delta}_1(\omega_1,\omega_2) & \mathbf{0} & \cdots & \cdots & & \mathbf{0} \\ \mathbf{0} & \ddots & & & & \vdots \\ \vdots & & \boldsymbol{\Delta}_q(\omega_1,\omega_2) & & & \vdots \\ \vdots & & & \ddots & & \mathbf{0} \\ \mathbf{0} & \cdots & & \cdots & \mathbf{0} & (1-\omega_2)\mathbf{I}_{p-q} \end{bmatrix}$$

(3.2a)
where each 2×2 matrix $\boldsymbol{\Delta}_i(\omega_1, \omega_2)$ is given by

$$(3.2b) \quad \boldsymbol{\Delta}_i(\omega_1,\omega_2) = \begin{bmatrix} (1-\omega_1) & \omega_1 s_i \\ \omega_2(1-\omega_1)s_i & (1-\omega_2) + \omega_1\omega_2 s_i^2 \end{bmatrix} \quad i = 1,2,\ldots,q$$

where s_i are the singular values of (2.2). With virtually no change in argument given in Theorem (3.1), we have proved the following:

Theorem 3.1 *Given matrices* $\mathbf{B}_J$ *of (1.9) and* L_{ω_1,ω_2} *of (3.1). Then eigenvalues* $\mu_i \in \sigma(\mathbf{B}_J)$ $\lambda_i \in \sigma(L_{\omega_1,\omega_2})$ *are linked by the functional relation*

(3.3) $$(1 - \omega_1 - \lambda_i)(1 - \omega_2 - \lambda_i) - \lambda_i \omega_1 \omega_2 \mu_i^2 = 0.$$

Moreover, eigenvalues and 2-norms of matrices L_{ω_1,ω_2} *and* $\mathbf{\Delta}(\omega_1,\omega_2)$ *of (3.2a), (3.2b) are related as follows:*

(3.4a) $\sigma(L_{\omega_1,\omega_2}) = \sigma(\mathbf{\Delta}(\omega_1,\omega_2))$

(3.4b) $\rho(L_{\omega_1,\omega_2}) = \rho(\mathbf{\Delta}(\omega_1,\omega_2)) = \max\limits_{1 \le i \le r} \|\rho(\mathbf{\Delta}_i(\omega_1,\omega_2))\|$

(3.4c) $\|L_{\omega_1,\omega_2}^k\|_2 = \|\mathbf{\Delta}^k(\omega_1,\omega_2)\|_2 = \max\limits_{1 \le i \le r} \|\mathbf{\Delta}_i^k(\omega_1,\omega_2)\|_2,$ *for all* k.

Remark: *When* $\omega_1 = \omega_2$ *in (3.3), we recover (1.11), (2.14), the well-known relation between* $\mu \in \sigma(\mathbf{B})$ *and* $\lambda \in \sigma(L_{\omega_1,\omega_2})$.

In a future paper, we shall see how some of these ideas extend to the case when matrix $\mathbf{A}$ is non-symmetric.

4   A Numerical Example[2]

We have computed the norms of the two-parameter iteration matrices $L_{\omega_1,\omega_2}^{50}$ where L_{ω_1,ω_2} is constructed from the tri-diagonal Jacobi iteration matrix which produces iteration matrix $\mathbf{B}_J'$ given by

$$\mathbf{B}_J' = \frac{1}{2} \begin{bmatrix} 0 & -1 & 0 & \ldots & 0 \\ -1 & 0 & \ddots & & 0 \\ & \ddots & \ddots & \ddots & \\ 0 & \ldots & \ddots & 0 & -1 \\ 0 & 0 & \ldots & -1 & 0 \end{bmatrix}, \qquad \rho(\mathbf{B}_J) = \cos(\pi/(n+1)).$$

The red-black ordering produces an orthogonally equivalent matrix with Property A as per (1.7). That is, for some orthogonal $\mathbf{Q}$, we have

$$\mathbf{A}_{n,n} = \mathbf{Q}\mathbf{A}_{n,n}'\mathbf{Q}^T = \mathbf{I} - \mathbf{B}_J$$

where B_J has form (2.3) and replaces the tri-diagonal B_J'. From the $\mathbf{B}_J$ with Property A, we construct the two-parameter family of SOR iteration matrices L_{ω_1,ω_2} given by (3.1). NOTE: When $\omega_1 = \omega_2 = \omega$, then two-parameter $L_{\omega,\omega}$ of (3.1) reduces to the usual one-parameter SOR iteration matrix L_ω.

---

[2]We thank Patrick Witting of Stanford University for developing important preliminary norm calculations for the one-parameter SOR. Also, Figure 7.1 is generated in PostScript by MATLAB software which we found to be invaluable.

Choosing matrix dimension $n = 100$ and power $k = 50$, notice what Figure 7.1 tells us—among all pairs of parameters $\omega_2, \omega_1 \in [0,2] \times [0,2]$, the norm $\|L^{50}_{\omega_1,\omega_2}\|$ is approximately minimized (optimal) for $(\omega_2, \omega_1) = (2.00, 0.6961)$. At the same time, for (ω_2, ω_1) running over the rectangular domain $[0,2] \times [0,2]$, we have

$$\|L^{50}_{\omega_b}\| = \|L^{50}_{\omega_b,\omega_b}\| = 8.643 \approx \|L^{50}_{\omega_1,\omega_2}\| = 9.508 = \texttt{Max}$$

so that the norm of the *optimal* iteration matrix $L^{50}_{\omega_b}$ is very near to the global maximum! In summation,

ω_2	ω_1	$\|L^{50}_{\omega_1,\omega_2}\|$
2.0000	0.6961	0.9508 (Min)
ω_b=1.9391	1.9391	8.6428
1.8397	2.0000	9.0471 (Max)

and

$$\rho(\mathbf{B}_J) = 0.9995, \qquad \rho(L_{\omega_b}) = 0.9391.$$

It is interesting to note that `Max` and $\|L^{50}_{\omega_b}\|$ have nearly equal values and occur very near to each other on the ω_2-ω_1 plane.

These computations and Figure 7.1 confirm the potential of using a mixed strategy to improve convergence. That is, for the first m iterations, say, construct vector iterates (1.2) using sub-optimal two-parameter SOR iteration matrix $\mathbf{B} = L_{\omega_1,\omega_2}$. This strategy is desirable as long as

(4.1) $$\|L^m_{\omega_1,\omega_2}\| < \|L^m_{\omega_b,\omega_b}\|.$$

When inequality (4.1) is violated for some $m > \bar{m}$, then for $k > \bar{m}$, continue calculation of the vector iterates (1.2) using the optimal $L_{\omega_b,\omega_b} = L_{\omega_b}$.

The determination of the *optimal* parameters ω_1, ω_2 for such a mixed strategy is still an open question. In a future paper, we will present results of further numerical experiments on the choice of optimal parameters and render comparisons with the Chebyshev semi-iterative method.

Acknowledgements

This work was written with partial support of NSF under grant DCR-8412314 (Golub) and from the IBM Bergen Scientific Research Centre, Bergen, Norway (de Pillis). The original manuscript was typed by the second author in LaTeX and reformatted by Lisa Laguna.

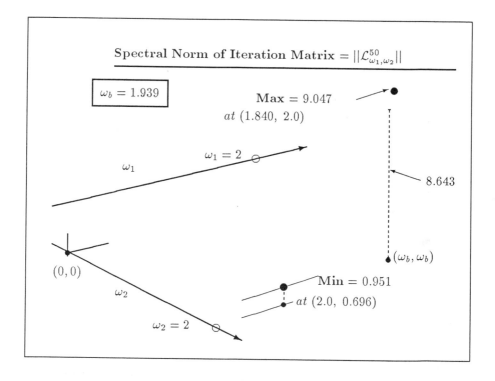

Figure 7.1 Spectral norm of iteration matrix $= \|L^{50}_{\omega_1,\omega_2}\|$

References

[1] Golub, G.H., and Charles Van Loan. *Matrix Computations, Edition 2*, The Johns Hopkins University Press, Baltimore, 1989.

[2] Lanczos, C. *Linear Differential Operators*, London, New York, Van Nostrand, 1961.

[3] Varga, R. *Matrix Iterative Analysis*, Prentice Hall, Englewood Cliffs, New Jersey, 1962.

[4] Young, D.M. *Iterative Solution of Large Linear Systems*, Academic Press, New York, 1971.

Appendix

Construction of Permutation Matrix P: A special case may illustrate the general construction of $\mathbf{P}$ in (2.11a)–(2.11b) used to reduce Γ_ω to block diagonal $\Delta(\omega)$. Accordingly, take $p = 3$ and $q = 2$ and let Γ_ω (2.10) and permutation matrix $\mathbf{P}$ be defined by

$$(A.1) \quad \Gamma_\omega = \begin{bmatrix} a_1 & 0 & 0 & b_1 & 0 \\ 0 & a_2 & 0 & 0 & b_2 \\ 0 & 0 & a_3 & 0 & 0 \\ c_1 & 0 & 0 & d_1 & 0 \\ 0 & c_2 & 0 & 0 & d_2 \end{bmatrix}, \quad \mathbf{P} : \begin{bmatrix} x_1 \\ x_2 \\ x_3 \\ y_1 \\ y_2 \end{bmatrix} \longrightarrow \begin{bmatrix} x_1 \\ y_1 \\ x_2 \\ y_2 \\ x_3 \end{bmatrix}.$$

Then the similarity transformation is block-diagonal

$$\mathbf{P}\Gamma_\omega\mathbf{P}^T = \begin{bmatrix} a_1 & b_1 & 0 & 0 & 0 \\ c_1 & d_1 & 0 & 0 & 0 \\ 0 & 0 & u_2 & b_2 & 0 \\ 0 & 0 & c_2 & d_2 & 0 \\ 0 & 0 & 0 & 0 & a_3 \end{bmatrix}$$

where each diagonal block is either a 2×2 or 1×1 scalar matrix.

For completeness, we state the *precise* generalization of the re-ordering given in (A.1) which "dove-tails" or "interlaces" longer p-vector $\mathbf{x}$ with shorter q-vector $\mathbf{y}$. Let vector $\mathbf{x} = [x_k] \in I\!R^{p+q}$. Then the permutation matrix $\mathbf{P}$ is defined by

$$[\mathbf{P}(\mathbf{x})]_k = [x_{\pi(k)}], \quad k = 1, 2, \ldots, p+q$$

where the permutation π on integers $\{1, 2, \ldots, (p+q)\}$ is given by

$$\pi : k \to 2k - 1 \quad \text{for} \quad 1 \le k \le q$$

$$\pi : k \to k + q \quad \text{for} \quad q + 1 \le k \le p$$

$$\pi : k \to 2(k - p) \quad \text{for} \quad p + 1 \le k \le p + q.$$

Chapter 8

Relaxation Parameters for the IQE Iterative Procedure for Solving Semi-Implicit Navier-Stokes Difference Equations

LOUIS A. HAGEMAN
Westinghouse-Bettis Atomic Power Laboratory

Dedicated to David M. Young, Jr., on the occasion of his sixty-fifth birthday.

Abstract

Numerical solutions of the time-dependent Navier-Stokes equations for a thermally expandable fluid require advancing the unknown flow variables in time by either an implicit, a semi-implicit, or an explicit procedure. Explicit methods require a modest amount of computational effort per time step, but require excessively small time-step lengths to ensure numerical stability. On the other hand, implicit and semi-implicit schemes allow the use of large time-step lengths, but suffer from increased computational complexities. Fully-implicit methods generate coupled systems of nonlinear equations that must be solved at each point in time, whereas semi-implicit procedures generate coupled systems of linear equations.

The focus of this paper is on the use of a recently formulated iterative procedure, called the IQE method, for solving the linear systems produced by a semi-implicit method. The IQE method requires the use of a relaxation parameter ω which must be properly chosen to obtain the greatest rate of convergence. Using heuristic arguments, we develop a numerical procedure for estimating near-optimal values of ω. Numerical test results are included.

121

1   Introduction

The system of equations produced by the semi-implicit method we use is of the form

(1.1)
$$\begin{bmatrix} \mathbf{Q} & -\mathbf{G} \\ \mathbf{G}^T & 0 \end{bmatrix} \begin{bmatrix} \mathbf{w} \\ \mathbf{p} \end{bmatrix} = \begin{bmatrix} \mathbf{s}_w \\ \mathbf{s}_p \end{bmatrix},$$

where $\mathbf{Q}$ is a nonsymmetric matrix. The facts that the coefficient matrix in (1.1) is nonsymmetric and has a zero diagonal block complicate any solution procedure. Either a direct method or an iterative method can be used to solve (1.1). Direct procedures are often cost ineffective for problems in more than one space dimension because of large storage and arithmetic requirements.[1] On the other hand, most iterative procedures used to solve systems such as (1.1) suffer from a lack of mathematical theory that ensures convergence and predicts convergence behavior. However, because of the versatility of iterative methods and because of their potential to solve large problems in reasonable computer times, it appears that the system (1.1) for complex two and three dimensional flow problems can be solved more effectively by a good iterative procedure than by a good direct procedure.

A family of iterative methods for solving the system (1.1) is defined in [2]. Preliminary analysis and numerical results for one member of this family, called the IQE method, is also given in [2]. The IQE method requires the use of a relaxation parameter ω which must be properly chosen to obtain the greatest rate of convergence. Using heuristic arguments, we develop in this paper a numerical procedure for estimating near-optimal values for ω.

2   The Continuous and Discrete Problems

The semi-implicit conservation equations for incompressible fluid flow are given here by

(2.1)
$$\frac{\partial u^{n+1}}{\partial x} + \frac{\partial v^{n+1}}{\partial y} = 0,$$

(2.2)
$$\frac{u^{n+1} - u^n}{\Delta t} + \frac{\partial u^n u^{n+1}}{\partial x} + \frac{\partial v^n u^{n+1}}{\partial y} - \nu \nabla^2 u^{n+1} = -\frac{\partial p^{n+1}}{\partial x} + g_x,$$

(2.3)
$$\frac{v^{n+1} - v^n}{\Delta t} + \frac{\partial u^n v^{n+1}}{\partial x} + \frac{\partial v^n v^{n+1}}{\partial y} - \nu \nabla^2 v^{n+1} = -\frac{\partial p^{n+1}}{\partial y} + g_y,$$

where the superscript n refers to the time point $t_n = n\Delta t$. Here u and v are the x and y direction velocity components, p is the ratio of pressure to constant density, ν is the kinematic viscosity coefficient, and g_x and g_y are accelerations due to external forces such as gravity.

---

[1]However, see Amit, Hall, and Porsching [1] for an effective direct solution procedure in two space dimensions.

For reasons of simplicity and concreteness, we assume that the flow region is rectangular and is subdivided into a union of rectangles whose side lengths in the x and y directions, respectively, are Δx and Δy. The MAC (Marker and Cell) scheme [4] is then used to obtain the discrete analogues to equations (2.1)–(2.3). For the MAC method, the discrete approximations for p are associated with mesh box centers while the discrete approximations for u and v are associated, respectively, with midpoints of the vertical and horizontal mesh sides. (See Fig. 8.1.) The MAC discrete analogues of equations (2.1)–(2.3) at time point $(n + 1)$ can be expressed in the matrix form

(2.4)
$$
\begin{bmatrix}
0 & \Delta y \mathbf{D}_x & \Delta x \mathbf{D}_y \\
-\dfrac{\Delta t}{\Delta x}\mathbf{D}_x^T & \mathbf{I} + \Delta t \mathbf{J}^u & 0 \\
-\dfrac{\Delta t}{\Delta y}\mathbf{D}_y^T & 0 & \mathbf{I} + \Delta t \mathbf{J}^v
\end{bmatrix}
\begin{bmatrix}
\mathbf{p}^{n+1} \\
\mathbf{u}^{n+1} \\
\mathbf{v}^{n+1}
\end{bmatrix}
=
\begin{bmatrix}
\mathbf{q}_c \\
\Delta t \mathbf{q}_u \\
\Delta t \mathbf{q}_v
\end{bmatrix}.
$$

Here, the matrices $\mathbf{D}_x$ and $\mathbf{D}_y$ denote, respectively, the discrete analogues of the operators $\Delta x(\partial/\partial x)$ and $\Delta y(\partial/\partial y)$. The superscript T denotes matrix transpose. The matrix-vector products $\mathbf{J}^u \mathbf{u}^{n+1}$ and $\mathbf{J}^v \mathbf{v}^{n+1}$ denote, respectively, the discrete analogues of the convection and the diffusion terms in (2.2) and (2.3). We assume that full upwinding [2] is used in the discretization of the convection terms. In what follows, we drop the superscript $(n + 1)$ on $\mathbf{u}, \mathbf{v}$, and $\mathbf{p}$.

As boundary conditions, we assume that the $y = 0$ boundary is a no-slip wall and that the top boundary, $y = y_J$, is either a no-slip or a free-slip wall. The $x = 0$ inlet boundary may be a combination of a no-slip wall for $0 \leq j \leq K$ and specified inlet values for $K < j \leq J$. (See Fig. 8.1) At the $x = x_I$ outlet boundary, the pressure values are specified.

Letting

(2.5) $\qquad \mathbf{C} \equiv (\Delta y \mathbf{D}_x, \ \Delta x \mathbf{D}_y), \quad \mathbf{G} \equiv \dfrac{1}{\Delta x \Delta y}\mathbf{C}^T, \quad \mathbf{w}^T \equiv (\mathbf{u}^T, \mathbf{v}^T),$

$$
\mathbf{q}^T \equiv (\mathbf{q}_u^T, \mathbf{q}_v^T), \quad \text{and} \quad \mathbf{J} \equiv \begin{bmatrix} \mathbf{J}^u & 0 \\ 0 & \mathbf{J}^v \end{bmatrix},
$$

we may express (2.4) alternatively in the form

(2.6)
$$
\begin{bmatrix} 0 & \mathbf{C} \\ -\Delta t \mathbf{G} & \mathbf{A} \end{bmatrix}
\begin{bmatrix} \mathbf{p} \\ \mathbf{w} \end{bmatrix}
=
\begin{bmatrix} \mathbf{q}_c \\ \mathbf{q}\Delta t \end{bmatrix},
$$

where $\mathbf{A} \equiv \mathbf{I} + \Delta t \mathbf{J}$. For each time point t_n, the semi-implicit time discretization scheme requires that a system of equations of the form (2.6) be solved for $\mathbf{p}$ and $\mathbf{w}$. For sufficiently small Δt, it is easy to show that $\mathbf{A}$ and $\mathbf{C}\mathbf{A}^{-1}\mathbf{G}$ are nonsingular. Thus, for sufficiently small Δt, the unique solution to (2.6) can be expressed in the

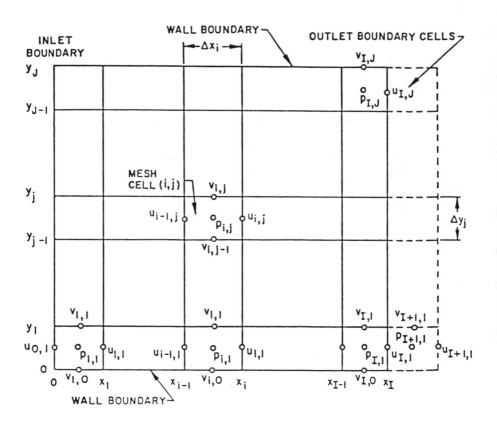

Figure 8.1 Placement of variables for MAC-type discretizations.

form

$$\mathbf{p} = \{\mathbf{CA^{-1}G}\}^{-1}\left\{\frac{1}{\Delta t}\mathbf{q}_c - \mathbf{CA^{-1}q}\right\}$$

(2.7)

$$\mathbf{w} = \Delta t\mathbf{A}^{-1}\{\mathbf{q} + \mathbf{Gp}\}.$$

Since $\mathbf{CA^{-1}G}$ is not sparse, it is not computationally feasible to obtain $\mathbf{p}$ and $\mathbf{w}$ using (2.7). In the next section, we define the IQE iterative method for solving the coupled system of equations (2.6).

3   The IQE Iterative Method

In what follows, we let the superscript (ℓ) denote the iteration count. The IQE iterative procedure involves a pressure vector $\mathbf{p}^{(\ell)}$ and two velocity vectors $\mathbf{w}^{(\ell)}$ and $\overset{*}{\mathbf{w}}{}^{(\ell)}$. The iterates $\mathbf{w}^{(\ell)}$ will satisfy a modified form of the momentum equations, while the iterates $\overset{*}{\mathbf{w}}{}^{(\ell)}$ will satisfy the continuity constraint; i.e., $\mathbf{C}\overset{*}{\mathbf{w}}{}^{(\ell)} = \mathbf{q}_c$. At convergence, we want $\mathbf{w} = \overset{*}{\mathbf{w}}$.

Let $\mathbf{Q} \equiv \mathbf{I} + \omega\Delta t\mathbf{J}$ and $\mathbf{A}_p \equiv \mathbf{CG}$. Then, given $\overset{*}{\mathbf{w}}{}^{(0)}$ and $\mathbf{p}^{(0)}$, the IQE method generates successive estimates for $\mathbf{w}$ and $\mathbf{p}$ by the process [2]

(3.1a)
$$\mathbf{Qw}^{(\ell+1)} = (1-\omega)\overset{*}{\mathbf{w}}{}^{(\ell)} + \omega\Delta t\left(\mathbf{Gp}^{(\ell)} + \mathbf{q}\right),$$

(3.1b)
$$\mathbf{A}_p\left(\mathbf{p}^{(\ell+1)} - \mathbf{p}^{(\ell)}\right) = -\frac{1}{\omega\Delta t}\left(\mathbf{Cw}^{(\ell+1)} - \mathbf{q}_c\right),$$

(3.1c)
$$\overset{*}{\mathbf{w}}{}^{(\ell+1)} = \mathbf{w}^{(\ell+1)} + \omega\Delta t\mathbf{G}\left(\mathbf{p}^{(\ell+1)} - \mathbf{p}^{(\ell)}\right).$$

The matrix $\mathbf{A}_p = \mathbf{CG}$ turns out to be the discrete analogue of the diffusion operator ∇^2 and is positive definite provided the pressure is fixed at some point.

The IQE procedure (3.1) can be derived easily. As before, let $\mathbf{A} = \mathbf{I} + \Delta t\mathbf{J}$. Then with $\overset{*}{\mathbf{w}}{}^{(\ell)}$ and $\mathbf{p}^{(\ell)}$ given, $\mathbf{w}^{(\ell+1)}$ is defined by

$$\mathbf{w}^{(\ell+1)} = \overset{*}{\mathbf{w}}{}^{(\ell)} - \omega\mathbf{Q}^{-1}\left[\mathbf{A}\overset{*}{\mathbf{w}}{}^{(\ell)} - \Delta t\mathbf{Gp}^{(\ell)} - \Delta t\mathbf{q}\right],$$

which when multiplied by $\mathbf{Q}$ gives (3.1a). The vectors $\mathbf{p}$ and $\overset{*}{\mathbf{w}}$ are updated by requiring $\overset{*}{\mathbf{w}}{}^{(\ell+1)}$ to satisfy the continuity constraint

(3.2)
$$\mathbf{C}\overset{*}{\mathbf{w}}{}^{(\ell+1)} = \mathbf{q}_c$$

and by requiring $\overset{*}{\mathbf{w}}{}^{(\ell+1)}$ and $\mathbf{p}^{(\ell+1)}$ to satisfy an equation analogous to (3.1a); namely that

(3.3)
$$\overset{*}{\mathbf{w}}{}^{(\ell+1)} + \omega\Delta t\mathbf{Jw}^{(\ell+1)} = (1-\omega)\overset{*}{\mathbf{w}}{}^{(\ell)} + \omega\Delta t\left(\mathbf{Gp}^{(\ell+1)} + \mathbf{q}\right).$$

Subtracting (3.1a) from (3.3) yields (3.1c) and the substitution of (3.1c) into (3.2) gives (3.1b).

For reasons of analysis, we eliminate $\overset{*}{\mathbf{w}}{}^{(\ell)}$ from the procedure (3.1). Combining (3.1b) with (3.1c) gives $\overset{*}{\mathbf{w}}{}^{(\ell)} = \left(\mathbf{I} - \mathbf{GA}_p^{-1}\mathbf{C}\right)\mathbf{w}^{(\ell)} + \mathbf{GA}_p^{-1}\mathbf{q}_c$, which when substituted into (3.1a) yields the equations

$$
\begin{bmatrix} \mathbf{Q} & \mathbf{0} \\ \dfrac{1}{\omega\Delta t}A_p^{-1}\mathbf{C} & \mathbf{I} \end{bmatrix}\begin{bmatrix} \mathbf{w}^{(\ell+1)} \\ \mathbf{p}^{(\ell+1)} \end{bmatrix} = \begin{bmatrix} (1-\omega)(\mathbf{I}-\mathbf{P}) & \omega\Delta t\mathbf{G} \\ \mathbf{0} & \mathbf{I} \end{bmatrix}\begin{bmatrix} \mathbf{w}^{(\ell)} \\ \mathbf{p}^{(\ell)} \end{bmatrix}
$$

(3.4)
$$
+ \begin{bmatrix} (1-\omega)\mathbf{GA}_p^{-1}\mathbf{q}_c + \omega\Delta t\mathbf{q} \\ \dfrac{1}{\omega\Delta t}A_p^{-1}\mathbf{q}_c \end{bmatrix}
$$

where $\mathbf{P} \equiv \mathbf{GA}_p^{-1}\mathbf{C}$.

Let the pseudo-residual vectors $\boldsymbol{\delta}$ and error vectors $\boldsymbol{\epsilon}$ associated with the IQE method be defined by

$$
\delta\mathbf{p}^{(\ell)} \equiv \mathbf{p}^{(\ell+1)} - \mathbf{p}^{(\ell)}, \quad \delta\mathbf{w}^{(\ell)} \equiv \mathbf{w}^{(\ell+1)} - \mathbf{w}^{(\ell)},
$$

(3.5)
$$
\epsilon_p^{(\ell)} \equiv \mathbf{p}^{(\ell)} - \mathbf{p}, \quad \text{and} \quad \epsilon_w^{(\ell)} \equiv \mathbf{w}^{(\ell)} - \mathbf{w},
$$

where $\{\mathbf{p}, \mathbf{w}\}$ is the unique solution to the system (2.6). Then using (3.4) together with the fact that $\{\mathbf{p}, \mathbf{w}\}$ also satisfies (3.4), we obtain for $\ell = 1, 2, \ldots$

(3.6)
$$
\begin{bmatrix} \epsilon_w^{(\ell+1)} \\ \epsilon_p^{(\ell+1)} \end{bmatrix} = \mathbf{S}_\omega \begin{bmatrix} \epsilon_w^{(\ell)} \\ \epsilon_p^{(\ell)} \end{bmatrix}, \quad \begin{bmatrix} \delta\mathbf{w}^{(\ell+1)} \\ \delta\mathbf{p}^{(\ell+1)} \end{bmatrix} = \mathbf{S}_\omega \begin{bmatrix} \delta\mathbf{w}^{(\ell)} \\ \delta\mathbf{p}^{(\ell)} \end{bmatrix},
$$

where

(3.7)
$$
\mathbf{S}_\omega = \begin{bmatrix} \mathbf{Q} & \mathbf{0} \\ \dfrac{1}{\omega\Delta t}A_p^{-1}\mathbf{C} & \mathbf{I} \end{bmatrix}^{-1}\begin{bmatrix} (1-\omega)(\mathbf{I}-\mathbf{P}) & \omega\Delta t\mathbf{G} \\ \mathbf{0} & \mathbf{I} \end{bmatrix}
$$

is the iteration matrix associated with the IQE method. We note that the error vector $\overset{*}{\epsilon}{}_w^{(\ell)} \equiv \overset{*}{\mathbf{w}}{}^{(\ell)} - \mathbf{w}$ satisfies $\overset{*}{\epsilon}{}_w^{(\ell)} = (\mathbf{I} - \mathbf{P})\epsilon_w^{(\ell)}$.

It follows from (3.6) that the IQE method for arbitrary guess vectors $\overset{*}{\mathbf{w}}{}^{(0)}$ and $\mathbf{p}^{(0)}$ will converge [3] if and only if the spectral radius,[2] $\rho(\mathbf{S}_\omega)$, of the matrix $\mathbf{S}_\omega$ is less than unity. Moreover, the smaller the value of $\rho(\mathbf{S}_\omega)$, the faster the convergence. We define the optimal value $\bar{\omega}$ for ω to be that ω which minimizes $\rho(\mathbf{S}_\omega)$.

If the matrix $\mathbf{J}$ is positive real,[3] it can be shown [2] that the IQE method converges for $\omega = 1.0$. However, no general convergence proofs are known if

---

[2] If λ_j is an eigenvalue of $\mathbf{S}_\omega$, then $\rho(\mathbf{S}_\omega) \equiv \max_j |\lambda_j|$.

[3] A matrix $\mathbf{H}$ is said to be positive real if $(\mathbf{H} + \mathbf{H}^T)$ is positive definite. When the flow velocities $\mathbf{u}^n$ and $\mathbf{v}^n$ from the previous time-step satisfy the continuity equation and when the mesh subdivisions are uniform, it can be shown [2] that $\mathbf{J}$ is positive real.

$\omega \neq 1$ and no prescription is known for $\bar{\omega}$. In spite of this void in rigor, numerical results indicate that the IQE method is an effective solution procedure provided a reasonable value of ω is used. In the next section we give a numerical procedure, based on heuristic arguments, for determining near-optimal values for ω.

4 The Calculation of ω

An eigenset $\{\lambda, \mathbf{w}, \mathbf{p}\}$ of the matrix $\mathbf{S}_\omega$ satisfies the pair of coupled equations

$$(4.1) \qquad \lambda \mathbf{Q}\mathbf{w} = (1-\omega)(\mathbf{I} - \mathbf{P})\mathbf{w} + \omega \Delta t \mathbf{G}\mathbf{p}; \qquad \frac{\lambda}{\omega \Delta t} \mathbf{A}_p^{-1} \mathbf{C}\mathbf{w} = (1-\lambda)\mathbf{p}.$$

The matrix $\mathbf{P} \equiv \mathbf{G}\mathbf{A}_p^{-1}\mathbf{C}$ is symmetric and idempotent, i.e., $\mathbf{P}^2 = \mathbf{P}$. Thus, the eigenvalues of $\mathbf{P}$ equal zero or unity. Moreover, if

$$(4.2) \qquad \kappa(\mathbf{P}) \equiv \{\mathbf{y} : \ \mathbf{y} = \mathbf{P}\mathbf{y}\} \quad \text{and} \quad \eta(\mathbf{P}) \equiv \{\mathbf{z} : \ \mathbf{0} = \mathbf{P}\mathbf{z}\},$$

then it can be shown [2] that for any velocity vector $\mathbf{w}$, there exists unique vectors $\mathbf{y}\epsilon\kappa(\mathbf{P})$ and $\mathbf{z}\epsilon\eta(\mathbf{P})$ such that $\mathbf{w} = \mathbf{y} + \mathbf{z}$, where the inner product $(\mathbf{y}, \mathbf{z}) = 0$. This splitting, $\mathbf{w} = \mathbf{y} + \mathbf{z}$, will be used throughout this section.

With $\omega > 0$, it can be shown [2] that no eigenvalue λ of $\mathbf{S}_\omega$ equals unity. Thus, $\mathbf{p} = [\lambda/\omega \Delta t(1-\lambda)]\mathbf{A}_p^{-1}\mathbf{C}\mathbf{w}$, which when substituted into (4.1) gives

$$(4.3) \qquad \lambda \mathbf{Q}\mathbf{w} = (1-\omega)\mathbf{z} + \frac{\lambda}{1-\lambda}\mathbf{y}.$$

Multiplying (4.3) by the conjugate transpose of $\mathbf{w}$, which we denote by $\mathbf{w}^H$, then gives

$$(4.4) \qquad \lambda = \frac{(1-\omega)\mathbf{z}^H\mathbf{z} + [\lambda/(1-\lambda)]\mathbf{y}^H\mathbf{y}}{1 + \omega \Delta t \mathbf{w}^H \mathbf{J}\mathbf{w}}.$$

Here and in subsequent discussions we assume that $\mathbf{w}$ is normalized such that $1 = \mathbf{w}^H\mathbf{w} = \mathbf{y}^H\mathbf{y} + \mathbf{z}^H\mathbf{z}$. In what follows, unless stated otherwise, *we assume that the eigenvalue being analyzed is real* and, moreover, that $\rho(\mathbf{S}_\omega)$ is an eigenvalue of $\mathbf{S}_\omega$. We use a series of Remarks to indicate the behavior of $\rho(\mathbf{S}_\omega)$ as a function of ω.

Remark 1: Let $\mathbf{w} = \mathbf{y} + \mathbf{z}$ be the eigenvector component associated with the eigenvalue $\rho(\mathbf{S}_\omega)$. If $\mathbf{z}^T\mathbf{z}$ is small relative to $\mathbf{y}^T\mathbf{y}$, then

$$(4.5) \qquad \rho(\mathbf{S}_\omega) \approx \frac{\omega \Delta t \mu_y}{1 + \omega \Delta t \mu_y}$$

where μ_y is the largest eigenvalue of $\mathbf{J}$ with associated eigenvector $\mathbf{y}\epsilon\kappa(\mathbf{P})$.
Comment: If $\mathbf{z}$ is small relative to $\mathbf{y}$, then $\omega \Delta t \mathbf{J}\mathbf{y} \approx [\lambda/(1-\lambda)]\mathbf{y}$ from (4.3). From this it follows that λ is maximized when $\mathbf{y}$ is an eigenvector of $\mathbf{J}$ with eigenvalue μ_y.

Remark 2: Let $\mathbf{w} = \mathbf{y} + \mathbf{z}$ be the eigenvector component associated with the eigenvalue $\rho(\mathbf{S}_\omega)$. If $\mathbf{y}^T\mathbf{y}$ is small relative to $\mathbf{z}^T\mathbf{z}$, then

$$(4.6) \qquad\qquad \rho(\mathbf{S}_\omega) \approx \frac{1 - \omega}{1 + \omega\Delta t\mu_z},$$

where μ_z is the smallest (algebraically) eigenvalue of $\mathbf{J}$ with associated eigenvector $\mathbf{z}\epsilon\eta(\mathbf{P})$.

Comment: If $\mathbf{y}^T\mathbf{y}$ is small relative to $\mathbf{z}^T\mathbf{z}$, then $\omega\Delta t\mathbf{Jz} \approx \left[\frac{(1-\omega)-\lambda}{\lambda}\right]\mathbf{z}$ from (4.3), which implies that λ is maximized when $\mathbf{z}$ is an eigenvector of $\mathbf{J}$ with eigenvalue μ_z.

Remark 3: Let $\omega = 1.0$. If $\mathbf{w}$ is any $\mathbf{w}$-eigenvector component of $\mathbf{S}_{\omega=1}$ with associated eigenvalue λ, then

$$(4.7) \qquad \text{(a)} \quad \Delta t\mathbf{Jw} = -\mathbf{z} + \frac{\lambda}{1-\lambda}\mathbf{y} \quad \text{and (b)} \quad \lambda = 1 - \frac{\mathbf{y}^T\mathbf{y}}{1 + \Delta t\mathbf{w}^T\mathbf{Jw}}.$$

Moreover,

$$(4.8) \qquad\qquad \lim_{\Delta t\to 0} \lambda = 0, \quad \text{and} \quad \lim_{\Delta t\to 0} \mathbf{w} = \mathbf{y}.$$

Comment: The equations of (4.7) follow easily from (4.3) and (4.4). Multiplying (4.7a) by $\mathbf{z}^T$ and $\mathbf{y}^T$ give, respectively, the limits on $\mathbf{w}$ and λ in (4.8). Note that (4.8) implies that as Δt is decreased, the convergence rate of the IQE method increases and becomes less sensitive to the value used for ω. These properties are useful in the solution of transient problems which require smaller time-steps for accuracy reasons.

Remark 4: As before, let $\bar{\omega}$ denote that value of ω which minimizes $\rho(\mathbf{S}_\omega)$. Numerical results indicate that there exist small positive numbers ε_y and ε_z such that approximation (4.5) is valid for $\bar{\omega} + \varepsilon_y < \omega \le 1$ and that approximation (4.6) is valid for $0 < \omega \le \bar{\omega} - \varepsilon_z$.

Comment: In Figure 8.2, the graph of $\rho(\mathbf{S}_\omega)$ vs. ω is given for Problem (2.a) of Figure 8.3. A time-step of $\Delta t = .2$ was used. The values of $\rho(\mathbf{S}_\omega)$ were determined numerically. Also given in Figure 8.2 are graphs of the functions

$$(4.9) \qquad\qquad \rho_y \equiv \frac{\omega\Delta t\mu_y}{1 + \omega\Delta t\mu_y} \quad \text{and} \quad \rho_z \equiv \frac{1 - \omega}{1 + \omega\Delta t\mu_z}$$

with $\mu_y = 762.0$ and $\mu_z = 12.5$. Note that approximation (4.5) appears to be valid for all $\omega > \bar{\omega}$ and that $\bar{\omega}$ is close to the intersection of the ρ_y and ρ_z curves. Thus, if μ_y and μ_z were known, a good approximation for $\bar{\omega}$ could be obtained. A numerical procedure (given below) has been developed to estimate μ_y numerically from iteration data obtained from the IQE method with $\omega = 1.0$. The value for μ_y in Fig. 8.2 was obtained in this manner. No efficient procedure is known at this time for estimating μ_z. In Figure 8.2, μ_z was obtained by equating $\rho(\mathbf{S}_\omega)$ and ρ_z at $\omega = 0.1$. Knowing only μ_y, however, one can calculate that ω, say ω_γ, such that $\rho(\mathbf{S}_{\omega_\gamma}) = 1 - \omega_\gamma$. This is what we do; i.e., ω_λ is estimated instead of $\bar{\omega}$.

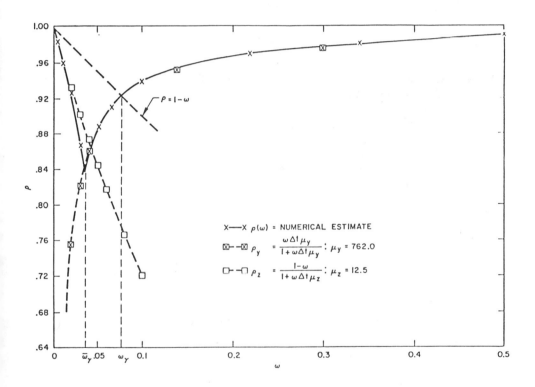

Figure 8.2 Spectral radius ρ vs. ω: Problem (2.a) with $\Delta t = .2$.

Remark 5: If approximation (4.5) is valid, then

$$(4.10) \qquad \omega_\gamma = \frac{\sqrt{\Delta t \mu_y + .25} - .5}{\Delta t \mu_y}$$

is that ω such that $\rho(\mathbf{S}\omega_\gamma) = 1 - \omega_\gamma$.

Comment: Setting $\rho(\mathbf{S}\omega_\gamma) = 1 - \omega_\gamma$ in (4.5) leads to the quadratic equation $\Delta t \mu_y \omega_\gamma^2 + \omega_\gamma - 1 = 0$. The positive solution of this quadratic is the ω_γ of (4.10).

Remark 6: Let $\omega = 1.0$ and let $\rho(\mathbf{S}_{\omega=1})$ be estimated numerically from iteration data of the IQE method. If Approximation (4.5) is valid, then μ_y can be approximated by

$$(4.11) \qquad \mu_y \approx \frac{\rho(\mathbf{S}_{\omega=1})}{\Delta t \left[1 - \rho(\mathbf{S}_{\omega=1})\right]}.$$

Comment: Equation (4.11) follows easily from (4.5). Since the convergence rate of the IQE method can be slow when $\omega = 1.0$, some care must be exercised in the estimation of $\rho(\mathbf{S}_{\omega=1})$. To ensure sufficiently rapid convergence, we pick a Δt such that $\rho(\mathbf{S}_{\omega=1})$ is approximately 0.9. Such a Δt is obtained by solving (4.11) for Δt after setting ρ to 0.9 and replacing μ_y by its upper bound $\hat{\sigma}$. If R_i is the i-th row sum of the elements, in absolute value, of the matrix $(\mathbf{J} + \mathbf{J}^T)/2$, then it can be shown that $\mu_y \leq \max_i R_i \equiv \hat{\sigma}$. Thus, to calculate $\rho(\mathbf{S}_{\omega=1})$ for use in (4.11), we use a time-step $\Delta t \leq \hat{\Delta} t$, where

$$(4.12) \qquad \hat{\Delta} t = \frac{9}{\hat{\sigma}}.$$

REMARK 7: The general procedure we use to estimate ω for the IQE iterations at time-step n is the following:

1. For the initial time-steps, $n = 1, 2, \ldots, n_1$, use $\omega = \hat{\omega}_\gamma$, where $\hat{\omega}_\gamma$ is calculated from (4.10) using $\hat{\sigma}$ instead of μ_y.

2. On time-step $n_1 + 1$, use $\Delta t_{n_1+1} = \min\{\Delta t, \hat{\Delta} t\}$, where $\hat{\Delta} t$ is given by (4.12). With $\omega = 1$, determine $\rho(\mathbf{S}_{\omega=1})$ numerically from the data of the IQE iterations. Determine μ_y from (4.11) and the ω_γ from (4.10).

3. Use $\omega = \omega_\gamma$ on succeeding time-steps $n = n_1 + 2, n_1 + 3, \ldots$ provided (a) Δt_n does not change, and provided (b) $\rho(\mathbf{S}_\omega)_n \leq 1 - \omega_\gamma$.

4. If (a) is not satisfied, i.e., if $\Delta t_n \neq \Delta t_{n-1}$, recompute ω_γ using (4.10).

5. If (b) is not satisfied, go to 2.

Comment: Since $\rho(\mathbf{S}_\omega) \leq 1 - \omega$ for any ω satisfying $0 < \omega \leq \omega_\gamma$, the ω strategy given above will not detect the case when $\omega < \bar{\omega}$. Although not included here, a test (c) for detecting this condition is needed in step 3 above.

We remark that any rigorous justification concerning a prescription for the optimal value of ω must include analysis of the complex eigenvalues of $\mathbf{S}_\omega$.

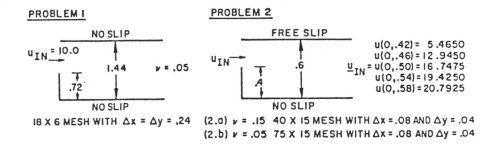

Figure 8.3 Description of test problems.

5 Numerical Results

In this section, we describe results of numerical experiments that substantiate the assumptions of the previous section. We will give numerical data for the estimation of μ_y and ω_γ using step 2 of Remark 7 and numerical data for the behavior of the IQE method when $\omega = \omega_\gamma$ is used.

The test problems that we consider are given in Fig. 8.3.[4] Any wall at the inlet is assumed to be noslip. Where the inlet u is specified, the inlet v is assumed to satisfy a free-slip condition.

We use the residual vector $\delta\mathbf{w}^{(\ell)} \equiv \mathbf{w}^{(\ell+1)} - \mathbf{w}^{(\ell)}$ to estimate $\rho(\mathbf{S}_\omega)$ and to terminate the IQE iterations. Specifically, with

$$(5.1) \qquad \rho\left(\boldsymbol{\delta}^{(\ell)}\right) = \frac{\|\delta\mathbf{w}^{(\ell)}\|}{\|\delta\mathbf{w}^{(\ell-1)}\|} \quad \text{and} \quad \mathrm{ER}^{(\ell)} \equiv \frac{\|\delta\mathbf{w}^{(\ell)}\|}{\|\delta\mathbf{w}^{(0)}\|},$$

we take $\rho\left(\boldsymbol{\delta}^{(\ell)}\right)$ as an approximation of $\rho(\mathbf{S}_\omega)$ and $\mathrm{ER}^{(\ell)}$ as an approximation to the ratio $\|\boldsymbol{\varepsilon}_w^{(\ell)}\|/\|\boldsymbol{\varepsilon}_w^{(0)}\|$. Here, $\boldsymbol{\varepsilon}_w^{(\ell)}$ is defined by (3.5). In the numerical data given, we also list a crude numerical estimate for $[(\mathbf{z}^{(\ell)})^T(\mathbf{z}^{(\ell)})]/[(\mathbf{y}^{(\ell)})^T(\mathbf{y}^{(\ell)})]$, where $\mathbf{y}^{(\ell)}$ and $\mathbf{z}^{(\ell)}$ satisfy $\boldsymbol{\varepsilon}_w^{(\ell)} = \mathbf{z}^{(\ell)} + \mathbf{y}^{(\ell)}$. The method used to obtain this estimate is given in [2].

In Table 8.1, we give the data required to estimate μ_y in step 2 of Remark 7. The time increment $\hat{\Delta}t$ is determined by (4.12), whereas Δt is the time-step used in the calculation. The column headed by ℓ gives the number of IQE iterations carried out.

The data given in Tables 8.2–8.4 illustrate the behavior of the IQE method as a function of Δt and ω. The value of ω_γ given here is obtained from (4.10) using μ_y

---

[4]For the test problems of Fig. 8.3, the units of the variables specified are in feet and seconds.

PROBLEM	$\hat{\sigma}$	$\hat{\Delta t}$	Δt	ℓ	ER^ℓ	$\rho\left(\delta^{(\ell)}\right)$	μ_y	$z^T z / y^T y$
1	216	.042	.050	65	$.9 \times 10^{-6}$	.895	170.5	$.2 \times 10^{-2}$
(2.a)	965	.009	.010	49	$.3 \times 10^{-4}$	.884	762.0	$.3 \times 10^{-3}$
(2.b)	644	.014	.014	41	$.4 \times 10^{-3}$	.864	453.8	$.3 \times 10^{-2}$

Table 8.1 Iteration data used in the calculation of μ_y.

Δt	ω_γ	ω	ℓ	$\text{ER}^{(\ell)}$	$\rho\left(\delta^{(\ell)}\right)$	$z^T z / y^T y$
		.200	50	$.9 \times 10^{-6}$	.873	$.8 \times 10^{-2}$
.02	.157	.175	46	$.9 \times 10^{-6}$	.857	$.3 \times 10^{-1}$
		.157	46	$.9 \times 10^{-6}$	.845	$.4 \times 10^{1}$
		.100	71	$.9 \times 10^{-5}$	.910	$.3 \times 10^{-1}$
.60	.094	.050	61	$.9 \times 10^{-5}$	.815	$.1 \times 10^{5}$
1.20	.068	.068	97	$.9 \times 10^{-6}$	.934	$.6 \times 10^{-2}$
		.020	296	$.9 \times 10^{-6}$	.980	$.7 \times 10^{-2}$
14.20	.020	.015	245	$.9 \times 10^{-6}$	.974	$.6 \times 10^{-2}$

Table 8.2 Iteration behavior of the IQE method for Problem 1.

Δt	ω_γ	ω	ℓ	$\text{ER}^{(\ell)}$	$\rho\left(\delta^{(\ell)}\right)$	$z^T z / y^T y$
		.1000	124	$.2 \times 10^{-5}$	.94	$.7 \times 10^{-3}$
.2	.0778	.0778	89	$.4 \times 10^{-5}$	.92	$.7 \times 10^{-3}$
		.0350	93	$.9 \times 10^{-6}$	.83	$.7 \times 10^{4}$
		.0600	95	$.9 \times 10^{-5}$	.950	$.2 \times 10^{-2}$
.4	.0557	.0557	99	$.4 \times 10^{-5}$	.947	$.9 \times 10^{-3}$
		.0300	53	$.9 \times 10^{-5}$	.865	$.1 \times 10^{3}$
		.100	250	$.9 \times 10^{-4}$	.983	$.5 \times 10^{-1}$
1.2	.033	.033	224	$.2 \times 10^{-5}$	.968	$.1 \times 10^{-3}$
		.010	59	$.9 \times 10^{-5}$	.875	$.4 \times 10^{3}$
		.0131	237	$.3 \times 10^{-3}$	.990	$.3 \times 10^{-1}$
14.2	.0096	.0096	239	$.2 \times 10^{-3}$	.987	$.2 \times 10^{-1}$

Table 8.3 Iteration behavior of the IQE method for Problem (2.a).

Δt	ω_γ	ω	ℓ	$ER^{(\ell)}$	$\rho\left(\boldsymbol{\delta}^{(\ell)}\right)$	$\mathbf{z}^T\mathbf{z}/\mathbf{y}^T\mathbf{y}$
.05	.189	.189	50	$.9 \times 10^{-6}$	.818	$.2 \times 10^3$
		.250	68	$.9 \times 10^{-5}$	.953	$.2 \times 10^{-1}$
.20	.100	.100	62	$.1 \times 10^{-5}$	.902	$.5 \times 10^{-1}$
		.085	56	$.9 \times 10^{-5}$	.880	$.3 \times 10^3$
		.060	114	$.9 \times 10^{-5}$	.969	$.6 \times 10^{-2}$
1.20	.042	.042	107	$.2 \times 10^{-5}$	.957	$.5 \times 10^{-2}$
		.035	76	$.8 \times 10^{-5}$	.944	$.8 \times 10^0$
3.60	.024	.024	157	$.3 \times 10^{-5}$	.974	$.7 \times 10^{-2}$

Table 8.4 Iteration behavior of the IQE method for Problem (2.b).

Δt	ω	ℓ	$ER^{(\ell)}$	$\rho\left(\boldsymbol{\delta}^{(\ell)}\right)$	$\mathbf{z}^T\mathbf{z}/\mathbf{y}^T\mathbf{y}$
.4000	1.0	275	$.3 \times 10^{-3}$	.993	$.6 \times 10^{-1}$
.200	1.0	238	$.4 \times 10^{-3}$	.990	$.6 \times 10^{-1}$
.0200	1.0	115	$.3 \times 10^{-5}$	.938	$.5 \times 10^{-3}$
.0100	1.0	49	$.8 \times 10^{-5}$	.884	$.3 \times 10^{-3}$
.0010	1.0	14	$.4 \times 10^{-7}$	.359	$.8 \times 10^{-4}$
.0001	1.0	7	$.6 \times 10^{-7}$	.039	$.1 \times 10^{-5}$

Table 8.5 Behavior of the IQE method when $\omega = 1.0$ as a function of Δt.

from Table 8.1. The column headed by ω gives the value of ω used in the calculation. Note that, indeed, $\rho(\boldsymbol{\delta}^{(\ell)}) \approx 1 - \omega_\gamma$ when $\omega = \omega_\gamma$ is used.

In Table 8.5, we give $\rho(\mathbf{S})$ and $\mathbf{z}^T\mathbf{z}/\mathbf{y}^T\mathbf{y}$ with $\omega = 1.0$ for various values of Δt. Note that both $\rho(\mathbf{S})$ and $\mathbf{z}^T\mathbf{z}/\mathbf{y}^T\mathbf{y}$ decrease as Δt is decreased.

The numerical data given in this section supports the assumptions made in Sec. 4 for the class of problems considered here. In addition, the ω strategy given in Remark 7 has also worked well for the more general thermally expandable fluid problems similar to those considered by Porsching [5].

Acknowledgements

The author would like to thank Dr. J.H. Anderson for his many constructive comments concerning the contents of this paper. Also sincere thanks to Joan Nichol who typed the original paper in NBI and to Lisa Laguna who retyped it in LaTeX. This work was supported by the U.S. Department of Energy under Contract DE-AC11-76PN00014.

References

[1] Amit, R., C.A. Hall, and T.A. Porsching. "An application of network theory to the solution of implicit Navier-Stokes difference equations," *J. of Computational Physics* **40** (1980) 183–201.

[2] Hageman, L.A. "A Family of Iterative Procedures for Solving the Semi-implicit Navier-Stokes Difference Equations," Bettis Atomic Power Laboratory Report WAPD-TM-1525, Pittsburgh, PA, 1982.

[3] Hageman, L.A., and D.M. Young. *Applied Iterative Methods*, Academic Press, New York, 1982.

[4] Hirt, C.W., B.D. Nichols, and N.C. Romero. "SOLA—A Numerical Solution Algorithm for Transient Fluid Flows," Report LA-5852, Los Alamos Scientific Laboratory, Los Alamos, New Mexico, 1976.

[5] Porsching, T.A. "A finite difference method for thermally expandable fluid transients," *Nucl. Sci. and Eng.* **64** (1972) 177–186.

Chapter 9

Hodie Approximation of Boundary Conditions

ROBERT E. LYNCH

Purdue University

Dedicated to David M. Young, Jr., on the occasion of his sixty-fifth birthday.

Abstract

A method is described for automatically computing coefficients of difference approximations to second-order linear elliptic differential equations on general domains with general linear boundary conditions.

1 Introduction

We describe a new method for incorporating boundary conditions in 9-point finite difference approximations of the differential equation in the linear elliptic boundary value problem

$$(1.1) \quad \mathbf{L}[u] = Au_{xx} + 2Bu_{xy} + Cu_{yy} + Du_x + Eu_y + Fu = G, \quad \text{on} \quad \Omega,$$

$$(1.2) \quad \mathbf{B}[u] = eu_n + fu = g, \quad e^2 + f^2 > 0, \quad \text{on} \quad \Gamma.$$

In these expressions, Ω is a two-dimensional domain with piecewise smooth boundary Γ; $A, ..., G, e, ..., g$, are smooth functions of x and y with $AC > B^2$; and u_n denotes the outward directed normal derivative.

To construct a difference approximation to the solution of (1.1)–(1.2), a rectangular mesh is placed over Ω. At each mesh point (x_i, y_j) in the *interior* of Ω, an approximation $U_{i,j} = U(x_i, y_j)$ to $u(x_i, y_j)$ is obtained. These $U_{i,j}$ are the *unknowns* of the approximating difference equation (ΔE). In contrast to usual difference approximations of (1.2), in the new discretization described below, there are *no unknowns at boundary mesh points* nor are there 'extra' equations approximating the boundary condition equation (1.2). Instead, the boundary conditions (1.2) are included in the ΔEs approximating (1.1).

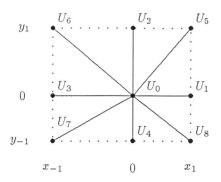

Figure 9.1

2   Approximation 'Away from the Boundary'

Before explaining how boundary conditions are treated (in Sec. 4), we review some features of Hodie[1] approximations to the differential equation (DE) (1.1) at mesh points 'away from the boundary'. We say that an interior mesh point $P_{i,j} = (x_i, y_j)$ is *away from the boundary* when each of the 8 neighboring mesh points

(2.1) $P_{i+p,j+q} = (x_{i+p}, y_{j+q}),$ $|p| = 1$ and/or $|q| = 1,$

and the points on the 8 line segments with endpoints $P_{i,j}$ and $P_{i+p,j+q}$, are in (the open set) Ω. No generality is lost by taking $P_{i,j}$ as the origin; then it is 'away from the boundary' when the points on the solid line segments in Fig. 9.1 are in Ω. The point $P_{0,0}$ is called the *central point* of this 'stencil' of 9 points.

Consider approximations to (1.1) which equate a linear combination of 9 values of the unknowns $U_{i,j}$ to some linear combination of values of G, the right side of the DE (1.1). To simplify the notation, we use a single subscript for these unknowns, as illustrated in Fig. 9.1: U_0 is the value at the central stencil point and the values at the eight neighboring mesh points are denoted by U_k, $k = 1, 2, \ldots, 8$. Any such approximation has the form

(2.2)
$$\sum_{k=0}^{8} \alpha_k U_k = \sum_{\ell=0}^{L} \beta_\ell G(\xi_\ell, \eta_\ell).$$

The usual difference approximations use a *single* value of G at $(0,0)$ in the approximation (2.2), thus making $L = 0$. However, use of more values of G can increase the accuracy. The points at which G is evaluated are called *evaluation points*, denoted by (ξ_ℓ, η_ℓ); they are all chosen in the rectangle $[x_{-1}, x_1] \times [y_{-1}, y_1]$. Typically, we take 5, 9, or 13 evaluation points.

---

[1] [6, 7]; the acronym is for 'High Order Difference approximations via Identity Expansions'.

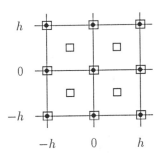

Figure 9.2

Example 1. Consider the Poisson equation

$$(2.3) \qquad \nabla^2 u = u_{xx} + u_{yy} = G.$$

At the origin the usual 9-point approximation to the Laplacian on a square mesh with spacing h, as in Fig. 9.2, is (see, for example, [1], p. 87)

$$(2.4) \qquad \mathbf{N}_h[u]_0 = \frac{1}{6h^2}\left[-20u_0 + 4\sum_{k=1}^{4} u_k + \sum_{k=5}^{8} u_k\right].$$

For sufficiently smooth u, expansion in Taylor series gives

$$(2.5) \qquad \mathbf{N}_h[u] = \nabla^2 u + \frac{h^2}{12}\nabla^4 u + \frac{h^4}{360}\left[\nabla^6 u + 2\frac{\partial^4}{\partial x^2 \partial y^2}\nabla^2 u\right] + \mathbf{R}_h[u],$$

where the 'remainder' is

$$(2.6) \qquad \mathbf{R}_h[u] = \frac{h^6}{60480}\left[3\nabla^8 u + 16\frac{\partial^4}{\partial x^2 \partial y^2}\nabla^4 u + 20\frac{\partial^8}{\partial x^4 \partial y^4}u\right] + O(h^8).$$

This shows that $\mathbf{N}_h[U]_0 = G_0$ approximates the Poisson equation with only $O(h^2)$ accuracy unless $\nabla^2 u = G = 0$, whereas the Mehrstellenverfahren[2]

$$(2.7) \qquad \mathbf{N}_h[U]_0 = G_0 + \frac{h^2}{12}\nabla^2 G_0 + \frac{h^4}{360}\left[\nabla^4 G_0 + 2\frac{\partial^4}{\partial x^2 \partial y^2}G_0\right]$$

has $O(h^6)$ accuracy. It also follows that higher than $O(h^6)$ accuracy is unattainable with 9-point approximations because the term proportional to h^6 in (2.6) cannot be expressed in terms of $G = \nabla^2 u$.

Appropriate divided difference approximations can be used to express the right side of (2.7) as an $O(h^6)$ accurate linear combination of values of G to get an

---

[2] [10]; see also [11], p. 991.

approximation of the form (2.2); [9] gives one involving values of G at mesh points; [6] give one involving 13 values of G at the 9 stencil point and at centroids of 4 mesh squares, indicated by little squares in Fig. 9.2; thus, with $\tilde{h} = h/2$:

$$
(2.8) \quad \mathbf{N}_h[u]_0 = \frac{1}{360} \Bigg[148 G_0 + 4 \sum_{k=1}^{4} G_k + \sum_{k=5}^{8} G_k
$$
$$
+ 48[G(\tilde{h}, \tilde{h}) + G(-\tilde{h}, \tilde{h}) + G(-\tilde{h}, -\tilde{h}) + G(\tilde{h}, -\tilde{h})] \Bigg] + O(h^6).
$$

Other evaluation points can be used; see [3] for a study of the effect of the location of evaluation points on the accuracy of the approximation for the Poisson equation.

For very special cases, such as the Poisson equation and a square mesh, one can easily derive simple formulas for the coefficients in the ΔE (2.2); e.g., (2.6) gives the α's and (2.8) the β's for a particular set of evaluation points. For the general case of variable coefficients or for a general distribution of stencil points, formulas for these coefficients are very complicated. See, for example the formulas for Mehrstellenverfahren in [10].

In [6] and [7], a method for *automating* the construction of 9-point approximations (2.2) of (1.1) was proposed which requires less computation than evaluation of *formulas* for these coefficients; e.g., with divided differences from the Mehrstellenverfahren formulas in [10].

Given the coefficients A, B, ..., F of the differential operator and the evaluation points, the basic idea of this Hodie method is to have a *computer compute* the coefficients α's and β's of the ΔE (2.2) by *solving a linear system*, and so that, with $M = L + 9$, the difference approximation is 'exact' on a given M-dimensional linear space S_M of polynomials. That is, if S_M is spanned by $s^{(1)}$, $s^{(2)}$, ..., $s^{(M)}$, then whenever a polynomial $p = \sum a_m s^{(m)}$ satisfies the DE (1.1), then it also satisfies the ΔE (2.2). In this situation, G is $\mathbf{L}[p]$, and the ΔE (2.2) is

$$
(2.9) \quad \sum_{k=0}^{8} \alpha_k p_k = \sum_{\ell=0}^{L} \beta_\ell \mathbf{L}[p](\xi_\ell, \eta_\ell).
$$

Normalizing the coefficients with

$$
(2.10) \quad \beta_0 = 1,
$$

one *computes* the coefficients α_k and the other β_ℓ by *solving the linear system*

$$
(2.11) \quad \sum_{k=0}^{8} \alpha_k (s^{(m)})_k - \sum_{\ell=1}^{L} \beta_\ell \mathbf{L}[s^{(m)}](\xi_\ell, \eta_\ell) = \mathbf{L}[s^{(m)}](0, 0),
$$
$$
m = 1, 2, \ldots, M = L + 9.
$$

For DEs (1.1) having constant coefficients, A, ..., F, on a uniform rectangular mesh, the coefficients α_k, β_ℓ, are computed at all points away from the boundary by solving (2.11) once. In general (2.11) must be solved once for each interior mesh point; i.e., it is solved $O(h^{-2})$ times for variable coefficient DEs. (For operations counts of Hodie and a comparison with collocation with Hermite bicubics, see [7]. Technical details about an ELLPACK module, HODIEG, which constructs a discretization of (1.1)–(1.2) for general domains and boundary conditions is given in [4].)

After computing the coefficients of all of the ΔEs, there results a system of $O(h^{-2})$ linear algebraic equations; after indexing the equations and unknowns, this system can be written as the vector equation

$$(2.12) \qquad\qquad \mathbf{AU} = \mathbf{b}.$$

When this is solved by a direct method and h is sufficiently small, the time spent in solving $\mathbf{AU} = \mathbf{b}$ dominates the discretization time, i.e., the time for solving (2.11) $O(h^{-2})$ times. Some experimental results are given in [2], [7], and [8, Chaps. 10–11].

Example 2. Any basis for the 9-dimensional space of biquadratic polynomials and a single evaluation point at the origin, leads to the usual divided difference approximation of (1.1). To conserve space, we display only two of the coefficients which solve (2.11); setting $x_1 = h$, $x_{-1} = -\theta h$, $y_1 = k$, and $y_{-1} = -\tau k$, we obtain

$$(2.13)$$

$$\alpha_1 = \frac{2(A\theta k + Bh\tau[1-\theta]) + D\theta\tau hk}{\theta(\tau+1)h^2 k},$$

$$\alpha_5 = \frac{2B\theta\tau}{(1+\theta)(1+\tau)hk},$$

where the functions A, B, and D are evaluate at the origin.

Special choices of basis polynomials make (2.11) easy to solve. For example, the biquadratic polynomial

$$(2.14) \qquad\qquad s(x,y) = \frac{(x+\theta h)x(y+\tau k)(k-y)}{(1+\theta)h^2\tau k^2}$$

is zero at 8 of the 9 stencil points and is unity at the ninth. When used as a basis polynomial, the corresponding equation in (2.11) reduces to

$$(2.15) \qquad \alpha_1 = [As_{xx} + 2Bs_{xy} + Cs_{yy} + Ds_x + Es_y + Fs](0,0),$$

whence the first formula in (2.13).

Example 3. One can get higher-order accuracy, by increasing the dimension of the space. Moreover, one can choose the basis polynomials to simplify the solution of (2.11). In particular, the cubic and quartic basis polynomials

$$(2.16)$$

$$s^{(10)}(x,y) = (x - x_{-1})x(x - x_1), \qquad s^{(12)}(x,y) = x\, s^{(10)}(x,y),$$

$$s^{(11)}(x,y) = (y - y_{-1})y(y - y_1), \qquad s^{(13)}(x,y) = y\, s^{(11)}(x,y),$$

are *zero* at the 9 stencil points in Fig. 9.1. Equation (2.11) reduces to four equations for the four unknown β's:

$$(2.17) \qquad -\sum_{\ell=1}^{L} \beta_\ell \mathbf{L}[s^{(m)}](\xi_\ell, \eta_\ell) = \mathbf{L}[s^{(m)}](0,0), \quad m = 10, 11, \ldots, 13.$$

The evaluation points can be taken as

$$(x_1, 0), \qquad (0, y_1), \qquad (x_{-1}, 0), \qquad (0, y_{-1}),$$

for $\ell = 1, 2, 3, 4$, respectively. After solving these four equations for $\beta_1, \beta_2, \ldots, \beta_4$, then any basis for the space of biquadratic polynomials can be used to determine the α's by solving

$$(2.18) \qquad \sum_{k=0}^{8} \alpha_k (s^{(m)})_k = \sum_{\ell=0}^{L} \beta_\ell \mathbf{L}[s^{(m)}](\xi_\ell, \eta_\ell), \quad m = 1, 2, \ldots, 9.$$

To make the approximation exact on 4th degree polynomials, one would like to augment the basis with a pair of polynomials so that $x^3 y$ and $x y^3$ are in the space S_{15}. However, the calculation can become delicate. For example, $x^3 y - x y^3$ is a harmonic polynomial and thus when $\mathbf{L} = \nabla^2$ and the mesh is square, then the system (2.11) is singular, but consistent.

3   Hodie as Interpolation

When the system (2.11) is nonsingular, then the resulting ΔE (2.2) approximating the DE (1.1) is also derivable from an *interpolant* in S_M. Indeed, the matrix of coefficients of the system (2.11) is the transpose of the coefficient matrix for the interpolation problem: Find $p \in S_M$ which satisfies the interpolation conditions

$$(3.1) \qquad \begin{aligned} p_k &= U_k, & k &= 0, 1, \ldots, 8, \\ \mathbf{L}[p](\xi_\ell, \eta_\ell) &= G(\xi_\ell, \eta_\ell), & \ell &= 1, \ldots, L. \end{aligned}$$

Then there is a dual (or 'cardinal') basis, $\{\sigma^{(1)}, \sigma^{(2)}, \ldots, \sigma^{(M)}\}$, for S_M satisfying

$$(3.2) \qquad \left. \begin{aligned} (\sigma^{(m)})_k &= \delta_{m,k+1}, \ k = 0, 1, \ldots, 8 \\ \mathbf{L}[\sigma^{(m)}](\xi_\ell, \eta_\ell) &= \delta_{m,\ell+9}, \ \ell = 1, 2 \ldots, L \end{aligned} \right\}, \ m = 1, 2, \ldots, M = L + 9,$$

where $\delta_{m,n}$ is the Kronecker delta function. The unique interpolant can then be written as

$$(3.3) \qquad p(x, y) = \sum_{k=0}^{8} U_k \, \sigma^{(k+1)}(x, y) + \sum_{\ell=1}^{L} G_\ell \, \sigma^{(\ell+9)}(x, y),$$

where $G_\ell = G(\xi_\ell, \eta_\ell)$. Apply $\mathbf{L}$, evaluate the result at the origin, and then set $G_0 = \mathbf{L}[p](0,0)$ to get

$$(3.4) \qquad G_0 = \sum_{k=0}^{8} \mathbf{L}[s^{(k+1)}](0,0)\, U_k + \sum_{\ell=1}^{L} \mathbf{L}[s^{(\ell+9)}](0,0)\, G_\ell.$$

By taking the basis polynomials $s^{(m)}$ in (2.11) to be the $\sigma^{(m)}$'s the algebraic system (2.11) reduces to

$$
\begin{aligned}
\alpha_k &= \mathbf{L}[s^{(k+1)}](0,0), & k &= 0,\, 1,\, \ldots,\, 8, \\
-\mathbf{L}[p](\xi_\ell, \eta_\ell)\,\beta_\ell &= \mathbf{L}[s^{(\ell+9)}](0,0), & \ell &= 1,\, 2,\, \ldots,\, L,
\end{aligned}
$$

(3.5)

so that (3.4) becomes (2.2). We formalize this rather obvious result as follows.

Theorem 3.1 *If the system* (2.11) *is nonsingular with solution* α_k, β_ℓ, *then*

$$\sum_{k=0}^{8} \alpha_k U_k - \sum_{\ell=1}^{L} \beta_\ell G(\xi_\ell, \eta_\ell)$$

is the value of $\mathbf{L}[p]$ *at the origin, where p solves the interpolation problem* (3.1).

4   Boundary Conditions

An interior mesh point $P_{i,j}$ is said to be *next to the boundary* if one or more of its 8 nearest neighboring mesh points $P_{i+p,j+q}$ [as in (2.1)] is in the exterior of the open domain Ω, or if a point on the line segment joining $P_{i,j}$ and $P_{i+p,j+q}$ is on the boundary Γ. In the 9-point stencil centered at $P_{i,j}$, the exterior point $P_{i+p,j+q}$ is replaced with the boundary point closest to $P_{i,j}$ on the line segment joining $P_{i,j}$ and $P_{i+p,j+q}$. Figure 9.3 illustrates such a configuration where $P_{i,j}$ is taken as the origin.

The 9-point stencil centered at an interior mesh point $P_{i,j}$ next to the boundary is taken as $P_{i,j}$, together with the boundary mesh points (in place of the corresponding $P_{i+p,j+q}$'s), and its nearest neighboring mesh points which are interior and joined to $P_{i,j}$ by a line segment of points in the interior of Ω.

In the usual finite difference approximations, it is customary to construct a ΔE centered at each interior mesh point next to the boundary and include in the ΔE *unknowns* at the boundary mesh points (e.g., for the configuration of Fig. 9.3, the unknowns U_3, U_6, U_7, are included in the ΔE).

Consequently, when Neumann or, more generally, mixed boundary conditions (1.2) are given, one must supplement the system with an additional ΔE which approximates the given boundary condition equation (1.2) for each boundary mesh

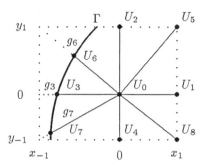

Figure 9.3

point where non-Dirichlet conditions apply. For the situation illustrated in Fig. 9.3 and a single evaluation point, one has

$$(4.1) \qquad \sum_{k=0}^{8} \alpha_k U_k = G(0,0), \quad \text{and} \quad \sum_{k=0}^{8} \zeta_{k,m} U_k = g_m, \quad m = 3, 6, 7,$$

where the ζ's are chosen to yield an approximation of the boundary condition (1.2). See, for example, [1], pp. 52–56 and the examples and the references given there.

Our new approximation for (1.1)–(1.2) exploits the observation in Sec. 3 that the resulting difference approximation is one obtained by *interpolation*. It applies interpolation not only away from the boundary, but also next to the boundary. Let I_0 denote the set of indices of those stencil points which are mesh points, and B_0 the indices of the boundary points. For the configuration in Fig. 9.3:

$$(4.2) \qquad I_0 = \{0, 1, 2, 4, 5, 8\}, \qquad B_0 = \{3, 6, 7\}.$$

Then our $\Delta\mathbf{E}$ approximation of (1.1)–(1.2) is derived from the following interpolation problem: Find $p \in S_M$ such that

$$p_k = U_k, \qquad k \in I_0, \qquad \text{[interior mesh points]}$$

$$(4.3) \qquad \mathbf{B}[p]_k = g_k, \qquad k \in B_0, \qquad \text{[boundary mesh points]}$$

$$\mathbf{L}[p](\xi_\ell, \eta_\ell) = G(\xi_\ell, \eta_\ell), \quad \ell = 1, 2, \ldots, L \quad \text{[evaluation points]}.$$

Apply $\mathbf{L}$ to p and evaluate the result at the origin. Write the result in terms of the interpolation data as

$$(4.4) \qquad \mathbf{L}[p](0,0) = \sum_{k \in I_0} \alpha_k U_k - \sum_{k \in B_0} \gamma_k g_k - \sum_{\ell=1}^{L} \beta_\ell G(\xi_\ell, \eta_\ell).$$

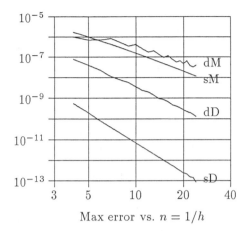

Max error vs. $n = 1/h$

Figure 9.4 (s) square, (s) disk, (D) Dirichlet, (M) mixed.

The value of the right side of (4.4) is then an approximate value of $G_0 = \mathbf{L}[u]_0$. By replacing the left side of (4.4) with G_0, one obtains a ΔE relating the unknowns U_k. By construction, any solution of this ΔE results in values of U_k, $k \in I_0$, which are values of the interpolating polynomial $p \in S_M$ which satisfy the conditions (4.3) and (by construction) the additional condition

$$\mathbf{L}[p](0,0) = G(0,0).$$

Rearrangement gives

$$(4.5) \qquad \sum_{k \in I_0} \alpha_k U_k = G(0,0) + \sum_{\ell=1}^{L} \beta_\ell G(\xi_\ell, \eta_\ell) + \sum_{k \in B_0} \gamma_k g_k.$$

As for ΔEs away from the boundary, to *compute* the values of the coefficients α, β, and γ, one *solves* a linear system, whose m-th equation, for $m = 1, 2, \ldots, M$, is

$$(4.6) \quad \sum_{k \in I_0} \alpha_k (s^{(m)})_k - \sum_{k \in B_0} \gamma_k \mathbf{B}[s^{(m)}]_k - \sum_{\ell=1}^{L} \beta_\ell \mathbf{L}[s^{(m)}](\xi_\ell, \eta_\ell) = \mathbf{L}[s^{(m)}](0,0),$$

To show the effect of the approximation of the boundary conditions, we solved the Laplace equation having solution $e^x \sin y$. Away from the boundary the accuracy is $O(h^6)$, so deviations from this are due to errors at the boundary. The graphs in Fig. 9.4 shows log-log plots of the max-error versus h for the unit square (s) $[0,1]^2$ and for the disk (d), in the unit square, when Dirichlet boundary conditions are applied (D) and when the mixed conditions (M) $u + u_n = g$ are applied. When the least squares approximation of the error by Ch^ν is obtained, the values of the exponents are (s) 5.99, (d) 4.49, for Dirichlet conditions and (s) 3.54, (d) 2.59, for the mixed conditions; see [2] for more experimental results.

5 Extension of $U_{i,j}$ to Ω

The computed values $U_{i,j} = U(x_i, y_j)$, obtained by solving the linear system of ΔEs, $\mathbf{AU} = \mathbf{b}$ give approximate values of $u(x_i, y_j)$ at each interior mesh point in Ω. To obtain an approximation to u at a point $(x, y) \in \Omega$ which is *not* a mesh point, we can use the value of a corresponding interpolant.

For example, suppose (x, y) is near the origin in Fig. 9.3. Having computed the approximate mesh values (U_k, $k = 0, 1, 2, 4, 5, 8$, in the figure), one can determine the interpolant (4.3) and take $p(x, y)$ as the approximation to $u(x, y)$. Similarly, for approximations of values of derivatives of u, or its integral, etc.

6 Indexing of Unknowns

Since the only unknowns, U, are values at interior mesh points, any ordering of mesh points in a *rectangular domain* containing the mesh on Ω imposes an ordering on these unknowns. Additional 'fictitious unknowns' can be added to the system, one for each mesh point in the rectangle and exterior to Ω; e.g., equations of the form $U_k = 0$ can be added to the system. There results an equation $\mathbf{AU} = \mathbf{b}$ whose matrix $\mathbf{A}$ has the *same structure* as the one corresponding to a problem on the rectangular domain. For example, in the 'natural ordering', this matrix has nonzero entries on exactly 9 diagonals. Such a property can be exploited when the system is solved with a vector computer (such as the Cyber 205), especially when iterative methods are used—in contrast, the matrix of the original system without the 'fictitious unknowns', has nonzero entries on many diagonals, a diagonal structure which does not lead to efficient processing with a vector computer.

7 Eigenproblems

Consider the eigenvalue problem

(7.1) $$(pu_x)_x + (pu_y)_y + Fu = -\lambda wu$$

rewritten in the form (1.1) with $G = -\lambda wu$, and homogeneous Dirichlet boundary conditions. The usual 9-point discretization with a single value of G yields a linear system

(7.2) $$\mathbf{AU} = -\mu\mathbf{WU},$$

where $\mathbf{W}$ is a diagonal matrix. With four evaluation points on mesh lines through the central stencil point (e.g., points labeled 1, 2, 3, and 4 in Figs. 9.1 and 9.3), the

ΔE (4.5) becomes

$$(7.3) \qquad \sum_{k \in I_0} \alpha_k U_k = \sum_{\ell=0}^{4} \beta_\ell G(\xi_\ell, \eta_\ell) = -\lambda \sum_{\ell=0}^{4} \beta_\ell w_\ell U_\ell.$$

However, in order that the right side approximates λwu (instead of a multiple of it), we must use the normalization $\sum_{\ell=0}^{4} \beta_\ell = 1$ instead of (2.10). Replacing λ with σ, we obtain the generalized eigenvalue problem

$$(7.4) \qquad \mathbf{AU} = -\sigma \mathbf{BU},$$

where $\mathbf{B}$ has at most 5 nonzero entries in each row. Because the Hodie scheme yields an approximation to the homogeneous DE, in which $G \equiv 0$, the matrix $\mathbf{A}$ in (7.4) is the same as $\mathbf{A}$ in (7.2).

As an example, consider the eigenproblem for the unit disk

$$(7.5) \qquad \nabla^2 u = -\lambda u \quad \text{on} \quad r < 1, \qquad u = 0 \quad \text{on} \quad r = 1.$$

Then $\lambda = j_{m,k}^2$, where $j_{m,k}$ is the k-th positive zero of the Bessel function J_m; the corresponding eigenfunction is

$$(7.6) \qquad \phi_{m,k}(r, \theta) = J_m(j_{m,k} r) \left\{ \begin{array}{c} \cos m\theta \\ \sin m\theta \end{array} \right\}.$$

For a square mesh with mesh length $h = 2/n$, away from the boundary the ΔEs are

$$\mathbf{N}_h[U]_0 = -\mu U_0$$

and

$$\mathbf{N}_h[U]_0 = -\frac{\sigma}{12} \left[8U_0 + \sum_{\ell=1}^{4} U_\ell \right],$$

where $\mathbf{N}_h$ is as in (2.4). Next to the boundary, these equations have different coefficients as explained in Sec. 4.

Table 9.1 lists percent errors in the approximations of the first few eigenvalues of (7.5) by the eigenvalues μ and σ. The second and third eigenvalues of the continuous problem are double. As expected, the errors in μ are very nearly $O(h^2)$ and $O(h^4)$ for σ.

Acknowledgements

The author thanks Garrett Birkhoff for many helpful suggestions and the Air Force Office of Scientific Research for partial support under grant 88–0243. This paper was supplied by the author in LaTeX and reformatted by Lisa Laguna.

	Percent error in μ			Percent error in σ		
	$n = 4$	8	10	4	8	10
$j_{0,1}{}^2 = 5.78319$	7.0	2.3	1.6	0.19	0.041	0.014
$j_{1,1}{}^2 = 14.6820$	20.6	5.8	4.0	1.62	0.096	0.052
$j_{1,1}{}^2 = 14.6820$	20.6	5.8	4.0	1.62	0.096	0.052
$j_{2,1}{}^2 = 26.3746$	42.4	11.3	7.4	16.17	0.735	0.306
$j_{2,1}{}^2 = 26.3746$	36.0	9.9	7.1	2.57	0.103	0.0014
$j_{0,2}{}^2 = 30.4713$	36.7	12.4	8.4	8.36	0.398	0.17

Table 9.1

References

[1] Birkhoff, G., and R. E. Lynch, *Numerical Solution of Elliptic Problems*, SIAM Publications, Philadelphia, 1984.

[2] Birkhoff, G., and R. E. Lynch, "ELLPACK and ITPACK as research tools for solving elliptic problems," Chapter 3, these Proceedings.

[3] Boisvert, R. F., "Optimal order compact discretizations of the Poisson equation," *SIAM J. Sci. Stat. Comp.* **2** (1981) 193–199.

[4] Lynch, R. E., "HODIEG: A new ELLPACK discretization module," Purdue Univ., Dept. Compt. Sci., Report CSD-TR-871, West Lafayette, IN, March 1989.

[5] Lynch, R. E., "$O(h^4)$ and $O(h^6)$ finite difference approximations to the Helmholtz equation in n-dimensions," in *Advances in Computer Methods for Partial Differential Equations* **V**, Proceedings of the Fifth IMACS International Symposium on Computer Methods for Partial Differential Equations, V. R. Vichnevetsky and R. S. Stepleman (eds.), (1984) 199–202.

[6] Lynch, R. E., and J. R. Rice, "High accuracy finite difference approximation to solutions of elliptic partial differential equations," Proc. Nat. Acad. Sci. 75 (1978), 2541–2544.

[7] Lynch, R. E., and J. R. Rice, "The Hodie method and its performance for solving elliptic partial differential equations," in *Recent Advances in Numerical Analysis*, C. de Boor and G. H. Golub (eds.), Academic Press, New York, 1978, 143–175.

[8] Rice, J. R., and R. F. Boisvert, *Solving Elliptic Problems Using ELLPACK*, Springer-Verlag, New York, 1984.

[9] Rosser, J. B., "Nine point difference solutions for Poisson's equations," *Comp. and Math. with Appls.* **1** (1975) 351–360.

[10] Young, D. M., and J. H. Dauwalder, "Discrete representations of partial differential operators," in *Errors in Digital Computation*, Vol. 2, L. B. Rall (ed.), Wiley, New York, 1965, 181–217.

[11] Young, D. M., and R. T. Gregory, *A Survey of Numerical Mathematics*, Vol. II, Addison-Wesley, New York, 1973. (Reprinted by Dover, 1988).

Chapter 10

Iterative Methods for Nonsymmetric Linear Systems

Wayne D. Joubert
University of Texas at Austin

and

Thomas A. Manteuffel
Los Alamos National Laboratory
   and University of Colorado, Denver

Dedicated to David M. Young, Jr., on the occasion of his sixty-fifth birthday.

Abstract

This paper will present a survey of polynomial methods for solving non-symmetric linear systems. An emphasis will be placed on generalizations of the classical conjugate gradient method for symmetric problems. To provide a framework for these methods, we will first outline a general theory of projection methods. Then, the more specific class of Krylov projection methods will be presented, including examples such as the classical conjugate gradient method, normal equations methods, the GCW method and the biconjugate gradient method. After this, truncated and restarted Krylov space methods will be discussed. An attempt will be made to expose open questions.

1   Introduction

In this paper we are concerned with the solution of systems of the form

$$(1.1) \qquad Au = b$$

where $A \in \mathbb{C}^{N \times N}$ is a large nonsingular matrix and b and u are column vectors.

149

When A is large and sparse, iterative methods are usually necessary in order to solve this problem efficiently. In particular, when A is self-adjoint with respect to some inner product, then the conjugate gradient (CG) method in many cases is a useful and practical solution technique. However, for general A, no general-purpose short-recurrence iterative technique, other than forming the normal equations, is currently available.

The purpose of this paper is to establish a theoretical framework for iterative methods for non-Hermitian problems. We will be concerned mainly with generalizations of the conjugate gradient method which are applicable to the case when A is not Hermitian positive definite (HPD). This framework will attempt to give a systematic structure to the landscape of iterative methods. We will also prove certain results such as convergence results for various classes of methods, and suggest some open questions which merit further study.

Iterative Methods. An *iterative method* is defined by an initial guess $u^{(0)}$ to the exact solution $\bar{u} = A^{-1}b$, followed by a sequence of further approximations $\{u^{(i)}\}_{i>0}$. The error is given by $e^{(i)} = \bar{u} - u^{(i)}$ and the residual by $r^{(i)} = b - Au^{(i)} = Ae^{(i)}$.

Typically the system $Au = b$ is derived from the equivalent problem

$$(1.2) \qquad\qquad \hat{A}\hat{u} = \hat{b}$$

by preconditioning: $A = Q_L^{-1}\hat{A}Q_R^{-1}$, $u = Q_R\hat{u}$, $b = Q_L^{-1}\hat{b}$, where $Q_L, Q_R \in \mathbb{C}^{N \times N}$, so that (1.1) becomes

$$(Q_L^{-1}\hat{A}Q_R^{-1})(Q_R\hat{u}) = (Q_L^{-1}\hat{b}).$$

We thus have the associated quantities for the system (1.2): $\hat{u}^{(n)} = Q_R^{-1}u^{(n)}$, $\hat{u} = \hat{A}^{-1}\hat{b}$, $\hat{r}^{(n)} = \hat{b} - \hat{A}\hat{u}^{(n)}$ and $\hat{e}^{(n)} = Q_R^{-1}e^{(n)}$. Note that the postconditioning Q_R^{-1} distinguishes $e^{(n)}$ from $\hat{e}^{(n)}$, resulting in differing norms for the two quantities. In many settings the postconditioning may be included within the preconditioning (see [1]).

For a given choice of $u^{(0)}$, we will say an that iterative method *converges* if $u^{(n)} \to \bar{u}$ in some norm as $n \to \infty$. We will say that the method experiences *finite termination* if for some d we have $u^{(n)} = \bar{u}$ for all $n \geq d$.

Polynomial Methods. A *polynomial method* is an iterative method satisfying

$$e^{(n)} = P_n(A)e^{(0)}$$

for some polynomial $P_n(z) = 1 - zQ_{n-1}(z)$ of degree no greater than n satisfying $P_n(0) = 1$. Equivalently,

$$u^{(n)} - u^{(0)} \in \mathbf{K}_n(r^{(0)}, A)$$

where $\mathbf{K}_n(v, A) = \mathrm{span}\{A^i v\}_{i=0}^{n-1}$ is the Krylov space spanned by the n vectors $A^i v$, $0 \leq i \leq n - 1$. Typically Q_{n-1} is chosen so that $Q_{n-1}(A)$ approximates A^{-1} in some sense.

It is possible to compute P_n in a number of ways. *Chebyshev-like* methods, for example, typically attempt to calculate P_n which in some sense is small on the spectrum of A. On the other hand, *conjugate gradient-like* methods employ certain inner products of vectors from the Krylov space to attempt to reduce the error.

In the next section, we will define a class of methods that covers nearly all the existing conjugate gradient-like methods.

2 Projection Methods

We define a *projection method* to be an iterative method satisfying the following relations. For any n and initial guess $u^{(0)}$ let $\mathbf{L}_n$ be an l_n-dimensional and $\mathbf{R}_n$ an r_n-dimensional subspace. Assuming that $u^{(n-1)}$ exists, we define $u^{(n)}$ to be a vector satisfying

(2.1) $$u^{(n)} - u^{(n-1)} \in \mathbf{R}_n, \qquad u^{(n)} - \bar{u} \perp \mathbf{L}_n.$$

Orthogonality is defined in the standard Euclidean sense, with inner products defined by $(x, y) = \sum x_i \bar{y}_i$ for vectors $x = [x_1, x_2, \ldots, x_N]^T$, $y = [y_1, y_2, \ldots, y_N]^T$. We note that such $u^{(n)}$ may not exist and may not be unique. In this setting $u^{(n-1)} + \mathbf{R}_n$ is called the *solution space*, and the orthogonality condition of (2.1) is called the *Petrov-Galerkin condition*.

We now consider the specific form such $u^{(n)}$ takes. Let L_n (R_n) be a $N \times l_n$ ($N \times r_n$) matrix whose columns form a basis for $\mathbf{L}_n$ ($\mathbf{R}_n$). Then for some $r_n \times 1$ vector α we have

(2.2) $$u^{(n)} = u^{(n-1)} + R_n \alpha$$

or

$$e^{(n)} = e^{(n-1)} - R_n \alpha.$$

The Petrov-Galerkin condition (2.1) yields

$$L_n^* e^{(n)} = L_n^* e^{(n-1)} - L_n^* R_n \alpha = 0.$$

If $L_n^* R_n$ is square and nonsingular then

(2.3) $$\alpha = (L_n^* R_n)^{-1} L_n^* e^{(n-1)}.$$

Thus,

(2.4) $$\begin{aligned} u^{(n)} &= u^{(n-1)} + R_n (L_n^* R_n)^{-1} L_n^* e^{(n-1)}, \\ e^{(n)} &= \left[I - R_n (L_n^* R_n)^{-1} L_n^* \right] e^{(n-1)}. \end{aligned}$$

Let

(2.5) $$P_n = I - R_n (L_n^* R_n)^{-1} L_n^*.$$

Clearly P_n is a projection; hence the name projection methods is appropriate.

We now examine some properties of projection methods. We will begin with some definitions and then prove several theorems.

It is desirable that $u^{(n)}$ exist and be unique for any n. Given a particular $u^{(n-1)}$, we will say that a given projection method *breaks down* at step n if $u^{(n)}$ as defined in (2.1) does not exist or is not unique. It is clear that if $l_n < r_n$ and $u^{(n)}$ satisfying (2.1) exists, then such $u^{(n)}$ cannot be unique. On the other hand, when $l_n = r_n$ we have the following result.

Theorem 2.1 (Existence/Uniqueness) *Suppose $l_n = r_n > 0$. Then breakdown occurs at step n if and only if $L_n^* R_n$ is singular.*

Proof: The proof follows from the discussion above. ∎

Corollary 2.2 (Finite Termination) *Suppose for a particular $e^{(0)}$ breakdown never occurs for any step of the projection method, and $u^{(n-1)} \neq \bar{u}$ implies* $\dim \mathbf{R}_n \overset{>}{\neq} \dim \mathbf{R}_{n-1}$ *for any n. Then convergence is attained within N steps:* $u^{(d)} = \bar{u}$ *for some $d \leq N$.*

Proof: Suppose breakdown does not occur at step n. Then by uniqueness $u^{(n)} = \bar{u}$ if and only if $e^{(n-1)} \in \mathbf{R}_n$. Since $\dim \mathbf{R}_n$ is strictly increasing, $e^{(d-1)} \in \mathbf{R}_d$ for some $d \leq N$. ∎

Note that whenever $l_n = r_n \neq 0$, then it is possible to construct B_n such that $L_n = B_n^* R_n$. In particular, we define B_n by $B_n^* = L_n (R_n^* R_n)^{-1} R_n^*$. Thus (2.5) becomes

$$(2.6) \qquad P_n = I - R_n (R_n^* B_n R_n)^{-1} R_n^* B_n.$$

The matrix B_n embodies the relation between L_n and R_n. If B_n is *definite* on the subspace $\mathbf{R}_n$, in the sense that $v^* B_n v \neq 0$ for all $v \in \mathbf{R}_n \setminus \{0\}$, then it follows that $L_n^* R_n = R_n^* B_n R_n$ is nonsingular and breakdown will not occur at step n. This motivates the following theorem.

Theorem 2.3 (Boundedness) *Suppose $l_n = r_n \neq 0$ and $L_n = B_n^* R_n$ for every n, where $\mathbf{L}_n$, $\mathbf{R}_n$ and B_n may depend on $e^{(0)}$. Suppose, furthermore, that there exists c_n independent of $e^{(0)}$ such that*

$$0 < ||B_n|| \cdot ||x||^2 \leq c_n \cdot |x^* B_n x|$$

for every $e^{(0)}$ and $x \in \mathbf{R}_n \setminus \{0\}$. Then breakdown is impossible at any step and

$$||u^{(n)}|| \leq ||u^{(n-1)}|| + c_n ||e^{(n-1)}||.$$

Proof: If $x = R_n v$, then $x^* B_n x = v^* L_n^* R_n v$. By hypothesis,

$$\frac{v^* L_n^* R_n v}{v^* v} \cdot \frac{||v||^2}{||R_n v||^2} = \frac{|x^* B_n x|}{||x||^2} \geq \frac{||B_n||}{c_n} > 0.$$

This implies $L_n^* R_n$ is definite, thus nonsingular, implying no breakdown at this step.
Continuing, we have

$$u^{(n)} = u^{(n-1)} + R_n(R_n^* B_n R_n)^{-1} R_n^* B_n e^{(n-1)}.$$

Now let $\tilde{R}_n = R_n Q$, where Q is square and $\tilde{R}_n^* \tilde{R}_n = I$. Then it is easily seen that

$$u^{(n)} = u^{(n-1)} + \tilde{R}_n(\tilde{R}_n^* B_n \tilde{R}_n)^{-1} \tilde{R}_n^* B_n e^{(n-1)}.$$

Then

$$||u^{(n)}|| \le ||u^{(n-1)}|| + ||(\tilde{R}_n^* B_n \tilde{R}_n)^{-1}|| \cdot ||\tilde{R}_n^* B_n e^{(n-1)}||.$$

We seek an upper bound for $||u^{(n)}||$. We have

$$
\begin{aligned}
||(\tilde{R}_n^* B_n \tilde{R}_n)^{-1}||^2 &= 1/\min_{||w||=1} ||\tilde{R}_n^* B_n \tilde{R}_n w||^2 \\
&= 1/\min_{||w||=1} w^* \tilde{R}_n^* B_n^* \tilde{R}_n \left[ww^* + (I - ww^*) \right] \tilde{R}_n^* B_n \tilde{R}_n w \\
&= 1/\min_{||w||=1} \left[|w^* \tilde{R}_n^* B_n \tilde{R}_n w|^2 + ||(I - ww^*)\tilde{R}_n^* B_n \tilde{R}_n w||^2 \right] \\
&\le 1/\min_{||w||=1} |w^* \tilde{R}_n^* B_n \tilde{R}_n w|^2 \le 1/\min_{x \in R_n, ||x||=1} |x^* B_n x|^2.
\end{aligned}
$$

The last inequality follows from letting $x = \tilde{R}_n w$ and noting that $||\tilde{R}_n w|| = ||w||$.
We have, therefore,

$$||u^{(n)}|| \le ||u^{(n-1)}|| + \sup_{x \in R_n \setminus \{0\}} \frac{||x||^2}{|x^* B_n x|} \cdot ||B_n|| \cdot ||e^{(n-1)}||.$$

The hypothesis on B_n yields the result.   ∎

Balanced Projection Methods

We say a projection method is *balanced* (a BPM) if there exists a fixed square
matrix B such that $L_n = B^* R_n$ for all n and $e^{(0)}$.

Consider the sesquilinear form $(B\cdot, \cdot)$. The Petrov-Galerkin condition (3) is now
equivalent to the condition

$$(2.7) \qquad\qquad (Be^{(n)}, z) = 0 \qquad \forall z \in \mathbf{R}_n.$$

If B is *definite*, that is, $x^* B x \ne 0$ for all $x \ne 0$, then the hypothesis of Theorem 2.3
is satisfied. Recall that a real matrix B is definite if and only if either its Hermitian
part $B_H \equiv H(B) \equiv (B + B^*)/2$ or its negative $-H(B)$ is Hermitian positive definite
(HPD) (see [12]).

We have then the following

Corollary 2.4 (Boundedness) *Let B be definite. Then a BPM based on B will not break down at any step. Furthermore, let*

$$\rho_{\min}(B) = \inf_{x \neq 0} \frac{|x^* B x|}{x^* x}.$$

Then

$$\|u^{(n)}\| \leq \|u^{(n-1)}\| + \frac{\|B\|}{\rho_{\min}(B)} \|e^{(n-1)}\|.$$

Proof: The proof is a direct application of Theorem 2.3.   ■

Theorem 2.5 (Uniform Boundedness) *Suppose the hypotheses of Corollary 2.4 are satisfied, and furthermore $\mathbf{R}_n \subseteq \mathbf{R}_{n+1}$ for every n. Then the following bound holds for every n:*

$$\|u^{(n)}\| \leq \|u^{(0)}\| + \frac{\|B\|}{\rho_{\min}(B)} \|e^{(0)}\|.$$

Proof: The proof is similar to the proof of Theorem 2.3 and follows from the fact that such $u^{(n)}$ may be equivalently defined by

$$u^{(n)} = u^{(0)} + R_n (R_n^* B R_n)^{-1} R_n^* B e^{(0)}.$$

                                                                                                ■

If a BPM satisfies the further criterion that the matrix B is HPD, then the projector

$$P_n = I - R_n (R_n^* B R_n)^{-1} R_n^* B$$

is orthogonal with respect to the B inner product, defined by $(\cdot, \cdot)_B \equiv (B\cdot, \cdot)$. Let $\|v\|_B = (Bv, v)^{1/2}$, the B norm of a vector v. Then we have

Theorem 2.6 (Minimization) *Suppose we have a BPM based on a matrix B which is HPD. Then $u^{(n)}$ is the unique minimizer of $\|e^{(n)}\|_B$ over the solution space, and $\|e^{(n)}\|_B \leq \|e^{(n-1)}\|_B$.*

Proof: Now $e^{(n)}$ is the B-orthogonal projection of $e^{(n-1)}$ onto the complement of $\mathbf{R}_n$, and

$$\begin{aligned} e^{(n)*} B e^{(n)} &= e^{(n-1)*} B e^{(n-1)} - e^{(n-1)*} B R_n (R_n^* B R_n)^{-1} R_n^* B e^{(n-1)} \\ &\leq e^{(n-1)*} B e^{(n-1)} \end{aligned}$$

with R_n as above. Suppose $e' = \bar{u} - u'$ is the error arising from u', another element of the solution space. Then for some w,

$$e' = e^{(n)} + R_n w.$$

Using the properties of the projector yields

$$\begin{aligned} e'^* B e' &= e^{(n)*} B e^{(n)} + w^* R_n^* B R_n w \\ &> e^{(n)*} B e^{(n)} \text{ for } w \neq 0, \end{aligned}$$

showing $u^{(n)}$ gives the unique minimum. ∎

We also have the following sufficient condition for convergence, generalized from a theorem in [3]:

Theorem 2.7 (Convergence) *Suppose we have a BPM with B HPD. Suppose also that $Z = BA^{-1}$ is definite and $r^{(n-1)} \in \mathbf{R}_n$ for every n. Then there exists $\epsilon > 0$ independent of $e^{(0)}$ such that for all n,*

$$||e^{(n)}||_B \leq (1 - \epsilon)||e^{(n-1)}||_B.$$

Proof: Since B is HPD, the B norm of $e^{(n)}$ is minimized over the affine space $e^{(n-1)} + \mathbf{R}_n$. Let

$$\tilde{e}^{(n)} = e^{(n-1)} - \lambda A e^{(n-1)},$$

where λ is defined by the condition $B\tilde{e}^{(n)} \perp Ae^{(n-1)} = r^{(n-1)}$; note that since this causes a minimization over a smaller space, we must have

$$||e^{(n)}||_B \leq ||\tilde{e}^{(n)}||_B.$$

We obtain

$$\lambda = \frac{(Ae^{(n-1)})^* B(e^{(n-1)})}{(Ae^{(n-1)})^* B(Ae^{(n-1)})}.$$

Through the projection property we obtain

$$||\tilde{e}^{(n)}||_B^2 = ||e^{(n-1)}||_B^2 - \frac{|e^{(n-1)*} B Ae^{(n-1)}|^2}{(Ae^{(n-1)})^* B(Ae^{(n-1)})}.$$

Dropping superscripts and letting $Ae = r$,

$$\begin{aligned} \frac{||\tilde{e}||_B^2}{||e||_B^2} &= 1 - \frac{|e^* BAe|^2}{(e^* Be)(e^* A^* BAe)} \\ &= 1 - \frac{|r^* A^{-*} Br|^2}{(r^* A^{-*} BA^{-1}r)(r^* Br)} \\ &= 1 - \frac{|r^* Zr|^2}{(r^* A^{-*} BA^{-1}r)(r^* Br)} \\ &\leq 1 - \frac{\rho_{\min}(Z)^2}{||A^{-1}||^2 \cdot ||B||^2} \end{aligned}$$

where

$$\rho_{\min}(M) = \min_{v \neq 0} \frac{|v^* M v|}{v^* v}$$

for any matrix M, so that M is definite if and only if $\rho_{\min}(M) \neq 0$.   ∎

We remark that examples may be constructed of BPM's for which B is not HPD and the projection method diverges, i.e., $||e^{(n)}||_{B_H} \to \infty$, where $B_H = (B + B^*)/2$ is the Hermitian part of B (see [11]). However, if $r^{(n-1)} \in \mathbf{R}_n$, $Z = BA^{-1}$ is definite and the skew-Hermitian part $B_N \equiv (B - B^*)/2$ is sufficiently small, then convergence is assured. The following theorem gives a specific bound for this case of the form

$$||e^{(n)}||^2_{B_H} \leq (1 - \epsilon + K)||e^{(n-1)}||^2_{B_H},$$

where $\epsilon > 0$ and $K \geq 0$ is a constant which is small when B_N is small.

Theorem 2.8 (Convergence) *Consider a BPM based on B with the decomposition $B = B_H + B_N$ into its Hermitian and skew-Hermitian parts. If $r^{(n-1)} \in \mathbf{R}_n$, then*

$$\frac{||e^{(n)}||^2_{B_H}}{||e^{(n-1)}||^2_{B_H}} \leq 1 - \frac{(\rho_{\min}(BA^{-1}))^2}{||A^{-1}||^2 \cdot ||B_H||^2 (1 + ||B_N||^2 \cdot ||B_H^{-1}||/||B_H||)}$$

$$+ 2||B_H^{-1}||^{3/2}||B_N||(||B_H||^{1/2} + ||B_H^{-1}||^{1/2} \cdot ||B_N||).$$

Proof: See [11].   ∎

3   Krylov Projection Methods

We will now begin to consider projection methods which are also polynomial methods.

A *(full) Krylov projection method* (KPM) is a balanced projection method satisfying $\mathbf{R}_n = \mathbf{K}_n(r^{(0)}, A)$. That is,

$$u^{(n)} - u^{(0)} \in \mathbf{K}_n(r^{(0)}, A), \qquad Be^{(n)} \perp \mathbf{K}_n(r^{(0)}, A).$$

The above theorems applied to Krylov projection methods yield the following very desirable properties:

- *Finite Termination.* If there is no breakdown at step n, then $u^{(n)} = \bar{u}$ if and only if $n \geq d(r^{(0)}, A)$. Here $d(v, A)$, the *degree* of a vector v with respect to A, is given by $d(v, A) = \min\{\deg P : P \text{ monic}, P(A)v = 0\}$. The *degree* of a matrix A is given by $d(A) = \min\{\deg P : P \text{ monic}, P(A) = 0\}$.

Class	B property
Idealized Generalized Conjugate Gradient (IGCG) Method [21]	B positive real
Orthogonal Error Method (OEM) [5]	B definite
Hermitian Krylov Projection Method (HKPM)	B Hermitian
Conjugate Gradient Method (CGM) [4]	B HPD

Table 10.1

- *No Breakdown for B Definite.* For fixed A and B, the KPM will not break down for any $r^{(0)}$ if and only if B is definite. (If B is not definite, breakdown is caused at step 1 if $r^{(0)} \neq 0$ is such that $r^{(0)*}Br^{(0)} = 0$.)

- *Bounded Iterates for B Definite.* If B is definite and $u^{(0)}$ and $e^{(0)}$ are constrained to be in some bounded set, then the size of $u^{(n)}$ is bounded independent of n.

- *Minimization for B HPD.* If B is HPD, then $u^{(n)}$ is the unique element of $u^{(0)} + \mathbf{K}_n(r^{(0)}, A)$ to minimize $||e^{(n)}||_B$.

The choice of B and its associated sesquilinear form $(B\cdot, \cdot)$ is very strategic in terms of the properties of the method as well as the feasibility of implementing the method by some practical algorithm. Table 10.1 defines important classes of KPM's, based on the properties of B.

Most notable of these classes is the class of conjugate gradient methods (CGM), which possess a minimization property over the entire Krylov space due to B being HPD. Some examples of conjugate gradient methods are the classical conjugate gradient method (see [9]) ($B - A$ HPD), the preconditioned conjugate gradient method ($A = Q^{-1}\hat{A}$, $B = \hat{A}$, with Q and $\hat{A}$ HPD), and the conjugate residual method ($B = A^*A$, A Hermitian). (For a complete discussion of conjugate gradient methods, see [1].)

We now examine some common features of Krylov projection methods. Suppose no breakdown occurs up to step n. From (2.2) we have

$$(3.1) \qquad u^{(n+1)} = u^{(n)} + \alpha_n p^{(n)}, \quad p^{(n)} \in \mathbf{K}_{n+1}(r^{(0)}, A)$$

where α_n is a scalar. This yields

$$e^{(n+1)} = e^{(n)} - \alpha_n p^{(n)}.$$

The orthogonality condition indicates that $e^{(n+1)} \perp B^*\mathbf{K}_{n+1}(r^{(0)}, A)$. Similarly, we have that $e^{(n)} \perp B^*\mathbf{K}_n(r^{(0)}, A)$. Thus, if such $\alpha_n p^{(n)}$ exists and $\alpha_n \neq 0$ then

$$(3.2a) \qquad p^{(n)} \in \mathbf{K}_{n+1}(r^{(0)}, A),$$

$$(3.2b) \qquad p^{(n)} \perp B^*\mathbf{K}_n(r^{(0)}, A).$$

An easy induction argument can be used to show that such a $p^{(n)}$ exists uniquely up to scaling if and only if $p^{(j)*}Bp^{(j)} \neq 0$, $j < n$. Furthermore, if $p^{(n)}$ exists uniquely, then it may be chosen nonzero if and only if $n < d(r^{(0)}, A)$. Since $p^{(n)} \in \mathbf{K}_{n+1}(r^{(0)}, A)$, the orthogonality condition yields

(3.3) $$\alpha_n = \frac{p^{(n)*}Be^{(n)}}{p^{(n)*}Bp^{(n)}}$$

which exists uniquely if and only if $p^{(n)*}Bp^{(n)} \neq 0$. We sum up this discussion in the following theorem.

Theorem 3.1 *Suppose the Krylov projection method based on A and B with initial residual $r^{(0)}$ does not break down for steps $i = 1, 2, \ldots, n$, and $u^{(n)} \neq \bar{u}$. Then the KPM does not break down at step $n + 1$ if and only if the unique (up to scaling) vector $p^{(n)}$ satisfying (3.2a,3.2b) has the property $p^{(n)*}Bp^{(n)} \neq 0$.*

Proof: The proof proceeds by induction.   At each step the vector $p^{(i)} \in \mathbf{K}_{i+1}(r^{(0)}, A)$ such that $p^{(i)} \perp B^*\mathbf{K}_i(r^{(0)}, A)$ exists uniquely if and only if $p^{(j)*}Bp^{(j)} \neq 0$ for $j < i$. Then, α_i exists uniquely if and only if $p^{(i)*}Bp^{(i)} \neq 0$. ∎

We remark that it is possible to skip over several steps. That is, it may be possible to find $u^{(n+s)}$ satisfying

$$u^{(n+s)} - u^{(n)} \in \mathbf{K}_{n+s}(r^{(0)}, A)$$

and

$$e^{(n+s)} \perp B^*\mathbf{K}_{n+s}(r^{(0)}, A)$$

even if $p^{(n)*}Bp^{(n)} = 0$. This is the essence of the work on hyperbolic pairs given in [13]. For a further discussion of these ideas, see [11].

It is clear that in the absence of breakdown the direction vectors $\{p^{(i)}\}_{i=0}^{n}$ form a basis for the space $\mathbf{K}_{n+1}(r^{(0)}, A)$ and possess the property $p^{(j)*}Bp^{(i)} = 0$ for $j < i$. In the next section we discuss several algorithms for constructing these vectors.

3.1   Computational Schemes for Krylov Projection Methods

We will now examine practical algorithms for computing the KPM iterates. Practical algorithms hinge on the formation of a basis for the Krylov space which is well-conditioned as well as efficiently computed.

We will now introduce some further definitions. An *adjoint* (or *right adjoint*) of an arbitrary matrix A with respect to a matrix B is a matrix $A^\dagger$ satisfying $(Bu, A^\dagger v) = (BAu, v)$ for all u, v. If B is nonsingular, then $A^\dagger$ is given uniquely by $(BAB^{-1})^*$. The matrix A is said to be *B-self-adjoint* if $BA = A^*B$ (see [10]).

It should be noted that A may also have a *left* adjoint with respect to B, a matrix A^x satisfying $(BA^x u, v) = (Bu, Av)$ for every u and v. When B is nonsingular, we have $A^x = B^{-1}A^*B$. Furthermore, when B is HPD, the adjoints are equal: $A^x = A^\dagger$; they are also equal if B is definite and A is B-normal (see [5]). Here we will be concerned only with $A^\dagger$, the right adjoint.

For a given matrix B, a matrix A is *B-normal* if $BA = q(A)^*B$ for some polynomial q. The *normal degree* of A with respect to B, given by $n(A) = n(A, B)$, is the lowest degree of any such polynomial. If $s = n(A, B)$, then A will be called *normal(s)* with respect to B or *B-normal(s)*.

When $B = I$, this definition of normal coincides with the classical definition of normal, that $AA^* = A^*A$. In fact, it is shown in [5] that for B definite the following are equivalent:

[1] A and $A^\dagger$ commute.

[2] A and $A^\dagger$ have the same complete set of B-orthogonal eigenvectors.

[3] $A^\dagger = q(A)$ for some polynomial q.

This equivalence does not hold for general nonsingular A and B, however.

We now present three principal algorithms for the computation of the KPM iterates. They are the Orthodir, Orthomin and Orthores algorithms (see [21]). The formulas for these KPM algorithms are given in Figures 10.1–10.3. Of these algorithms, the Orthomin algorithm is most widely used and in the case of $B = A$ HPD embodies the common two-term form of the conjugate gradient method of Hestenes and Stiefel. In the formulas that follow and throughout we will make the definition $Z = BA^{-1}$.

Breakdown. If B is definite, then the Krylov projection method is guaranteed not to break down. However, even when the iterates $\{u^{(i)}\}$ are uniquely defined by the KPM, it may be possible that one of the KPM algorithms reaches an impasse before convergence is attained. In particular, we will say that *the algorithm breaks down at step n* if iterates $\{u^{(i)}\}_{i=0}^{n-1}$ have been computed and $e^{(n-1)} \neq 0$ but the algorithm is unable to compute $u^{(n)}$. It is important to note that breakdown of the algorithm may occur even if the projection method itself does not break down.

The Orthodir algorithm is guaranteed not to break down whenever the KPM iterates exist and are unique. This is assured if B is definite. On the other hand Orthomin and Orthores are guaranteed not to break down if and only if both B and Z are definite.

Computability. A necessary condition for these algorithms to be practical is that B is chosen so that the necessary inner products are computable in a practical way. In particular, we require that inner products of the form $(Q_L^{-*}ZQ_L^{-1}\cdot, \cdot)$ of vectors in the space $\mathbf{K}_n(\hat{r}^{(0)}, \hat{A}Q_R^{-1}Q_L^{-1})$ be computable. This would exclude, for instance, the general case of methods with $B = I$, though it is possible to implement $B = I$ methods if special choices of Q_L and Q_R are made (see [1]).

Economical Computation. In general the recursions for Orthodir, Orthomin and Orthores given above are not practical, since the work per iteration increases for every iteration. However, when A and B satisfy certain relationships, the formulas

ORTHODIR

$$q^{(0)} = r^{(0)},$$

$$q^{(n)} = Aq^{(n-1)} + \sum_{i=\theta(n)}^{n-1} \beta_{n,i} q^{(i)}, \qquad n > 0. \qquad (\theta(n) = 0 \ \forall n)$$

$$\beta_{n,i} = \frac{-q^{(i)*} B A q^{(n-1)} - \sum_{j=\theta(n)}^{i-1} \beta_{n,j} q^{(i)*} B q^{(j)}}{q^{(i)*} B q^{(i)}}.$$

$$u^{(n+1)} = u^{(n)} + \hat{\lambda}_n q^{(n)},$$
$$r^{(n+1)} = r^{(n)} - \hat{\lambda}_n A q^{(n)}.$$

$$\hat{\lambda}_n = \frac{q^{(n)*} Z r^{(n)}}{q^{(n)*} B q^{(n)}}.$$

If B Hermitian : $\qquad \beta_{n,i} = -\dfrac{q^{(i)*} B A q^{(n-1)}}{q^{(i)*} B q^{(i)}}$

If $n(A, B) = s$: $\qquad \beta_{n,i} = 0, \quad i < n - (s+1)$

Figure 10.1

ORTHOMIN

$$p^{(n)} = r^{(n)} + \sum_{i=\phi(n)}^{n-1} \alpha_{n,i} p^{(i)}, \qquad n \geq 0. \qquad (\phi(n) = 0 \ \forall n)$$

$$\alpha_{n,i} = \frac{-p^{(i)*} B r^{(n)} - \sum_{j=\phi(n)}^{i-1} \alpha_{n,j} p^{(i)*} B p^{(j)}}{p^{(i)*} B p^{(i)}}.$$

$$u^{(n+1)} = u^{(n)} + \lambda_n p^{(n)},$$
$$r^{(n+1)} = r^{(n)} - \lambda_n A p^{(n)}.$$

$$\lambda_n = \frac{p^{(n)*} Z r^{(n)}}{p^{(n)*} B p^{(n)}} = \frac{r^{(n)*} Z r^{(n)}}{p^{(n)*} B p^{(n)}}.$$

If B Hermitian : $\qquad \alpha_{n,i} = -\dfrac{p^{(i)*} B r^{(n)}}{p^{(i)*} B p^{(i)}}$

If $n(A, B) = s$: $\qquad \alpha_{n,i} = 0, \quad i < n - s$

Figure 10.2

ORTHORES

$$u^{(n+1)} = \lambda_n \left(r^{(n)} + \sum_{i=\theta(n+1)}^{n} \sigma_{n+1,i} u^{(i)} \right) \qquad (\theta(n) = 0 \ \forall n)$$

$$r^{(n+1)} = -\lambda_n \left(A r^{(n)} - \sum_{i=\theta(n+1)}^{n} \sigma_{n+1,i} r^{(i)} \right)$$

$$\lambda_n = \left(\sum_{i=\theta(n+1)}^{n} \sigma_{n+1,i} \right)^{-1}$$

$$\sigma_{n+1,i} = \frac{r^{(i)*} Z A r^{(n)} - \sum_{j=\theta(n+1)}^{i-1} \sigma_{n+1,j} r^{(i)*} Z r^{(j)}}{r^{(i)*} Z r^{(i)}}.$$

If Z Hermitian : $\qquad \sigma_{n+1,i} = \dfrac{r^{(i)*} Z A r^{(n)}}{r^{(i)*} Z r^{(i)}}$

If $n(A, B) = s$: $\qquad \sigma_{n+1,i} = 0, \quad i < n - s$

Figure 10.3

for these three algorithms may be simplified to give short recurrences while still computing the true KPM iterates. The following theorems characterize these cases. Their proofs are found in [4, 5] and [12].

Theorem 3.2 (Simplification) *For the Orthodir algorithm to yield* $\beta_{n,i} = 0$ *for all* $i < n - (s+1)$ *and* $n \leq d(r^{(0)}, A) - 1$ *for all* $r^{(0)}$, *it is sufficient that* $d(A) \leq s + 2$ *or that* A *be* B-*normal with* $n(A, B) \leq s$. *If* B *is definite, then the condition is also necessary.*

Proof: See [4, 5]. ∎

Theorem 3.3 (Simplification) *For the Orthomin and Orthores algorithms to yield* $\alpha_{n,i} = 0$ *and* $\sigma_{n+1,i} = 0$ *respectively for all* $i < n - s$ *and* $n \leq d(r^{(0)}, A) - 1$ *for all* $r^{(0)}$, *it is sufficient that* $d(A) \leq s + 1$ *or that* A *be* B-*normal with* $n(A, B) \leq s$. *If* B *and* Z *are definite, then the condition is also necessary.*

Proof: See [12]. ∎

Thus, if A is B-normal with $n(A, B) = s$ then, in the definition of the Orthodir algorithm, we may set $\theta(n) = \max(0, n - (s+1))$ and still obtain the KPM iterates.

In this case we may likewise set $\phi(n) = \max(0, n - s)$ in the Orthomin algorithm and $\theta(n + 1) = \max(0, n - s)$ in the Orthores algorithm and still obtain the KPM iterates. The simplifications may also be accomplished when $d(A) \leq s+1$; however, this case is not very useful, since $d(A)$ is usually large for practical problems, which would require s to be large also.

The above simplifications of formulas for the KPM algorithms holds for an important but restrictive class of matrices. The next two results help define this class.

Theorem 3.4 (B-Normal(s) Matrices) *Suppose that B is definite and A is B-normal. If $s = n(A, B) > 1$, then $s \geq \sqrt{d(A)}$.*

Proof: The proof is a simple generalization of a result in [4]; see also [5] and [12]. ∎

Since $d(A)$ is typically large, this case is not practical. On the other hand,

Theorem 3.5 (B-Normal(1) Matrices) *Suppose that B is definite and A is B-normal. If $n(A, B) = 1$, then A is of the form*

$$e^{i\theta}(riI + G),$$

where r and θ are real and G is B-self-adjoint.

Proof: This proof is also a simple generalization of a result given in [4]; see also [5] and [12]. ∎

Note that A B-normal(1) with B definite requires that the spectrum of A be contained in a line in the complex plane (see [5]). Furthermore, B must be such that the necessary inner products are feasibly computable.

The above theorems show that when B is definite, short recurrences occur naturally for an important but restrictive class of matrices. Alternatively, one can artificially truncate the recursions in the case of more general matrices. However, the full orthogonality and optimality conditions are lost and with them finite termination. More will be said about truncated methods below.

If on the other hand B is allowed to be indefinite, short recurrences may be used to compute the full KPM iterates for a larger class of matrices A. However, breakdown may occur and the iterates may become unbounded. An important example of such a method is the biconjugate gradient method described in the next section.

3.2 Examples of Krylov Projection Method

Within the class of Krylov projection methods are three important subclasses, defined by restrictions applied to B. In this section we will consider particular methods in each of these subclasses.

Conjugate Gradient Methods. These methods are characterized by choosing B to be HPD. We have already mentioned several well-known methods in this class: the classical conjugate gradient method of Hestenes and Stiefel; the preconditioned conjugate gradient method; and the conjugate residual method. One further class of methods is the class of normal equations methods, which includes the methods defined by $A = \hat{A}^*\hat{A}$ with $B = I$ or $B = A$. Here $\hat{A}$ itself may be the result of a preconditioning. These cases are examined in detail in [1].

Orthogonal Error Methods. These methods are characterized by choosing B to be definite. A classic example of an OEM is the GCW method [2, 20]. Here we let $A = Q^{-1}\hat{A}$, $Q = \hat{A}_H \equiv (\hat{A} + \hat{A}^*)/2$ and $B = \hat{A}$, assuming $\hat{A}$ definite and $\hat{A}_H$ nonsingular. It is easily verified that the B-adjoint, $A^\dagger$, is a degree-1 polynomial in A. Since $(B\cdot, \cdot)$ is not in general an inner product, no obvious minimization property is satisfied. However, it is shown in [8] that when $\hat{A}$ is real, the even iterates of the GCW method are the same as the iterates of the CG method defined by $B = \hat{A}_H$ and $A = \hat{A}_H^{-1}\hat{A}^*\hat{A}_H^{-1}\hat{A}$.

Hermitian Krylov Projection Methods. These methods are characterized by restricting B to be Hermitian. Relaxing the constraint that B be definite is a drastic measure; the resulting methods may break down, and the iterates may be unbounded.

A classic example of an HKPM is the SYMMLQ method of [15], for which $B = A$ where A is Hermitian but possibly indefinite. Another example of an HKPM is the Lanczos or biconjugate gradient (BCG) method (Fig. 10.4). It is shown in [10] that when A, $u^{(0)}$, b, $\tilde{u}^{(0)}$ and $\tilde{b}$ are restricted to be real, then the biconjugate gradient iterates may be obtained by applying the Orthomin algorithm to the double system

$$\textcircled{A}\;\textcircled{u} = \textcircled{b}$$

with

$$\textcircled{A} = \begin{pmatrix} A & \\ & A^* \end{pmatrix}, \quad \textcircled{u} = \begin{pmatrix} u \\ \tilde{u} \end{pmatrix}, \quad \textcircled{b} = \begin{pmatrix} b \\ \tilde{b} \end{pmatrix}$$

and

$$\textcircled{Z} = \begin{pmatrix} & I \\ I & \end{pmatrix}, \quad \textcircled{B} = \textcircled{Z}\,\textcircled{A} = \begin{pmatrix} & A^* \\ A & \end{pmatrix}.$$

The resulting iterates are

$$\textcircled{u}^{(n)} = \begin{pmatrix} u^{(n)} \\ \tilde{u}^{(n)} \end{pmatrix}$$

where $u^{(n)}$ is the same as the BCG iterate. Here $\tilde{r}^{(0)} \equiv \tilde{b} - A^*\tilde{u}^{(0)}$ is some arbitrary vector, typically set to $\tilde{Z}^* r^{(0)}$ for some fixed matrix $\tilde{Z}$, e.g. $\tilde{Z} = I$. The BCG method may also be seen as a polynomial projection method with $\mathbf{R}_n = \mathbf{K}_n(r^{(0)}, A)$ and $\mathbf{L}_n = \mathbf{K}_n(\tilde{r}^{(0)}, A^*)$.

THE BCG METHOD

$$p^{(0)} = r^{(0)}$$

$$\tilde{p}^{(0)} = \tilde{r}^{(0)}$$

$$p^{(n)} = r^{(n)} + \alpha_n p^{(n-1)}, \quad n > 0.$$

$$\tilde{p}^{(n)} = \tilde{r}^{(n)} + \alpha_n^* \tilde{p}^{(n-1)}, \quad n > 0.$$

$$\alpha_n = \frac{\tilde{r}^{(n)*} r^{(n)}}{\tilde{r}^{(n-1)*} r^{(n-1)}}$$

$$u^{(n+1)} = u^{(n)} + \lambda_n p^{(n)},$$

$$r^{(n+1)} = r^{(n)} - \lambda_n A p^{(n)},$$

$$\tilde{r}^{(n+1)} = \tilde{r}^{(n)} - \lambda_n^* A^* \tilde{p}^{(n)}.$$

$$\lambda_n = \frac{\tilde{r}^{(n)*} r^{(n)}}{\tilde{p}^{(n)*} A p^{(n)}}$$

Figure 10.4

The BCG method may break down unless certain very special conditions are satisfied, such as A HPD and $\tilde{r}^{(0)} = r^{(0)}$. Furthermore, the residuals may get arbitrarily large before convergence. However, for certain classes of problems, the set of initial residuals $r^{(0)}$ for which the BCG method breaks down is only a set of measure zero in the space of all real vectors [11]. It remains an open question of how likely it is that the method may "nearly" break down, in the sense that the residual becomes unacceptably large during the iteration process. Further work must be done to understand the convergence properties of BCG.

4 Semi-Krylov Projection Methods

Methods such as the normal equations methods, the GCW method and the BCG method attempt to embed the original problem (1.1) into a problem satisfying the B-normal(1) condition for some B. Another means to surmounting the problem of long recurrences is to define projection methods which use smaller subspaces than the whole Krylov space. This is the idea behind various truncated and restarted methods. These methods are more economical than the full-space methods; however, in general they do not possess certain desirable properties of KPM's, such as finite termination and minimization over the entire Krylov space when B is HPD.

In particular, we will define a *semi-Krylov projection method* (SKPM) to be a projection method utilizing a portion of the Krylov space: $\mathbf{R}_n \subseteq \mathbf{K}_n(r^{(0)}, A)$ and $\mathbf{L}_n \subseteq B^*\mathbf{K}_n(r^{(0)}, A)$ for some matrix B. If an SKPM is also balanced (i.e., $\mathbf{L}_n = B^*\mathbf{R}_n$), we will say it is a *balanced* semi-Krylov projection method (BSKPM).

Some examples of SKPM's are the truncated Orthodir, Orthomin and Orthores algorithms. These are obtained from the above-stated KPM algorithms by a simple redefinition of the index functions θ and ϕ: Orthodir(s) ($\theta(n) = \max(0, n - s)$), Orthomin($s$) ($\phi(n) = \max(0, n - s)$), and Orthores($s$) ($\theta(n + 1) = \max(0, n - s)$). Other truncated methods are documented in [16], including the Axelsson Least Squares ($B = A^*A$) and Axelsson Galerkin ($B = A$) methods. Restarted methods also fall into this category, including the GCR($s - 1$) [3] and GMRES(s) [17] methods. These two restarted methods both utilize $B = A^*A$ and restart the full KPM every s steps. The GCR($s - 1$) method, which is restarted Orthomin, is guaranteed not to break down for A definite; GMRES(s) will never break down but may stagnate for general nonsingular A.

Truncated and restarted methods do not in general possess all the useful KPM properties. However, certain desirable properties are possessed by some of the methods:

- *Behavior near the normal(1) case.* It is desirable that if A happens to be B-normal(1), then an adequate part of the Krylov space is saved in order to give the same iterate as the full KPM method. Likewise, by continuity, if A is a perturbation from a B-normal(1) matrix, it is hoped that nearly the KPM result will be given. This property is possessed by some truncated methods, e.g., Orthomin(s) with $s \geq 1$ and A nearly B-normal(1).

- *Minimization property.* It is hoped that the error $e^{(n)}$ is a minimization over some space, so that the error in some norm is nonincreasing.

- *Minimization over some Krylov space.* If each $e^{(n)}$ is minimized over an appropriate Krylov space, then certain error bounds may be invoked for the error, even though it is not a minimization over the entire Krylov space. In particular, we seek a balanced method with B HPD and $u^{(n)} \in u^{(n-k)} + \mathbf{K}_k(r^{(n-k)}, A)$ with k as large as possible. This property is possessed by restarted methods such as GCR($s - 1$) and GMRES(s).

4.1 Balanced SKPM's: Truncated/Restarted Methods

Some truncated methods of the SKPM category satisfy the additional condition that they are balanced, in the sense that $\mathbf{L}_n = B^*\mathbf{R}_n$ for a fixed map B. Included are the truncated Orthodir and Orthomin methods, GCR($s - 1$), GMRES(s), and the two methods of Axelsson mentioned above.

When B is HPD and $Z = B^{-1}A$ is definite, the Orthomin algorithm with the function $\phi(n)$ defined as an arbitrary nonnegative monotone function is guaranteed to converge (Theorem 2.7). Moreover, when A is definite, the restarted methods GCR($s - 1$) and GMRES(s) (for which $B = A^*A$) will always converge (see [3], [17]).

Many open questions remain concerning these methods, such as necessary and sufficient conditions for $r^{(0)}$ and A for which the methods converge, as well as sharp error bounds. A few things are known, however.

We make the following definitions. Let $K_n(v, A) = (\,v \quad Av \quad A^2 v \quad \ldots \quad A^{n-1}v\,)$, a matrix whose columns span $\mathbf{K}_n(v, A)$. Let $r^{(s)}$ be defined as a function of $r^{(0)}$ to be the residual at step s of the KPM with $B = A^* A$ applied to the initial residual $r^{(0)}$. Then let

$$
\begin{aligned}
\psi_s(A) \;=\; & \max_{r^{(0)} \neq 0} \frac{||r^{(s)}||}{||r^{(0)}||} = \max_{v \neq 0} \min_{P_s(0)=1} \frac{||P_s(A)v||}{||v||} \\[2mm]
=\; & \max_{r:d(r,A) \geq s} \frac{1}{||r||} \big\{ ||r||^2 - r^* A K_s(r, A) \\[2mm]
& \cdot [(A K_s(r, A))^* (A K_s(r, A))]^{-1} (A K_s(r, A))^* r \big\}^{1/2}
\end{aligned}
$$

and let

$$
\varphi_s(A) \;=\; \min_{P_s(0)=1} ||P_s(A)|| = \min_{P_s(0)=1} \max_{v \neq 0} \frac{||P_s(A)v||}{||v||}.
$$

We will say a matrix A is *s-definite* if $\psi_s(A) < 1$. One can show (see [11]) that s-definiteness is a necessary and sufficient condition for a restarted method such as GMRES(s) to converge (i.e., not stagnate) for all $r^{(0)}$.

We note that for any n and A:

$$
\psi_n(A) = \max_{v \neq 0} \min_{P_n(0)=1} \frac{||P_n(A)v||}{||v||} \leq \max_{v \neq 0} \min_{P_n(0)=1} ||P_n(A)|| = \varphi_n(A)
$$

Thus GMRES(s) satisfies

$$
\frac{||r^{(ms+s)}||}{||r^{(ms)}||} \leq \psi_s(A) \leq \varphi_s(A).
$$

Also, $\varphi_s(A) \leq \mathrm{cond}(P) \cdot \varphi_s(J)$, where $A = PJP^{-1}$ is a Jordan decomposition. It is known from complex analysis that for arbitrary nonsingular diagonal J, $\varphi_s(J)$ may be made arbitrarily small by s sufficiently large. Thus a bound may be established for GMRES(s) implying convergence for s sufficiently large, whenever A is diagonalizable. Computable and sharp bounds are still needed.

4.2   Balanced SKPM's: Generalized Minimal Error Methods

As a further class of BSKPM's, we now consider a class of generalizations of the minimal error method of [7] and [6]. Given B, let $A^\dagger$ be the B-adjoint of A. We define the iteration

$$
u^{(n)} - u^{(0)} \in A^\dagger \mathbf{K}_n(r^{(0)}, A), \qquad e^{(n)} \perp B^* A^\dagger \mathbf{K}_n(r^{(0)}, A).
$$

THE MINIMAL ERROR METHOD

$$p^{(0)} = Ar^{(0)},$$

$$p^{(n)} = Ap^{(n-1)} + \beta_{n,n-1}p^{(n-1)} + \beta_{n,n-2}p^{(n-2)},$$

$$\beta_{n,i} = -\frac{p^{(i)*}Ap^{(n-1)}}{p^{(i)*}p^{(i)}},$$

$$u^{(n+1)} = u^{(n)} + \hat{\lambda}_n p^{(n)},$$

$$r^{(n+1)} = r^{(n)} - \hat{\lambda}_n Ap^{(n)},$$

$$\hat{\lambda}_n = \begin{cases} r^{(0)*}r^{(0)} / p^{(0)*}p^{(0)}, & n = 0, \\ p^{(n-1)*}r^{(n)} / p^{(n)*}p^{(n)}, & \text{else.} \end{cases}$$

Figure 10.5

This defines a balanced projection method with $\mathbf{R}_n = A^\dagger \mathbf{K}_n(r^{(0)}, A)$. If, furthermore, A is B-normal(1), we have a BSKPM and the method has a three-term Orthodir-like recurrence. It will be further noted that we also have a computable method in the case when $B = I$. If $A = A^*$, we obtain Fridman's minimal error method (Fig. 10.5).

For $A = A^*$ this method has the benefit of minimizing the 2-norm of the error $e^{(n)}$, which is useful when A is ill-conditioned or indefinite. However, convergence is hindered by numerical problems as well as the fact that $\mathbf{R}_n$ is not the entire Krylov space $\mathbf{K}_{n+1}(r^{(0)}, A)$; in fact, $r^{(n)} = P_{n+1}(A)r^{(0)}$ where $P_{n+1}(0) = 1$ and $P'_{n+1}(0) = 0$ (see [11]).

5 Non-polynomial Projection Methods

The class of methods we have herein called "projection methods" includes methods which are not, strictly speaking, polynomial methods. One example is the block conjugate gradient method [14] for which A is HPD and:

$$u^{(n)} - u^{(0)} \in \text{span} \bigcup_{i=1}^k \mathbf{K}_n(r_i^{(0)}, A), \qquad e^{(n)} \perp \text{span} \bigcup_{i=1}^k A^* \mathbf{K}_n(r_i^{(0)}, A)$$

for some set of vectors $\{r_i^{(0)}\}$, where $r_\ell^{(0)} = r^{(0)}$ for some ℓ.

The methods USYMLQ and USYMQR [18] are also non-polynomial projection methods. These methods are defined respectively by

$$u^{(n)} - u^{(0)} \in \tilde{\mathbf{K}}_n(r^{(0)}, A^*), \qquad e^{(n)} \perp A^* \tilde{\mathbf{K}}_n(r^{(0)}, A),$$

THE CGS METHOD

$$p^{(0)} = f^{(0)} = r^{(0)}$$

$$u^{(n+1)} = u^{(n)} + \lambda_n(f^{(n)} + h^{(n+1)})$$

$$r^{(n+1)} = r^{(n)} - \lambda_n A(f^{(n)} + h^{(n+1)})$$

$$h^{(n+1)} = f^{(n)} - \lambda_n A p^{(n)}$$

$$f^{(n+1)} = r^{(n+1)} + \alpha_{n+1} h^{(n+1)}$$

$$p^{(n+1)} = f^{(n+1)} + \alpha_{n+1}(h^{(n+1)} + \alpha_{n+1} p^{(n)})$$

$$\lambda_n = \frac{\tilde{r}^{(0)*} r^{(n)}}{\tilde{r}^{(0)*} A p^{(n)}}, \qquad \alpha_{n+1} = \frac{\tilde{r}^{(0)*} r^{(n+1)}}{\tilde{r}^{(0)*} r^{(n)}}.$$

Figure 10.6

$$u^{(n)} - u^{(0)} \in \tilde{K}_n(r^{(0)}, A^*), \qquad e^{(n)} \perp A^* A \tilde{K}_n(r^{(0)}, A^*).$$

Here $\tilde{K}_n$ indicates the quasi-Krylov space given as the linear span of the n vectors

$$\tilde{K}_n(v, A) = \mathrm{span}\{v, Av, AA^*v, AA^*Av, \ldots\}.$$

These methods are not in general based on a Krylov space. They do, on the other hand, possess an economical 3-term recurrence. The USYMQR method is a balanced projection method with $B = A^*A$ HPD, thus minimizing the 2-norm of the residual. The USYMLQ method can break down, however. For A not nearly Hermitian, these methods in practice appear to perform comparably to the normal equations.

6   Non-projection Polynomial Methods

Finally, we give an example of a conjugate gradient-like polynomial method which is not known to fit into the above setting of projection methods: the conjugate gradient squared (CGS) method of [19] (Fig. 10.6). This method is closely related to the BCG method: in particular, if the BCG residuals are given by

$$r^{(n)} = r^{(n)}_{BCG} = P_n(A)r^{(0)},$$

then the CGS residuals are given by

$$r^{(n)} = r^{(n)}_{CGS} = (P_n(A))^2 r^{(0)}.$$

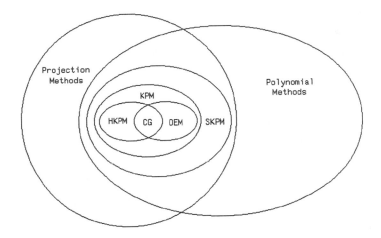

Figure 10.7 The landscape of iterative solvers.

Thus if $P_n(A)$ is a contraction, then its square is even more so; on the other hand, if $P_n(A)$ amplifies the error, as can happen in near-breakdown situations, the square can be even worse. In short, the CGS method shares a number of the convergence properties of the BCG method.

7   Conclusion

In this paper, we have suggested a structure for classifying and analyzing conjugate gradient-like iterative methods. The structure we have presented involves an "onion" of increasingly strong hypotheses. (See Fig. 10.7 for the landscape of iterative solvers.) The strengthening of these assumptions on the method, including the imposition of stronger properties on the generalized inner product matrix B, brings forth attractive features from the methods, such as assured convergence, minimization properties and error bounds. However, stronger assumptions also limit the domain of applicability of the methods to increasingly more specialized A. Further research is needed in order to study the implications of relaxing each of the assumptions, in the hope of better understanding the convergence properties of known and new methods. Further discussion of these issues may be found in [11].

Acknowledgements

The work of the first author was supported, in part, by the National Science Foundation NSF Grant DCR-8518722, and by the Department of Energy, through Grant

DE-FG05-87ER25048, with the University of Texas at Austin. The work of the second author was supported in part by the National Science Foundation, through Grant DMS-8704169, with Los Alamos National Laboratory. This paper was typed by the first author in TEX and reformatted in LATEX by Lisa Laguna.

References

[1] Ashby, Steven F., Thomas A. Manteuffel, and Paul E. Saylor. "A Taxonomy for Conjugate Gradient Methods," Lawrence Livermore National Laboratory Report UCRL-98508, March 1988, submitted to *SIAM Journal on Numerical Analysis*.

[2] Concus, Paul, and Gene H. Golub. "A Generalized Conjugate Gradient Method for Nonsymmetric Systems of Linear Equations," *Lecture Notes in Economics and Mathematical Systems 134*, R. Glowinski and J. L. Lions, eds., Springer Verlag, New York, 1976, 56–65.

[3] Eisenstat, Stanley C., Howard C. Elman, and Martin H. Schultz. "Variational iterative methods for nonsymmetric systems of linear equations," *SIAM Journal on Numerical Analysis* **2** (1983) 345–357.

[4] Faber, Vance, and Thomas Manteuffel. "Necessary and sufficient conditions for the existence of a conjugate gradient method," *SIAM Journal on Numerical Analysis* **21** (1984) 352–362.

[5] Faber, Vance, and Thomas Manteuffel. "Orthogonal error methods," *SIAM Journal on Numerical Analysis*, vol. 24, no. 1 (1987) 170–187.

[6] Fletcher, R. "Conjugate Gradient Methods for Indefinite Systems," *Numerical Analysis Dundee 1975*, G. A. Watson ed., New York: Springer, Lecture Notes in Mathematics, no. 506, 1976, 73–89.

[7] Fridman, V.M. "The method of minimum iterations with minimum errors for a system of linear algebraic equations with a symmetrical matrix," *USSR Computational Math. and Math. Phys.*, **2** (1963) 362–363.

[8] Hageman, L.A., Franklin T. Luk, and David M. Young. "On the equivalence of certain iterative acceleration methods," *SIAM Journal on Numerical Analysis* **17** (1980) 852–73.

[9] Hestenes, M.R., and E.L. Stiefel. "Methods of conjugate gradients for solving linear systems," *J. Res. Nat. Bur. Standards* **49** (1952) 409–436.

[10] Jea, Kang C., and David M. Young. "On the simplification of generalized conjugate-gradient methods for nonsymmetrizable linear systems," *Linear Algebra and its Applications* **52/53** (1983) 399–417.

[11] Joubert, Wayne D. "Iterative Methods for Nonsymmetric Systems of Linear Equations," Report, Center for Numerical Analysis, The University of Texas at Austin, in preparation.

[12] Joubert, Wayne D., and David M. Young. "Necessary and sufficient conditions for the simplification of generalized conjugate gradient algorithms," *Linear Algebra and its Applications,* **88/89** (1987) 449–485.

[13] Luenberger, D.G. "Hyperbolic pairs in the method of conjugate gradients," *SIAM Journal on Numerical Analysis* **17** (1969) 1263–1267.

[14] O'Leary, Dianne P. "The block conjugate gradient algorithm and related methods," *Linear Algebra and its Applications* **29** (1980) 293–322.

[15] Paige, C.C., and M.A. Saunders. "Solution of sparse indefinite systems of linear equations," *SIAM Journal on Numerical Analysis* **12** (1975) 617–629.

[16] Saad, Youcef, and Martin H. Schultz. "Conjugate gradient-like algorithms for solving nonsymmetric linear systems," *Mathematics of Computation* **44** (1985) 417–424.

[17] Saad, Youcef, and Martin H. Schultz. "GMRES: A generalized minimal residual algorithm for solving nonsymmetric linear systems," *SIAM Journal on Scientific and Statistical Computing* **7** (1986) 856–869.

[18] Saunders, M.A., H.D. Simon, and E.L. Yip. "Two conjugate-gradient-type methods for sparse unsymmetric linear equations," *SIAM Journal on Numerical Analysis* **25** (1988) 927–940.

[19] Sonneveld, Peter "CGS, a Fast Lanczos-type Solver for Nonsymmetric Linear Systems," *SIAM J. Sci. Stat. Comput.* **10** (1989) 35–52.

[20] Widlund, Olof. "A Lanczos method for a class of nonsymmetric systems of linear equations," *SIAM Journal on Numerical Analysis* **15** (1978) 801–812.

[21] Young, David M., and Kang C. Jea. "Generalized conjugate-gradient acceleration of nonsymmetrizable iterative methods," *Linear Algebra and its Applications* **34** (1980) 159–194.

Chapter 11

Solution of Three-Dimensional Generalized Poisson Equations on Vector Computers

DAVID L. HARRAR II AND JAMES M. ORTEGA
University of Virginia

Dedicated to David M. Young, Jr., on the occasion of his sixty-fifth birthday.

Abstract

We consider the solution of equations of the form $\nabla(K\nabla u) = f$. Discretization is by finite differences with variable grid spacing, and diagonal similarity transformations symmetrize the coefficient matrix. The conjugate gradient method with polynomial preconditioning based on the SSOR iteration is used in conjunction with the red/black ordering in order to give long vector lengths. Numerical results on a Cyber 205 and a CRAY-2 are given for a particular test problem, and for up to 250,000 unknowns. Timings are also given for the reduced system conjugate gradient method and these are slightly better than the best polynomial preconditioning times.

1 Introduction

We consider the three-dimensional Poisson-type equation

$$(1.1) \qquad \nabla(K\nabla u) = f$$

where K and f are given functions of the three spatial variables, x, y, and z. The domain is a rectangular parallelepiped, and Dirichlet boundary conditions are assumed on all surfaces.

Equation (1.1) is discretized by standard seven-point finite differences with, in general, non-uniform spacing. This discretization gives a nonsymmetric coefficient matrix, but it may be symmetrized by diagonal similarity transformations. We then use a preconditioned conjugate gradient method (Concus et al. [6]) with a polynomial preconditioner based on symmetric successive over-relaxation (SSOR) (Adams [2]). A red/black reordering of the unknowns is utilized in order to vectorize the SSOR iteration and the Eisenstat modification (Eisenstat et al. [8]) is used to save computation. It is known (see, e.g., Adams [1], Ashcraft and Grimes [3], Elman and Agron [9], Melhem [19], and Poole and Ortega [21]) that the use of red/black or multicolor orderings with SSOR or incomplete Choleski preconditioning tends to degrade the rate of convergence when compared to the natural ordering. Thus, there is a trade off between the computation rate and the number of iterations. For a particular example problem arising in fluid dynamics, the red/black SSOR polynomial preconditioner gives a dramatic decrease in the computation time, compared with no preconditioning. We will see, however, that much of this decrease could have been achieved with just Jacobi preconditioning. Moreover, the reduced system approach is more efficient still.

Section 2 covers the discretization and the resulting linear system. In Sec. 3, we discuss the conjugate gradient algorithm and the preconditioner, as well as the implementation on vector computers. Section 4 reports on numerical experiments on a CDC Cyber 205 and a Cray-2. Results are given for problems with up to 250,000 unknowns.

2   Discretization

In the coordinate directions in which variable spacing is used, we use the usual differencing scheme for non-uniform spacing. Thus, for example, if there is non-uniform spacing in the z-direction, then $(Ku_z)_z$ at the i, j, k grid point is approximated by

$$
\begin{aligned}
[(Ku_z)_z]_{i,j,k} = &\frac{2}{\zeta_k(\zeta_{k+1} + \zeta_k)} K_{i,j,k-\frac{1}{2}} u_{i,j,k-1} \\
&- \frac{2}{(\zeta_{k+1} + \zeta_k)} \left(\frac{1}{\zeta_{k+1}} K_{i,j,k+\frac{1}{2}} + \frac{1}{\zeta_k} K_{i,j,k-\frac{1}{2}} \right) u_{i,j,k} \\
&+ \frac{2}{\zeta_{k+1}(\zeta_{k+1} + \zeta_k)} K_{i,j,k+\frac{1}{2}} u_{i,j,k+1}
\end{aligned}
$$

(2.1)

where $u_{ijk} = u(x_i, y_j, z_k)$ and

(2.2) $\zeta_k = z_k - z_{k-1}, \quad K_{i,j,k\pm\frac{1}{2}} = K(x_i, y_j, z_{k\pm\frac{1}{2}}), \quad z_{k+\frac{1}{2}} = (z_{k+1} + z_k)/2.$

Similar expressions would be used for non-uniform spacing in the x and/or y directions.

Using (2.1) and corresponding expressions for the derivatives with respect to x and y to discretize equation (1.1), we obtain the following linear system of equations where W, E, S, N, L, U, and C are used mnemonically for west, east, south, north, lower, upper, and center, respectively, and η depends on the grid spacing:

$$-W_{ijk}u_{i-1,j,k} - E_{ijk}u_{i+1,j,k} - S_{ijk}u_{i,j-1,k} - N_{ijk}u_{i,j+1,k}$$

(2.3)
$$- L_{ijk}u_{i,j,k-1} - U_{ijk}u_{i,j,k+1} + C_{ijk}u_{i,j,k} = -\eta_{i,j,k}^2 f_{i,j,k}.$$

The system (2.3) can be written as

(2.4)
$$\widehat{\mathbf{A}}\hat{\mathbf{u}} = \hat{\mathbf{b}}$$

where $\hat{\mathbf{u}}$ is the vector of unknowns, $\hat{\mathbf{b}}$ contains the forcing terms as well as boundary values, and the coefficient matrix $\widehat{\mathbf{A}}$ has seven non-zero diagonals.

Although we discretize $K(x, y, z)$ in (2.1) in such a way that symmetry of the coefficient matrix is preserved in the case of uniform spacing, with non-uniform spacing, $\widehat{\mathbf{A}}$ is not symmetric. We can restore symmetry by diagonal similarity transformations which depend only on the spacings. This is an extension of the well-known way to symmetrize a tridiagonal matrix by diagonal scaling. For example, if there is non-uniform spacing in the z-direction, the diagonal scaling matrix is

(2.5) $\mathbf{D}_z = \mathrm{diag}(d_1, d_1, \ldots, d_1, d_2, d_2, \ldots, d_2, \ldots, d_{N_z}, d_{N_z}, \ldots, d_{N_z})$

with $d_1 = 1$ and

(2.6)
$$\left(\frac{d_{k+1}}{d_k} \right) = \left(\frac{\zeta_{k+1} + \zeta_{k+2}}{\zeta_k + \zeta_{k+1}} \right)^{1/2}, \quad k = 1, 2, \ldots, N_z - 1.$$

There are N_z different elements in $\mathbf{D}_z$, each one corresponding to a different plane of interior grid points in the z-direction. If there were also non-uniform spacing in the x-direction, we would use the diagonal scaling matrix

(2.7) $\mathbf{D}_x = \mathrm{diag}(\hat{d}_1, \hat{d}_2, \ldots, \hat{d}_{N_x}, \hat{d}_1, \hat{d}_2, \ldots, \hat{d}_{N_x}, \ldots, \hat{d}_1, \hat{d}_2, \ldots, \hat{d}_{N_x})$

where $\hat{d}_1 = 1$, $\hat{\xi}_i = x_i - x_{i-1}$ and

(2.8)
$$\left(\frac{\hat{d}_{i+1}}{\hat{d}_i} \right) = \left(\frac{\xi_{i+1} + \xi_{i+2}}{\xi_i + \xi_{i+1}} \right)^{1/2}, \quad i = 1, 2, \ldots, N_x - 1.$$

There are N_x different elements in $\mathbf{D}_x$, again each one corresponding to a different plane of interior grid points. We note that the ordering of the elements of $\mathbf{D}_x$ is different than that of $\mathbf{D}_z$ as a result of the way in which the grid points are ordered.

The system (2.3) is then transformed to

(2.9)
$$\mathbf{A}\mathbf{u} = \mathbf{b}$$

where $\mathbf{A}$ is symmetric and

(2.10) $\mathbf{A} = \mathbf{D}_x\mathbf{D}_z\widehat{\mathbf{A}}\mathbf{D}_z^{-1}\mathbf{D}_x^{-1}, \quad \mathbf{u} = \mathbf{D}_x\mathbf{D}_z\hat{\mathbf{u}}, \quad \mathbf{b} = \mathbf{D}_x\mathbf{D}_z\hat{\mathbf{b}}.$

If uniform spacing is used in the x-direction, then $\mathbf{D}_x$ does not appear in (2.10). On the other hand, if non-uniform spacing were also used in the y direction, a third diagonal matrix $\mathbf{D}_y$ would be present.

The matrix $\mathbf{A}$ of (2.9) is also positive definite. This follows by showing that the matrix $\widehat{\mathbf{A}}$ of (2.4) is irreducibly diagonally dominant with positive main diagonal elements. In particular, the diagonal coefficient C_{ijk} in (2.3) is just the sum of the magnitudes of the off-diagonal coefficients in that row of the matrix when the corresponding grid point contains no boundary points as neighbors. [See (4.4), for a particular case.] If the grid point corresponding to C_{ijk} is connected to a boundary point, then C_{ijk} is larger than the sum of the off-diagonal elements. Hence, $\widehat{\mathbf{A}}$ is nonsingular and its Gerschgorin circles are all in the right-hand complex plane. Thus, the symmetric matrix $\mathbf{A}$, which is similar to $\widehat{\mathbf{A}}$, must have positive eigenvalues.

3   The SSOR Preconditioned Conjugate Gradient Method

We will approximate the solution to the system (2.9) by the conjugate gradient iteration preconditioned by the SSOR method. The preconditioned conjugate gradient (PCG) method can be written in the form

(3.1b) $\mathbf{M}\tilde{\mathbf{r}}^k = \mathbf{r}^k$

(3.1c) $\gamma_k = (\tilde{\mathbf{r}}^k, \mathbf{r}^k), \quad \beta_k = \gamma_k/\gamma_{k-1} \quad (\beta_0 = 0)$

(3.1d) $\mathbf{p}^k = \tilde{\mathbf{r}}^k + \beta_k \mathbf{p}^{k-1} \quad (\mathbf{p}^{-1} = 0)$

(3.1e) $\alpha_k = \gamma_k/(\mathbf{p}^k, \mathbf{A}\mathbf{p}^k)$

(3.1f) $\mathbf{r}^{k+1} = \mathbf{r}^k - \alpha_k \mathbf{A}\mathbf{p}^k$

(3.1g) $\mathbf{x}^{k+1} = \mathbf{x}^k + \alpha_k \mathbf{p}^k$

Here $(\mathbf{x},\mathbf{y}) = \mathbf{x}^T\mathbf{y}$ is the usual inner product, $\mathbf{r}^k = \mathbf{b} - \mathbf{A}\mathbf{x}^k$ is the residual at the k-th step and $k = 0, 1, \cdots$ with $\mathbf{x}^0$ assumed given.

Equation (3.1a) is the preconditioning step, and we will be interested in preconditioning matrices of the form

(3.2) $\mathbf{M} = \mathbf{P}(\mathbf{I} + \mathbf{H} + \cdots + \mathbf{H}^{m-1})^{-1}$

where $\mathbf{A} = \mathbf{P}-\mathbf{Q}$ is a splitting of $\mathbf{A}$ and $\mathbf{H} = \mathbf{P}^{-1}\mathbf{Q}$. The use of $\mathbf{M}$ in (3.1a) is then equivalent to carrying out the subsidiary iteration

(3.3) $\tilde{\mathbf{r}}^k_{i+1} = \mathbf{H}\tilde{\mathbf{r}}^k_i + \mathbf{P}^{-1}\mathbf{r}^k = \tilde{\mathbf{r}}^k_i + \mathbf{P}^{-1}(\mathbf{r}^k - \mathbf{A}\tilde{\mathbf{r}}^k_i), \quad i = 0, 1, \cdots, m-1, \quad \tilde{\mathbf{r}}^k_0 = 0$

for the equation $\mathbf{A}\tilde{\mathbf{r}} = \mathbf{r}^k$ and setting $\tilde{\mathbf{r}}^k = \tilde{\mathbf{r}}^k_m$. For SSOR preconditioning with relaxation parameter ω, we have

(3.4) $\mathbf{P} = (\widehat{\mathbf{D}} - \mathbf{L})\, \tilde{\mathbf{D}}^{-1}(\widehat{\mathbf{D}} - \mathbf{L}^T)$

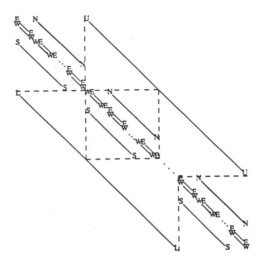

Figure 11.1 The structure of B after red/black reordering.

where $\mathbf{A} = \mathbf{D} - \mathbf{L} - \mathbf{L}^T$ with $\mathbf{D}$ diagonal, $\mathbf{L}$ strictly lower triangular, $\widehat{\mathbf{D}} = \omega^{-1}\mathbf{D}$, and $\tilde{\mathbf{D}} = 2\widehat{\mathbf{D}} - \mathbf{D}$. For the problem of Sec. 2, each SSOR iteration is carried out by two SOR sweeps through the grid points, one in the forward direction followed by one in the reverse order. Equation (3.3) is just the mathematical representation of m SSOR iterations. For further details see Ortega [20].

In order to vectorize the SSOR iteration, we use a red/black reordering of the unknowns (Young [23]). In this ordering, the coefficient matrix has the form

$$(3.5) \qquad\qquad \mathbf{A} = \begin{bmatrix} \mathbf{D}_R & \mathbf{B} \\ \mathbf{B}^T & \mathbf{D}_B \end{bmatrix}$$

where $\mathbf{D}_R$ and $\mathbf{D}_B$ are diagonal matrices, and $\mathbf{B}$ is shown in more detail in Figure 3.1. Since $\mathbf{A}$ is symmetric, we only need to store its main diagonal and the non-zero diagonals of $\mathbf{B}$. We note that the diagonal blocks in Figure 3.1 are not square when there is an odd number of grid points in each direction. Also, the two upper blocks have one more row than the two lower blocks, and the two left blocks have one more column than the two right blocks; this is because we have colored the first grid point red.

The computation in the algorithm (3.1) will be dominated by the preconditioning and by the matrix-vector multiply $\mathbf{Ap}^k$; hence, it is imperative that these be handled efficiently. We have stored the matrix by diagonals, and we calculate $\mathbf{Ap}^k$ using multiplication by diagonals (Madsen *et al.* [18]). This reduces the matrix-vector multiplication to a series of vector-vector multiplications and additions. The C,E, and W diagonals are all of length $O(N/2)$, where $N = N_x N_y N_z$, and the U and L diagonals are of length $(N - N_x N_y)/2$. Since the intermediate zeroes of the N and S diagonals are not stored, they are of length $N_x(N_y - 1)/2$.

In the red/black ordering, the SSOR iteration (3.3) (with relaxation parameter $\omega = 1$) can be carried out as

(3.6a) $\tilde{\mathbf{r}}_{R,i+1}^k = \mathbf{D}_R^{-1}[\mathbf{r}_R - \mathbf{B}\tilde{\mathbf{r}}_{B,i}^k], \quad \tilde{\mathbf{r}}_{B,0}^k = 0$

(3.6b) $\tilde{\mathbf{r}}_{B,i+1}^k = \mathbf{D}_B^{-1}[\mathbf{r}_B - \mathbf{B}^T\tilde{\mathbf{r}}_{R,i+1}^k].$

The iteration (3.6) can be vectorized very efficiently since $\mathbf{D}_R^{-1}$ and $\mathbf{D}_B^{-1}$ are diagonal and vectors are multiplied by $\mathbf{B}$ or $\mathbf{B}^T$, using matrix-vector multiplication by diagonals. Considerable computation can be saved by using the Eisenstat modification (Eisenstat [7]). An extension of the original one-step Eisenstat procedure to m SSOR preconditioning steps was given in Eisenstat et al. [8], and this has been used in our algorithm. However, the major saving is obtained in 1-step preconditioning and the larger m is the less the savings.

Finally, we address the choice of ω for SSOR. It is known that for the red/black ordering (3.5), the optimum value of ω for the SSOR iteration is $\omega = 1$ (Young [23]). This is also true for m-step SSOR preconditioning. (Harrar and Ortega [12]. See also Kuo and Chan [17] who have proved this result by Fourier analysis for the two dimensional Poisson equation and 1-step preconditioning.) However, for polynomial preconditioning, $\omega = 1$ is not necessarily the optimum, as an example given below shows. But for the large problem sizes of interest here, $\omega = 1$ does indeed seem to be the optimum for the polynomial preconditioners we have used.

For polynomial preconditioning, the matrix $\mathbf{M}$ of (3.2) becomes

(3.7) $\mathbf{M} = \mathbf{P}(\alpha_0\mathbf{I} + \alpha_1\mathbf{H} + \cdots + \alpha_{m-1}\mathbf{H}^{m-1})^{-1}$

and the modified residual vector $\tilde{\mathbf{r}} = \mathbf{M}^{-1}\mathbf{r}$ of (3.1a) is a linear combination of the SSOR iterates generated by (3.3). Since

(3.8) $\mathbf{M}^{-1}\mathbf{A} = (\alpha_0\,\mathbf{I} + \cdots + \alpha_{m-1}\mathbf{H}^{m-1})(\mathbf{I} - \mathbf{H})$

the l_2–condition number of the preconditioned matrix is

(3.9) $$\frac{\max_i |p_m(\lambda_i)|}{\min_i |p_m(\lambda_i)|}$$

where $\lambda_1, \lambda_2, \ldots, \lambda_n$ are the eigenvalues of $\mathbf{H}$ and

(3.10) $p_m(\lambda) = (\alpha_0 + \alpha_1\lambda + \cdots + \alpha_{m-1}\lambda^{m-1})(1 - \lambda).$

To obtain the polynomials p_m, we follow the papers by Johnson et al. [13] and Adams [2] and minimize the integral

(3.11) $$\int_0^1 [p_m(\lambda) - 1]^2 d\lambda.$$

(See Saad [22] for other ways to obtain the polynomials.) The first few of these polynomials are given explicitly by

(3.12a) $p_2(\lambda) = (1 + 5\lambda)(1 - \lambda)$

(3.12b) $p_3(\lambda) = (1 - 2\lambda + 7\lambda^2)(1 - \lambda)$

(3.12c) $p_4(\lambda) = (1 + 7\lambda - 24.5\lambda^2 + 31.5\lambda^3)(1 - \lambda)$

(3.12d) $p_5(\lambda) = (1 - 4\lambda + 36\lambda^2 - 84\lambda^3 + 66\lambda^4)(1 - \lambda)$

(3.12e) $p_6(\lambda) = (1 + 9\lambda - 81\lambda^2 + 319\lambda^3 - 506\lambda^4 + 286\lambda^5)(1 - \lambda)$

We have scaled these so that $p_m(0) = \alpha_0 = 1$. This does not affect the ratio (3.9).

We first show that $\omega = 1$ is not always the optimum for these polynomials. For the 2-dimensional Poisson equation on the unit square, with Dirichlet boundary conditions and discretized by 5-point finite differences, the eigenvalues of the Jacobi iteration matrix are well-known (Young [23]) to be

$$(3.13) \qquad \frac{1}{2}\left(\cos\frac{k\pi}{N+1} + \cos\frac{j\pi}{N+1}\right), \quad j, k = 1, 2, \ldots, N.$$

This is true for either the natural or red/black ordering of the grid points since the corresponding Jacobi iteration matrices are permutationally similar. By using (3.13) for $N = 3$, the condition numbers (3.9) were computed for various ω and the following results obtained:

$$m = 2: \ \omega_{\mathrm{opt}} \doteq 1.50 \quad m = 3: \ \omega_{\mathrm{opt}} \doteq 1.59, \quad m = 4: \ \omega_{\mathrm{opt}} \doteq 1.39.$$

Hence, $\omega = 1$ is not optimum in general. However, we next indicate, at least for the polynomials (3.12), that $\omega = 1$ may be optimum in practical situations. Let $0 \le \lambda_1 \le \cdots \le \lambda_n$ be the eigenvalues of the SSOR iteration matrix H. The reason for the previous counterexample is that for small problems λ_n is not particularly close to 1, and therefore, the condition number (3.9) may be determined by local minima of p_m. On the other hand, if λ_n is close to 1, then we expect the condition number to be largely determined by λ_n: the larger λ_n, the closer to zero is $p_m(\lambda_n)$ and the larger the condition number. In this case, we wish to minimize λ_n as a function of ω and, in the setting of our problem, $\omega = 1$ will achieve this. Unfortunately, we have not been able to prove, under reasonable assumptions, an optimality result. But numerical experiments have indicated that $\omega = 1$ is optimum, at least for our sample problems, and thus $\omega = 1$ has been used in all of the results of the next section.

4   Numerical Results

In this section, we present the results of numerical experiments. The particular problem used for these experiments was motivated by our collaboration with the

High-Speed Aerodynamics Division at NASA's Langley Research Center. A time-splitting scheme presented in Krist and Zang [16] for the time-dependent Navier-Stokes equations requires the solution of a Helmholtz-type equation at each time step. The solution of the Poisson-type equation (1.1) embodies the major difficulties in solving this Helmholtz-type equation and the initial goal of this work was to obtain fast solution techniques for (1.1) on vector computers.

The function K used in our computations was suggested by Zang [24] and is

$$(4.1) \qquad K(x,y,z) = 1 + e^{-5z}\left(1 + 0.1\mathrm{Re}\left(e^{2i(x+y)}\right)\right)$$

where Re denotes the real part. In the time splitting scheme, the function f is calculated at the intermediate time step and is related to the divergence of the velocity field, but for our test cases, we chose f so that the exact solution of (1.1) is $x^2 + y^2 + z^2$. We use variable spacing in the z-direction so as to concentrate grid points near the lower surface of the domain where gradients are especially large, and variable spacing in the x-direction to concentrate grid points near the outflow region. These non-uniform spacings are based on cosine distributions of the grid points:

$$(4.2) \qquad z_k = 1 - \cos\left[\frac{\pi k}{2(N_z + 1)}\right], \qquad k = 0, 1, \dots, N_z + 1$$

$$(4.3) \qquad x_i = 3\cos\left[\frac{\pi(N_x + 1 - i)}{2(N_x + 1)}\right], \qquad i = 0, 1, \dots, N_x + 1$$

where N_x and N_z are the number of interior grid points in the x- and z-directions, respectively. The factor 3 appears in Equation (4.3) since the domain is elongated in the x-direction by a factor of three. With $\xi_i = x_i - x_{i-1}$, $\zeta_k = z_k - z_{k-1}$ and η the constant grid spacing in the y-direction, we let

$$\sigma_i = \frac{2\eta^2}{\xi_i + \xi_{i+1}}, \qquad \tau_k = \frac{2\eta^2}{\zeta_k + \zeta_{k+1}}.$$

Then the coefficients in (2.3) are given by

$$(4.4a) \qquad W_{ijk} = \frac{\sigma_i}{\xi_i}K_{i-\frac{1}{2},j,k}, \qquad E_{ijk} = \frac{\sigma_i}{\xi_{i+1}}K_{i+\frac{1}{2},j,k}$$

$$(4.4b) \qquad S_{ijk} = K_{i,j-\frac{1}{2},k}, \qquad N_{ijk} = K_{i,j+\frac{1}{2},k}$$

$$(4.4c) \qquad L_{ijk} = \frac{\tau_k}{\zeta_k}K_{i,j,k-\frac{1}{2}}, \qquad U_{ijk} = \frac{\tau_k}{\zeta_{k+1}}K_{i,j,k+\frac{1}{2}}$$

$$(4.4d) \qquad \begin{aligned} C_{ijk} &= \sigma_i\left[\frac{1}{\xi_i}K_{i-\frac{1}{2},j,k} + \frac{1}{\xi_{i+1}}K_{i+\frac{1}{2},j,k}\right] \\ &\quad + [K_{i,j-\frac{1}{2},k} + K_{i,j+\frac{1}{2},k}] + \tau_k\left[\frac{1}{\zeta_k}K_{i,j,k-\frac{1}{2}} + \frac{1}{\zeta_{k+1}}K_{i,j,k+\frac{1}{2}}\right]. \end{aligned}$$

Our experiments were run on a CDC Cyber 205 and a Cray-2. The CDC Cyber 205 at the NASA Langley Research Center has two pipelines and a cycle time of 20ns (nanoseconds). This gives a maximum megaflop rate of 100 for addition and multiplication and 200 for linked triads (operations of the form $\mathbf{z} = \alpha\,\mathbf{x} + \mathbf{y}$ with α a scalar). These maximum megaflop rates, of course, are not attainable but can be approached by using long vector lengths.

The Cray-2 at the NASA Ames Research Center has four processors, but our experiments used only a single processor. Each processor has separate pipelines for addition and multiplication, but unlike the Cray X-MP, chaining is not allowed. The cycle time is 4.1ns giving a maximum theoretical megaflop rate of 488 per processor if both pipelines are working simultaneously. It is a register-to-register machine with eight 64-element registers. There is only one path from main memory to the registers so that one may not load and store at the same time. The main memory has 256 million words, and each processor has a local memory of 16 thousand words; access of these local memories is substantially faster than that of the main memory (45ns as compared to about 205ns, assuming no main memory conflicts).

Four problem sizes were used with seven, fifteen, thirty-one, and sixty-three grid points in each direction; these grids correspond to 343, 3375, 29791, and 250047 unknowns, respectively. We first show some results for a code for the Cyber 205 which does not use the Eisenstat modification. Table 11.1 gives the number of iterations, and corresponding times in seconds. In all of the tables of this section, m is the number of preconditioning steps and $m = 0$ corresponds to conjugate gradient with no preconditioning.

The convergence criterion used for the results of Table 11.1 was

$$(4.5) \qquad (\mathbf{r}^{k+1},\ \mathbf{D}^{-1}\mathbf{r}^{k+1}) < \epsilon = 10^{-12}$$

where $\mathbf{D}$ is the main diagonal of $\mathbf{A}$. This is a rather stringent condition as evidenced by the fact that the computed solutions were accurate, on the average, to 9 or 10 decimal places. Norms of the initial residuals ranged from $O(10^6)$ for the smallest

	$7^3 = 343$		$15^3 = 3375$		$31^3 = 29791$		$63^3 = 250047$	
m	ITER	TIME	ITER	TIME	ITER	TIME	ITER	TIME
0	96	0.0321	375	0.614	1494	18.4	6108	615.
1	19	0.0115	39	0.111	80	1.67	161	27.5
2	11	0.0095	21	0.084	44	1.26	89	20.7
3	8	0.0092	16	0.082	32	1.16	66	19.4
4	8	0.0112	12	0.076	24	1.06	50	17.8
5	7	0.0117	10	0.076	20	1.05	42	17.6
6	7	0.0135	8	0.071	17	1.03	35	16.9

Table 11.1 Polynomial preconditioning on Cyber 205.

problem to $O(10^{12})$ for the largest.

Note that each time we double the number of grid points in each direction, the number of preconditioned iterations increases by about a factor of 2, for all values of m, and the number of unpreconditioned iterations increases by about a factor of 4. This suggests that we can predict the number of iterations necessary for larger problems. For example, for a $127 \times 127 \times 127$ problem, we would predict about 180 iterations for two preconditioning steps and about 25000 iterations for no preconditioning.

Clearly, the use of the preconditioner gives a dramatic decrease in both the number of iterations and the time, increasingly so for the larger problem sizes. However, this is somewhat misleading as to the efficacy of the SSOR preconditioner. In Table 11.2, we give results when the coefficient matrix has been scaled by

$$(4.6) \qquad \tilde{\mathbf{A}} = \mathbf{D}^{-1/2}\mathbf{A}\mathbf{D}^{-1/2}$$

where $\mathbf{D}$ is the main diagonal of $\mathbf{A}$. Thus, the diagonal elements of $\tilde{\mathbf{A}}$ are all unity. The times in Table 11.2 include the time to do the scaling (4.6). These times were 0.00024, 0.0015, 0.012, and 0.10 seconds for the four problem sizes.

Note that the big difference in Tables 11.1 and 11.2 is in the number of iterations and time with no preconditioning. This is due to the diagonal scaling (4.6) which is equivalent to one-step Jacobi preconditioning (See, e.g., Ortega [20]). Hence, this Jacobi preconditioning accounts for the major part of the improvement in the number of iterations and SSOR preconditioning gives only a relatively small additional improvement.

The convergence test for the results of Table 11.2 is

$$(4.7) \qquad (\tilde{\mathbf{r}}, \tilde{\mathbf{r}}) < 10^{-12}$$

where $\tilde{\mathbf{r}} = \tilde{\mathbf{b}} - \tilde{\mathbf{A}}\tilde{\mathbf{u}}$ is the residual corresponding to the scaled system $\tilde{\mathbf{A}}\tilde{\mathbf{u}} = \tilde{\mathbf{b}}$ with $\tilde{\mathbf{u}} = \mathbf{D}^{1/2}\mathbf{u}$ and $\tilde{\mathbf{b}} = \mathbf{D}^{-1/2}\mathbf{b}$. Since $\tilde{\mathbf{r}} = \mathbf{D}^{-1/2}\mathbf{r}$, this is exactly the same test

	$7^3 = 343$		$15^3 = 3375$		$31^3 = 29791$		$63^3 = 250047$	
m	ITER	TIME	ITER	TIME	ITER	TIME	ITER	TIME
0	37	0.0083	78	0.072	159	1.02	322	16.5
1	19	0.0085	39	0.069	80	0.933	162	15.0
2	11	0.0075	22	0.057	45	0.746	91	11.9
3	8	0.0074	16	0.055	33	0.721	67	11.3
4	8	0.0090	12	0.051	25	0.666	51	10.6
5	8	0.0106	10	0.053	22	0.697	44	10.8
6	7	0.0110	9	0.054	17	0.631	36	10.2

Table 11.2 Polynomial preconditioning with diagonal scaling on Cyber 205.

as (4.5). Still it seems curious that the number of iterations with preconditioning should be almost exactly the same in both tables. This is a consequence of the following result.

Theorem 4.1 *Let* $A_1 = P_1 - Q_1$ *be a symmetric positive definite matrix and for some nonsingular matrix* B *define*

$$(4.8) \qquad A_2 = BA_1B^T, \quad P_2 = BP_1B^T, \quad Q_2 = BQ_1B^T$$

and

$$R_i = (\alpha_0 I + \alpha_1 H_i + \cdots + \alpha_{m-1}H_i^{m-1})P_i^{-1}, \quad i = 1, 2$$

where $H_i = P_i^{-1}Q_i$ *and the* α_i *are scalars. Then* R_1A_1 *and* $R_2 A_2$ *are similar.*

Proof: From (4.8), it follows immediately that $A_2 = P_2 - Q_2$ and

$$H_2 = (BP_1B^T)^{-1}BQ_1B^T = B^{-T}H_1B^T.$$

Then

$$\begin{aligned} R_2 &= \left[\alpha_0 B^{-T}B^T + \alpha_1 B^{-T}H_1B^T + \cdots + \alpha_{m-1}(B^{-T}H_1B^T)^{m-1}\right](BP_1B^T)^{-1} \\ &= B^{-T}\left[\alpha_0 I + \alpha_1 H_1 + \cdots + \alpha_{m-1}H_1^{m-1}\right]P_1^{-1}B^{-1} \\ &= B^{-T}R_1B^{-1}. \end{aligned}$$

Thus

$$R_2A_2 = (B^{-T}R_1B^{-1})(BA_1B^T) = B^{-T}R_1A_1B^T$$

which was to be proved. ∎

Since it is the eigenvalues of R_iA_i that determine the condition number of the matrix preconditioned by the polynomial preconditioner, a consequence of the theorem is that the preconditioned conjugate gradient rate of convergence, at least as determined by the condition number, is the same for both A_1 and A_2 provided that the splitting changes by the same congruence. This is indeed the case for SSOR preconditioning under diagonal scaling of the coefficient matrix as the following corollary shows.

Corollary 4.1 *Let* $A = D - L - L^T$ *be symmetric positive definite. Then the condition number of the matrix obtained from* A *by SSOR polynomial preconditioning is the same as the condition number of the matrix obtained from* $\tilde{A} = D^{-1/2}AD^{-1/2}$ *by SSOR polynomial preconditioning.*

Proof: By Theorem 4.1, it suffices to show that (4.8) holds with $A_1 = A$, $A_2 = \tilde{A}$, and $B = D^{-1/2}$. The SSOR splitting of A_1 (see, e.g., Ortega [20]) is

$$P_1 = \alpha(D - \omega L)D^{-1}(D - \omega L^T)$$

$$Q_1 = \alpha[(1 - \omega)D + \omega L]D^{-1}[(1 - \omega)D + \omega L^T]$$

where $\alpha = [\omega(2 - \omega)]^{-1}$. The corresponding P_2 for the scaled matrix $\tilde{A}$ is

$$\mathbf{P}_2 = \alpha[\mathbf{I} - \omega\mathbf{D}^{-1/2}\mathbf{L}\mathbf{D}^{-1/2}][\mathbf{I} - \omega\mathbf{D}^{-1/2}\mathbf{L}^T\mathbf{D}^{-1/2}]$$

$$= \alpha\mathbf{D}^{-1/2}(\mathbf{D} - \omega\mathbf{L})\mathbf{D}^{-1}(\mathbf{D} - \omega\mathbf{L}^T)\mathbf{D}^{-1/2}$$

$$= \mathbf{D}^{-1/2}\mathbf{P}_1\mathbf{D}^{-1/2}.$$

Similarly, it is easy to verify that $\mathbf{Q}_2 = \mathbf{D}^{-1/2}\mathbf{Q}_1\mathbf{D}^{-1/2}$. Hence (4.8) is verified. ∎

Note that although the number of iterations with preconditioning is almost exactly the same in Tables 11.1 and 11.2, the times in Table 11.2 are considerably lower. This is due in part to the time saved by having the main diagonal of $\tilde{\mathbf{A}}$ equal to the identity, which saves time in carrying out the matrix-vector multiply and the convergence test (4.5), and in part to the code for Table 11.2 being more efficient. For example, the code for Table 11.1 is for general ω, while that for Table 11.2 takes advantage of $\omega = 1$.

Next, we give in Table 11.3 the results when the Eisenstat modification is added. As in Table 11.2, the time to do the scaling (4.6) is included.

The iterations in Tables 11.3 and 11.2 should be exactly the same but occasionally differ slightly due to, presumably, a different pattern of rounding errors. The times for $m = 0$ are also the same since the Eisenstat modification affects only the preconditioning. Note that the major benefit of the Eisenstat modification occurs for $m = 1$. As m increases, the savings decrease as predicted by the results in Eisenstat et al. [8]. Even though the largest saving results from the diagonal scaling, SSOR polynomial preconditioning with the Eisenstat modification saves almost another factor of two in time on the largest problem sizes.

Note that there is an optimal number of preconditioning steps, depending upon the problem size. For example, on a $15 \times 15 \times 15$ grid, the optimal number is 2. Although with additional preconditioning steps the number of iterations goes down, there is enough extra work being done in each preconditioning step to cause the

	$7^3 = 343$		$15^3 = 3375$		$31^3 = 29791$		$63^3 = 250047$	
m	ITER	TIME	ITER	TIME	ITER	TIME	ITER	TIME
0	37	0.0083	78	0.072	159	1.020	322	16.5
1	19	0.0059	39	0.046	80	0.631	162	10.1
2	11	0.0061	22	0.044	45	0.571	91	8.97
3	8	0.0065	16	0.045	33	0.577	67	9.02
4	8	0.0082	12	0.044	25	0.561	51	8.73
5	7	0.0089	10	0.046	21	0.575	43	8.93
6	7	0.0104	9	0.049	17	0.553	36	8.82

Table 11.3 Polynomial preconditioning with diagonal scaling and Eisenstat modification on Cyber 205.

$7^3 = 343$		$15^3 = 3375$		$31^3 = 29791$	
m	uni	var	uni	var	uni	var
0	40	37 (37)	84	79 (78)	170	158 (159)
1	21	19 (19)	42	40 (40)	86	80 (80)
2	11	11 (11)	23	22 (22)	47	45 (45)
3	9	8 (8)	17	16 (16)	35	33 (33)

Table 11.4 Iterations for Poisson equation.

total time to increase. The optimal number of steps increases as the problem size increases. We used polynomials of degree up to 10 on the two larger problems and found that the optimum occurred for $m = 6$ for the $31 \times 31 \times 31$ grid. For the largest problem, the optimum occurred at $m = 7$ with a time essentially equal to that obtained for $m = 4$. However, most of the decrease in both time and iterations occurs for $m = 1$.

In order to obtain some information as to the role the function K of equation (1.1) plays in the rate of convergence, we give in Table 11.4 the number of iterations for $K \equiv 1$ (i.e., Poisson's equation) for all but the largest problem size. Results for both variable spacing (the columns headed **var**) as well as for uniform spacing in all directions (the columns headed **uni**) are listed for the first few values of m. In parentheses are given the corresponding iteration counts from Table 11.2. These results show that K has essentially no effect on the rate of convergence, and the variable spacing actually decreases the number of iterations slightly. The convergence test for Table 11.4 is again (4.7), since the Poisson equations have also been diagonally scaled.

As mentioned earlier, the convergence test (4.5) is rather stringent. Table 11.5 gives the number of iterations required for smaller values of the tolerance ϵ. These results are for the code with diagonal scaling.

We next compare the polynomial preconditioning results with the reduced system approach. (See, for example, Axelsson and Gustafsson [4], Chandra [5], and Hageman and Varga [10].) Let

$$(4.9) \qquad \begin{bmatrix} \mathbf{I} & -\mathbf{C} \\ -\mathbf{C}^T & \mathbf{I} \end{bmatrix} \begin{bmatrix} \mathbf{u}_R \\ \mathbf{u}_B \end{bmatrix} = \begin{bmatrix} \mathbf{b}_R \\ \mathbf{b}_B \end{bmatrix}$$

be the system after the diagonal scaling (4.6). Then the reduced (Schur complement) system is

$$(4.10) \qquad (\mathbf{I} - \mathbf{C}^T\mathbf{C})\mathbf{u}_B = \mathbf{b}_B + \mathbf{C}^T\mathbf{b}_R.$$

ϵ	$7^3 = 343$		$15^3 = 3375$		$31^3 = 29791$		$63^3 = 250047$	
	m=0	m=1	m=0	m=1	m=0	m=1	m=0	m=1
10^{-12}	37	19	78	39	159	80	322	162
10^{-8}	29	15	63	32	131	66	267	134
10^{-4}	20	11	44	23	94	48	195	98

Table 11.5 Effect of ϵ on number of iterations.

$7^3 = 343$		$15^3 = 3375$		$31^3 = 29791$		$63^3 = 250047$	
ITER	TIME	ITER	TIME	ITER	TIME	ITER	TIME
19	0.0046	39	0.035	80	0.460	162	7.34

Table 11.6 RS-CG method on Cyber 205.

After $\mathbf{u}_B$ is obtained from (4.10), $\mathbf{u}_R$ can be computed from

$$(4.11) \qquad\qquad \mathbf{u}_R = \mathbf{C}\mathbf{u}_B + \mathbf{b}_R.$$

Table 11.6 gives results for the conjugate gradient method applied to (4.10), which we call the RS-CG method following Hageman and Young [11]. The times in Table 11.6 include the scaling time to produce the system (4.9). The convergence test is $(\mathbf{r}_B, \mathbf{r}_B) < 10^{-12}$, where $\mathbf{r}_B$ is the residual for the system (4.10). This test is compatible with that of (4.7) for the system (4.9).

Note that the times in Table 11.6 are about 20 percent lower than the best times in Table 11.3. Note also that the number of iterations is exactly the same as for $m = 1$ in Table 11.3. This is no coincidence since conjugate gradient on the reduced system (4.10) is equivalent to using 1-step SSOR preconditioning on (4.9). (See Keyes and Gropp [14] for a much more general equivalence due to S. Eisenstat.) This follows from the observation that the 1-step SSOR preconditioned matrix for (4.9) is

$$\begin{bmatrix} \mathbf{I} & \mathbf{0} \\ -\mathbf{C}^T & \mathbf{I} \end{bmatrix}^{-1} \begin{bmatrix} \mathbf{I} & -\mathbf{C} \\ -\mathbf{C}^T & \mathbf{I} \end{bmatrix} \begin{bmatrix} \mathbf{I} & \mathbf{0} \\ -\mathbf{C}^T & \mathbf{I} \end{bmatrix}^{-T} = \begin{bmatrix} \mathbf{I} & \mathbf{0} \\ \mathbf{0} & \mathbf{I} - \mathbf{C}^T \mathbf{C} \end{bmatrix}$$

With the Eisenstat modification, the 1-step code for (4.9) requires a multiplication by $\mathbf{C}$ and $\mathbf{C}^T$ on each iteration, as does the RS-CG method. However, our code for (4.9) does not take full advantage of the possibility of working with vectors of half-length, as does the RS-CG code. This explains the superior performance of the latter method.

We next consider the Cray-2. Tables 11.7 and 11.8 give results for polynomial preconditioning and the RS-CG method, respectively. The times to do the diagonal

m	$7^3 = 343$ ITER	TIME	$15^3 = 3375$ ITER	TIME	$31^3 = 29791$ ITER	TIME	$63^3 = 250047$ ITER	TIME
0	37	0.0068	78	0.081	159	1.340	322	21.6
1	19	0.0045	39	0.049	80	0.787	162	13.0
2	11	0.0048	22	0.046	45	0.734	91	11.8
3	8	0.0052	16	0.047	33	0.741	67	11.9
4	8	0.0065	12	0.047	25	0.707	51	11.7
5	7	0.0073	10	0.049	21	0.751	43	12.0
6	7	0.0084	9	0.052	17	0.707	36	11.8

Table 11.7 Polynomial preconditioning with diagonal scaling and Eisenstat modification on Cray-2.

scaling are included in both tables; these were 0.0001, 0.0007, 0.0059, and 0.052 seconds for the four problem sizes. As with the Cyber 205, the RS-CG method is the best, beating polynomial preconditioning on the largest problem size by about 20 percent.

Note that the Cray-2 times for the smallest problem size are considerably better than the Cyber 205 times (Tables 11.3 and 11.6). However, for the larger problem sizes the Cyber 205 times are better, presumably due to the longer vector lengths, which benefit the 205 more. It was our experience that measured runtimes can vary by well over 10 percent from one run to the next on the Cray-2 because of its multi-user environment. The timings given in Tables 11.7 and 11.8 were averages for three separate runs.

We suspect that the Cray-2 codes could be tuned to perform somewhat better. They were written originally for the Cyber 205 and then compiled by CFT77 on the Cray-2. No attempt was made to reorganize the algorithms so as to be more efficient on the Cray-2. In particular, no explicit attempt was made to utilize local memory nor to keep intermediate results in the vector registers.

Table 11.9 shows megaflop rates for conjugate gradient with diagonal scaling, for $m = 1$ and $m = 6$ polynomial preconditioning with the Eisenstat modification, and for the RS-CG method. These rates are quite satisfactory for the Cyber 205, especially for the larger problem sizes for which we achieve over 110 megaflops. Part

$7^3 = 343$ ITER	TIME	$15^3 = 3375$ ITER	TIME	$31^3 = 29791$ ITER	TIME	$63^3 = 250047$ ITER	TIME
19	0.0041	39	0.042	80	0.600	162	9.90

Table 11.8 RS-CG method on Cray-2.

	$7^3 = 343$		$15^3 = 3375$		$31^3 = 29791$		$63^3 = 250047$	
	205	Cray-2	205	Cray-2	205	Cray-2	205	Cray-2
CG	35	42	84	73	107	80	112	84
m=1	36	42	88	75	114	84	118	86
m=6	26	30	72	61	99	72	105	75
RS	27	29	71	57	97	72	103	74

Table 11.9 MFLOP rates.

	$7 \times 7 \times 7$	$15 \times 15 \times 15$	$31 \times 31 \times 31$	$63 \times 63 \times 63$
Nat	11	17	26	38
RB	19	39	80	162

Table 11.10 Iterations for 1-step SSOR preconditioning.

of the reason for these good megaflop rates is that the conjugate gradient algorithm requires three linked triads at each iteration, which, as mentioned earlier in this section, have an asymptotic rate of 200 megaflops. For the Cray-2 the megaflop rates do not begin to approach the maximum theoretical rate of 488 megaflops.

Finally, Table 11.10 shows the number of iterations for 1-step SSOR preconditioning using the natural ordering, as well as the corresponding iterations from Table 11.2 for the red/black ordering. The optimum ω for the natural ordering was determined experimentally for the four problem sizes to be approximately 1.45, 1.65, 1.8, and 1.9. Clearly, the number of iterations is considerably less with the natural ordering. The times were much higher, but this was for a code in which the preconditioning portion was entirely serial (although the conjugate gradient portion was vectorized). We are currently planning to use the techniques in Ashcraft and Grimes [3] and Melhem [19], as well as other possibilities, to achieve at least some vectorization in the preconditioning.

5   Summary and Conclusions

For the largest problem size considered (250,000 unknowns) the conjugate gradient method requires over 6,000 iterations and 615 seconds of time on a Cyber 205. Diagonal scaling (1-step Jacobi preconditioning) reduces these requirements to 322 iterations and 16.5 seconds. Polynomial SSOR preconditioning with the red/black ordering and the Eisenstat modification gives further reductions to 36 iterations and under 9 seconds. However, still better results are obtained by using the RS-CG method. The RS-CG method is theoretically equivalent to 1-step SSOR preconditioning on the full system but has the computational advantage of working with vectors of only half-lengths. These results corroborate and complement those in Kincaid, Oppe and Young [15], who also concluded that the RS-CG method was the fastest among several methods tested. We expect that the RS-CG method will be still faster with suitable preconditioning of the reduced system. It is also possible that sufficient vectorization can be achieved with the natural ordering so that its better convergence rate can be utilized.

Acknowledgements

Research supported in part under a National Science Foundation Graduate Fellowship and in part by the National Aeronautics and Space Administration under Grant NAG-1-242. This paper was originally typed in troff by Beverly Martin and retyped in LATEX by Lisa Laguna.

Added in Proof

Professor S. Eisenstat has observed that the theorem and corollary on the condition number may be strengthened as follows. If u^k are the conjugate gradient iterates for $Au = b$ with preconditioning matrix M and $\tilde{u}^k$ are those for $\tilde{A}\tilde{u} = \tilde{b}$ with $\tilde{A} = BAB^T$ and preconditioning matrix $\tilde{M} = BMB^T$, and if $\tilde{u} = B^{-T}u^0$, then $\tilde{u}^k = B^{-T}\tilde{u}^k$ for all k.

References

[1] Adams, L. "Iterative Algorithms for Large Sparse Linear Systems on Parallel Computers," Ph.D. Thesis, Applied Mathematics, University of Virginia, 1982.

[2] Adams, L. "An M-Step Preconditioned Conjugate Gradient Method for Parallel Computation," Proc. 1983 Int. Conf. Par. Proc., 36-43, 1983.

[3] Ashcraft, C., and R. Grimes. "On Vectorizing Incomplete Factorization and SSOR Preconditioners, *SIAM J. Sci. Stat. Comput.* **9** (1988) 122-151.

[4] Axelsson, O., and I. Gustafsson. "On the Use of Preconditioned Conjugate Gradient Methods for Red-Black Ordered Five-Point Difference Schemes," *J. Comp. Phys.* **35** (1980) 284-289.

[5] Chandra, R. "Conjugate Gradient Methods for Partial Differential Equations," Comp. Sci. Rep. 129, Yale University, 1978.

[6] Concus, P., G. Golub, and D. O'Leary. "A Generalized Conjugate Gradient Method for the Numerical Solution of Elliptic Partial Differential Equations," in *Sparse Matrix Computations*, J. Bunch and D. Rose, eds., Academic Press, New York, 309-322, 1976.

[7] Eisenstat, S. "Efficient Implementation of a Class of Conjugate Gradient Methods," *SIAM J. Sci. Stat. Comput.* **2** (1981) 1–4.

[8] Eisenstat, S., J. Ortega, and C. Vaughan. "Efficient Polynomial Preconditioning for the Conjugate Gradient Method," Applied Math. Rep. RM-88-14, University of Virginia, 1988.

[9] Elman, H., and E. Agron. "Ordering Techniques for the Preconditioned Conjugate Gradient Method on Parallel Computers," Comp. Sci. Rep. Tr-88-53, University of Maryland, 1988.

[10] Hageman, L., and R. Varga. "Block Iterative Methods for Cyclically Reduced Matrix Equations," *Numer. Math.* **6** (1964) 106–119.

[11] Hageman, L. and D. Young. *Applied Iterative Methods*, Academic Press, New York, 1981.

[12] Harrar, D., and J. Ortega. "Optimum *m*-step SSOR Preconditioning," *J. Comput. Appl. Math.* **24** (1988) 191–198.

[13] Johnson, O., C. Micchelli, and G. Paul. "Polynomial Preconditioners for Conjugate Gradient Calculations," *SIAM J. Numer. Anal.* **20** (1983) 362–376.

[14] Keyes, D., and W. Gropp. "A Comparison of Domain Decomposition Techniques for Elliptic Partial Differential Equations and Their Parallel Implementation," *SIAM J. Sci. Stat. Comput.* **8** (1987) s166–s203.

[15] Kincaid, D., T. Oppe, and D. Young. "Vector Computations for Sparse Linear Systems," *SIAM J. Alg. Disc. Meth.* **7** (1986) 99–112.

[16] Krist, S., and T. Zang. "Simulations of Transition and Turbulence on the Navier-Stokes Computer," AIAA Paper No. 87-1110, 1987.

[17] Kuo, C-C., and T. Chan. "Two-color Fourier Analysis of Iterative Algorithms for Elliptic Problems with Red/Black Ordering," CAM Rep. 88-15, Math. Dept., UCLA, 1988.

[18] Madsen, N., G. Rodrigue, and J. Karush. "Matrix Multiplication by Diagonals on a Vector/Parallel Processor," *Inf. Proc. Lett.* **5** (1976) 41–45.

[19] Melhem, R. "Towards Efficient Implementation of Preconditioned Conjugate Gradient Methods on Vector Supercomputers," *Int. J. Supercomput. Appl.* **1** (1987) 70–98.

[20] Ortega, J. *Introduction to Parallel and Vector Solution of Linear Systems*, Plenum Press, New York, 1988.

[21] Poole, E., and J. Ortega. "Multicolor ICCG Methods for Vector Computers," *SIAM J. Numer. Anal.* **24** (1987) 1394–1418.

[22] Saad, Y. "Practical Use of Polynomial Preconditioning for the Conjugate Gradient Method," *SIAM J. Sci. Stat. Comput.* **6** (1985) 865–882.

[23] Young, D. *Iterative Solution of Large Linear Systems*, Academic Press, New York, 1971.

[24] Zang, T. Private communication, 1987.

Chapter 12

Multi-Level Asynchronous Iteration for PDEs

D.C. MARINESCU AND J.R. RICE
Purdue University

Abstract

Consider solving a PDE on a domain D using a domain decomposition or Schwarz splitting approach. We decompose D several times, with each level having a finer (more accurate) decomposition than the preceding one. Iterative methods are initiated on each level so long as the higher level iterations have not converged. This is implemented in a parallel computing environment (such as a hypercube machine). The passing of information between levels is asynchronous. Parameters of such a method are (1) number of levels and number of domains per level, (2) number of processors per level, (3) choice of method on the subdomains of each level (and their convergence rates), and (4) the tolerances used to terminate higher level iterations. We report on both analytic and experimental studies of the effectiveness of such methods and of good values for the parameters.

1 Introduction

Modeling and analysis of synchronization in parallel computing raises difficult questions. Empirical data are largely unavailable, due to the present state of the art in the instrumentation of parallel systems. Only measurements related to the aggregate program behavior such as total execution time, processor utilization, etc., can be carried out routinely, while detailed data concerning communication and synchronization costs are usually unavailable. There are cases in parallel processing when synchronization cannot be avoided, for example in case of iteration techniques for partial differential equations (PDEs) [3], [5].

We have developed a non-deterministic model for parallel computation [1] which shows that in the general case the overhead associated with synchronization depends principally upon two factors, namely, the number of processing elements (PEs) running in parallel, and the actual distribution of the execution time on PEs. For particular distributions the overhead associated with synchronization is independent of the number of PEs running in parallel when this number is large, hence massive parallelism does not become prohibitively expensive solely due to synchronization. This is the case for the uniform distributions, when only the coefficient of variation of the distribution determines the synchronization overhead. For other distributions, like the exponential one, the synchronization overhead grows logarithmically in the number of PEs.

In a recent paper [4], we have analyzed some aspects of the behavior of a two level asynchronous algorithm for solving PDEs. The model developed in [1] has been specialized to this case and we explored several issues concerning the behavior of this algorithm on a hypercube machine. These issues include how to partition the hypercube, how to allocate processors to the two levels, the effects of synchronization, and how to relate the grids on the two levels. Our algorithm is designed to use less communication than is "natural" for such a computation because communication is so expensive on the present generation of hypercube machines. The general result of [4] is that the approach looks promising.

In this paper we extend the algorithm and its analysis to multiple levels (Sec. 2), study the effects of various iterations chosen within levels (Secs. 4 and 5), and provide a technique for efficiently embedding the multiple levels in the hypercube (Sec. 6).

2   Multiple Level Asynchronous PDE Algorithms

Consider the PDE problem $Lu = f$ on the domain $D = [0,1] \times [0,1]$ with Dirichlet boundary conditions. We first subdivide D to obtain *level 1* with $N(1)$ overlapping domains

$$D_k = [0,1] \times [(k-1)/N(1),\ k/N(1)], \qquad k = 1, 2, \ldots, N(1).$$

We then subdivide each D_k into $N(2)/N(1)$ domains

$$D_{kj} = [0,1] \times [(k-1)/N(1) + (j-1)\delta, (k-1)/N(1) + j\delta],$$

$$j = 1, 2, \ldots, N(2)/N(1),$$

where $\delta = 1/N(2)$. This determines *level 2* with $N(2)$ domains. We now formalize this with

Definition 1 (*Multi-level Structure*): A multi-level structure is an interconnection of M linear arrays $L_1, \ldots, L_M$ each representing a level of the structure. Each linear array is subdivided into subdomains. D_{ij} is the jth subdomain in level i. The number $N(p)$ of subdomains at level p is given by

$$N(p) = [R(1) \cdot R(2) \cdots R(j) \cdots R(p-1)]N(1)$$

for $1 \le p \le M - 1$, and $1 \le j \le p - 1$.

In the previous expression

- $N(i)$ is the number of subdomains at level i,
- $R(j)$ is the domain refinement factor at level j defined as
$R(j) = N(j+1)/N(j)$

Each subdomain D_{ij} at level i is connected to its neighbors $D_{i,j-1}$ and $D_{i,j+1}$, if $1 < j < N(i)$. $D_{i,1}$ and $D_{i,N(i)}$ are connected with only one neighbor $D_{i,2}$ and $D_{i,N(i)-1}$ respectively.

In addition a subdomain $D_{i,j}$ at level $i, 1 \leq j \leq N(i)$, is connected to f_i sub-domains at level $i+1$, for $1 \leq i \leq M-2$. The number f_i is called the *fan-out index*. An example of a multi-level structure is given in [3] in connection with a two level asynchronous algorithm for solving PDEs. Figure 12.1 shows this structure for $M = 2$, $N(0) = 4$, and $q_0 = 3$. The fan-out index is in this case $f_1 = 2$, a domain $D_{1,k}$ at level 1 has to communicate with two domains $D_{2,j}$ and $D_{2,j+1}$ at level 2.

We discretize the linear PDE on each level, using $nx(1)$ x-points on level 1 and $nx(p+1) = q(p)nx(p)$ x-points on level $p+1$. The y discretization is the same. An iteration method is then used to solve the resulting linear system on each level. We anticipate a parallel implementation of the iteration with $N(p)$ processors assigned to level p. We see that there are two nested structures here. First is that of the domains, each of which is to be assigned to a single processor. Second is that of the grids which determined the sizes of the linear systems for each domain (processor) and the error in discretizing the PDE. These structures may be visualized by the following parameters:

Level	Number of Domains	Domain Refinement Factor	Number of x-grid Points	Grid Size Ratio
1	$N(1)$	$R(1)$	$nx(1)$	$q(1)$
$\vdots$	$\vdots$	$\vdots$	$\vdots$	$\vdots$
p	$N(p)$	$R(p)$	$nx(p)$	$q(p)$

We have

$$N(p+1) = R(p)N(p), \quad nx(p+1) = q(p)nx(p).$$

For simplicity, we ignore for now the fact that the domains and grid sizes might not mesh nicely on the various levels.

In [4], we present a rationale for this multi-level iteration in the case of two levels. We note here merely its principal properties, namely

[1] The iterations on different levels operate asynchronously.

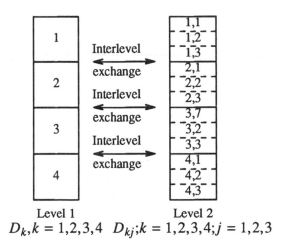

Level 1   Level 2

$$D_k, k = 1,2,3,4 \quad D_{kj}; k = 1,2,3,4; j = 1,2,3$$

Figure 12.1 A two level partition of D with $N(1)$ and $N(2)$ domains respectively. The algorithm has information passed asynchronously to and from level 2 along the boundaries of the domains of level 1.

[2] The iterations within levels are synchronized among the domains (processors) on that level.

[3] Information may be passed between levels along the domain boundaries.

[4] These are formulas to transfer information between different grid sizes (interpolate when going to a finer grid, smooth when going to a coarser grid).

The structure of this computation includes multi-level algorithms such as multigrid or nested iteration. It is practical even to have different "types" of iterations on different levels. At this point, we do not differentiate between all these possibilities.

We have analyzed this computation with models that assume rates of convergence for the linear system which range from those of Gauss-Seidel (slow) to SOR with optimum parameter (fast) to multigrid (very fast). We have also analyzed the effect of different discretizations, namely, second order and fourth order. See Sec. 4 for more details of the performance model and analysis.

3   A Unified Model of Parallel Computation

In [4], we have specialized the model of [1] to the case of a two level asynchronous iteration. We recall here the basic form of this model and extend it to multiple levels. We view a parallel computation as a sequence of w synchronization epochs each using I_i processors, and we express the expected execution time of a parallel computation as

$$T^* = E(T_c) = \alpha + w\beta + \sum_{i=1}^{w} \mu_i(1 + \Delta_i).$$

In this expression, α represents the cost due to control of the parallel computation. The communication costs are denoted by β. Note that β does reflect only the cost of communication which cannot be overlapped with computations. The final term is the execution time of the I_i processors active in any epoch i, whose execution times are independent, identically distributed random variables with mean μ_i, and variance σ_i. Now the average cost attributed to the load imbalance in epoch i is $\Delta_i = \Delta(C_i, Ii)$, with $C_i = \sigma_i/\mu_i$.

4 Model of Multi-Level Iteration On a Hypercube Machine

We now model the multi-level iteration algorithm and its implementation on a hypercube. The model is constructed in two phases. First, we study the algorithm and compare it with a standard, one level algorithm. We define the *algorithmic speed up* as the ratio of the execution time of the one level algorithm and the execution time of the multi-level algorithm on the same abstract machine.

Then we model the actual implementation of the algorithm with M levels on a parallel machine with a hypercube architecture, the NCUBE. We define the *actual speed up* as the ratio of the execution time on the NCUBE model when level M runs alone as compared with the execution time when all levels execute concurrently and lower levels feed information to the higher ones as prescribed by our algorithm.

For the algorithmic model, let us use the following notations:

S – algorithmic speed up,
m_B – number of iterations required by a one level PDE algorithm (using level M),
I_B – execution time per iteration for the one level algorithm on the hypothetical machine,
m_A – number of iterations required by the multiple level PDE algorithm,
I_A – execution time per iteration for the multiple level algorithm on the same hypothetical machine.

Then, the *algorithmic speed up* is

$$S = \frac{m_B I_B}{m_A I_A}.$$

As a first approximation, $I_A = I_B$ and S becomes m_B/m_B, the ratio of the number of iterations required.

The second model involves a more detailed account of computation, communication and synchronization at each level. To describe the model, the following notations are used to describe the machines and the computational structure.

t_a – time for arithmetic operations per iteration at a single
     grid point, a typical value for t_a is 5,
f_{sy} – synchronization factor (effect of synchronizing
     within levels),
t_c – communication cost per unit of data, and typical value is
     $t_c = 2$,
t_s – start-up time for a communication act, a typical value
     for t_s is 200,
N – total number of processors,
$N(p)$ – number of processors assigned to level p,
$R(p)$ – domain refinement factor,
M – total number of levels.

The model of the numerical aspects of the computation includes the error behavior parameters:

$$\text{ERR}(p) = \text{Discretization error on level } p$$
$$= [nx(0) + nx(p)]^{-\text{order}}$$

where

order $=$ order of the discretization error (usually 2, 3 or 4),
$nx(0) =$ number of x-points before any convergence toward
      the PDE solution starts,
$nx(p) =$ number of x-grid points at level p,
$q(p)$ $=$ grid refinement factor $nx(p+1) = q(p)nx(p)$.

We assume an initial guess for the PDE which has an error equal to 1.0. This model also includes the iteration behavior parameters

m_B $=$ iteration count required by the one level iteration,
$m(p,j) =$ count required by the iteration on level p in
       Phase j ($= 1$ or 2) of the iteration,
$r(p,j)$ $=$ convergence rate of the iteration on level p in
       Phase j of the iteration.

The iteration count m, convergence ratio r and discretization error ERR, are related by

$$r^m = \text{ERR}$$

The phases of the iteration on level p are defined as follows:

Phase 1: The error in the linear systems on level p exceeds the discretization error on level $p - 1$. The group of subdomains from a single level $p - 1$ domain are treated independently using boundary values transferred down from level $p - 1$. Thus $R(p - 1)$ subdomains are a unit.

Phase 2: The error in the linear systems on level p is less than the discretization error on level $p - 1$. All the subdomains on level p are then treated as a unit.

The convergence rates are assumed to depend on the number of x-points in the subdomains treated as a unit and we model this behavior by

$$r = 1 - 1/(nx)^k,$$

where nx is the number of x-points and k is a parameter of the iteration ($k = 2$ corresponds to Gauss-Seidel iteration, $k = 1$ to SOR and "faster" iterations may be approximately modeled by fractional values). The convergence rate formulas that result are

$$r(p, 1) = 1 - 1/(nx(p)/N(p))^k,$$
$$r(p, 2) = 1 - 1/nx(p)^k.$$

The number $m(p, 1)$ of iterations for Phase 1 is clearly more than (a) the total $m(p - 1, 1) + m(p - 1, 2)$ for level $p - 1$ adjusted by the factor $R(p - 1)/q(p - 1)$ relating (approximately) the time per iteration on each level. It is also plausible that $m(p, 1)$ is more than (b) the number required to reduce the error from the initial guess 1.0 to the discretization error level $ERR(p - 1)$. It is also clear that $m(p, 1)$ is less than the sum of these two numbers. We assume that $m(p, 1)$ is equal to the maximum of the two. Normally the latter is the larger but this is not always the case. Thus we have

$$m(p, 1) = \max[i_a, i_b],$$

where

$$i_a = [m(p - 1, 1) + m(p - 1, 2)]R(p - 1)/q(p - 1),$$
$$r(p, 1)^{i_b} = [nx(0) + nx(p - 1)]^{-\text{order}}.$$

We can now relate the iteration counts to the grids as follows:

$$r(p, 2)^{m(p,2)} = [(nx(0) + nx(p))/(nx(0) + nx(p - 1))]^{-\text{order}},$$

and obtain

$$i_b = \frac{-(\text{order})\log[nx(0) + nx(p - 1)]}{k\log[1 - N(p)/nx(p)]},$$
$$m(p, 1) = \max[i_a, i_b],$$
$$m(p, 2) = \frac{-(\text{order})\log[(nx(0) + nx(p))/(nx(0) + nx(p - 1))]}{k\log[1 - 1/nx(p)]}.$$

The same analysis leads to a simpler formula for m_B, namely,

$$m_B = -(\text{order})\log[nx(0) + nx(M)]/(k\log[1 - 1/nx(M)]).$$

The time T_B for the execution of the one level algorithm is the sum of T_B^a (the arithmetic time) and T_B^c (the communication time). Thus we have

$$
\begin{aligned}
T_B &= T_B^a + T_B^C \\
&= m_B \left[\frac{nx(M)^2}{N} \cdot t_a \cdot f_{sy} \right] \\
&\quad + m_B \left[(nx(M) + \log_2 N) \cdot 2t_c + (1 + \log_2 N) \cdot t_s \right].
\end{aligned}
$$

The time T_A for the execution of the multiple level algorithm is also the sum of T_A^a (the arithmetic time) and T_A^c (the communication time). Thus we have

$$T_A = T_A^a + T_A^c$$

$$= [m(M,1) + m(M,2)] \left[\frac{nx(M)^2}{N(M)} \cdot t_a \cdot f_{sy} \right]$$

$$+ [m(M,1) + m(M,2)][(nx(M) + \log_2 N(M)) \cdot 2t_c + (1 + \log_2 N(M)) \cdot t_s]$$

The value for $m(M,1)$, of course, depends on all the previous $m(p,j)$ values and these must be computed sequentially starting with $m(1,1) = 0$ and $m(1,2)$.

The *actual speedup* of the algorithm is defined as T_B/T_A.

5   Mapping Multi-Level Structures Onto a Hypercube

There are two broad classes of MIMD architectures, the shared memory, and the ensemble architecture. The latter consist of a large number of identical processors interconnected either by a network with fixed topology, or by a switch. In this type of architecture, all communication is done by message passing, there is no shared memory or global synchronization. In the following, we consider only ensemble architectures and interconnection networks with a fixed topology. The network topology is characterized by an interconnection graph $G' = (V', E')$ whose nodes represent the processors of the parallel machine and the edges correspond to the communication links. Examples of such machines are tree machines [8], different types of hypercubes [6, 7], and so on.

To use efficiently an ensemble architecture with a fixed interconnection topology, it is necessary to map a computation with a given topology to the machine architecture in an optimal way. Since communication and synchronization is done by message passing and communication is quite expensive, an optimal mapping should attempt to minimize the communication costs.

In this section we discuss first the issue of optimal mappings to ensemble architectures with a fixed topology and then we investigate mapping of multi-level structures onto a hypercube.

Definition 2 (*Embedding of a graph*): An embedding of a graph $G = (V, E)$ in a graph $G' = (V', E')$ is a one-to-one function $f : V \to V'$.

Definition 3 (*Expansion cost of an embedding*): The *expansion cost* of an embedding f is the ratio of the number of nodes in V' to the number of nodes in V.

Definition 4 (*Cost of an embedding*): The *cost* of an embedding f is the largest distance in G' between images of neighboring nodes in G. With $d(X, Y)$ the distance between X and Y, we have

$$C_f = \max_{(A,B) \in E} d(f(A), f(B))$$

Definition 5 (*Optimal mapping*): Given a computation with the computational graph G, an optimal mapping to a parallel machine with an interconnection graph G' is achieved by an embedding of G into G' with an expansion cost of one and with a cost of one.

Optimal mappings into hypercube architectures have been shown to exist for several computational graphs like: linear structures, multi-dimensional grids [6], as well as for some trees [8]. But it seems reasonable to expert that optimal mappings do not always exist. In the following, we discuss mapping with an expansion cost of one, but with costs larger than one. Such mappings will be called *non-optimal mappings*. The case of multi-level structures to be discussed later, provides such an example. In such a case, a sub-optimal mapping must be performed. To compare sub-optimal mappings, the concept of an effective cost of a mapping is defined.

Definition 6: (*Effective cost of a non-optimal mapping*) The effective cost of a non-optimal mapping of the computational graph G into a parallel machine with the interconnection graph G', is given by the following function of the embedding f of G into G'

$$C_f^{ef} = \sum_{\text{all}(A,B) \in E} (d(f(A), f(B)) - 1) \cdot i_{AB},$$

with $i_{A,B}$ = the intensity of traffic between the nodes A and B.

Note that the effective cost of a non-optimal mapping is for the function to be minimized by a sub-optimal mapping strategy.

Definition 7 (*Tearing*): Tearing an n-cube along the n-th direction is the process of separating the n-cube into two disjoint $(n-1)$ cubes, one obtained by considering all nodes whose n-th but is zero and the other one with all nodes whose n-th but is one. The two cubes will be denoted as $[n-1]^0$-cube and $[n-1]^1$-cube.

Proposition 1: *There exists a unique renumbering of the nodes of an n-cube such that any path of length 2^{n-1} is transformed into a path connecting all nodes of the $[n-1]^0$-cube.*

Proof: Consider a path of length 2^{n-1} connecting the nodes

$$(A_0, A_1, \ldots, A_i, \ldots, A_{2^{n-1}-1}).$$

Let us select the renumbering scheme that maps every node on the path given to a node of the $[n-1]^0$-cube as follows. A node with label A_i is mapped to a node with label B_i such that

$$B_i = G_n(i), \quad \text{for } 0 \le i \le 2^{n-1} - 1,$$

with $G_n(i)$ the Gray code of order n corresponding to integer i. Note that

$$G_n(i) < 2^{n-1}, \quad \text{if } i < 2^{n-1}.$$

This property of Gray nodes results immediately from the recursion defining the Gray codes. It follows that all B_i nodes are located on the $[n-1]^0$-cube. ∎

In order to complete the renumbering scheme, identify every node $\bar{B}_i = (1, b_{n-2}, \ldots, b_0)$ as the opposite of the node $B_i = (0, b_{n-2}, \ldots, b_0)$ and label it accordingly.

Definition 8 (*Opposite Paths*): Two paths of length ℓ, $\ell \le 2^{n-1}$ on an n-cube, $\pi^0 = (A_0, A_1, \ldots, A_{\ell-1})$, $\pi^1 = (B_0, B_1, \ldots, B_{\ell-1})$ are called opposite paths if and only if:

(a) All nodes A_i, $0 \le i \le \ell - 1$ are located on the $[n-1]^0$-cube and all nodes B_j, $0 \le j \le \ell - 1$ are located on the $[n-1]^1$-cube.

(b) The two paths contain pairs of opposite nodes, traversed in reverse order

$$B_j = A_{\ell-j}, \qquad 0 \le j \le \ell - 1.$$

Proposition 2: *A path of length 2^n on an n-cube can be decomposed into two opposite paths of length 2^{n-1}, one covering all nodes of the $[n-1]^0$-cube and the other covering all the nodes of the $[n-1]^1$-cube.*

Proof: A path of length 2^n on an n-cube can be constructed using Gray codes of order n, [3]. Note that there is a one-to-one correspondence between an integer $i < 2^n$ and its Gray code of order n, $G_n(i)$. The Gray codes of order n are generated recursively as

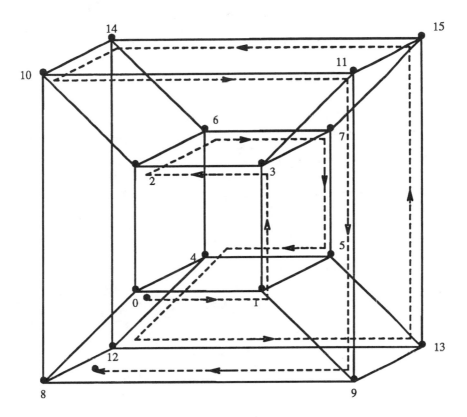

Figure 12.2 A hypercube of dimension 4 (16 nodes) showing the opposite paths of length 8. One is through nodes 0 to 7 (not in that order) and the other through nodes 8 to 15.

$$G_n = \{0G_{n-1}, 1G_{n-1}^R\}$$

with G_i^R the reverse of the i-tuple G_i. If $G_i = (g_{i-1}, g_{i-2}, \ldots, g_0)$ then
$G_i^R = (g_0, \ldots, g_{i-2}, g_{i-r})$.

From the definition of the Gray codes, it follows that

$$G_n(i) < 2^{n-1} \text{ if } i < 2^{n-1}$$

Hence the cycle determined by the Gray codes of order n covers first all the nodes
on the $[n-1]^0$-cube, then it covers the nodes of the $[n-1]^1$-cube. ∎

If we consider that the cycle starts at the node labeled 0, the inner path (on the
$[n-1]^0$-cube) runs to the node with label 2^{n-2} [since $G_n(2^{n-1}-1) = 2^{n-2}$] then
the outer path runs on the $[n-1]^1$-cube starting at the node with label $3 \cdot 2^{n-2}$
[since $G_n(2^{n-1}) = 2^{n-2} + 2^{n-1}$] and terminates at node $2^n - 1$.

Figure 12.2 illustrates this for $n = 4$. The inner path runs from the node labeled
0 to node 4, and the outer path runs from node 12 to node 8.

The mapping problem is to assign the subdomains of the multi-level structure to
the processors of the hypercube. The function f does this by assigning a subdomain
D_{ij} to a hypercube node A_k. Note that the indices i, j and k have the following
ranges

$$1 \leq i \leq M,$$
$$1 \leq j \leq N(i),$$
$$0 \leq k \leq 2^m,$$

and we have

$$N(1)N(2) \cdots N(M) = 2^m.$$

In case of multi-level structures, the effective cost of a non-optimal mapping has
two components, one due to the intra-level traffic, and a second one due to inter-
level traffic. If we assume equal traffic intensities, namely i_0 for the inter-level traffic
and βi_0 for the intra-level traffic, then the effective cost of a particular mapping
has two components

$$C_f^{ef} = C_f^{\text{intra}} + C_f^{\text{inter}},$$

with

$$C_f^{\text{intra}} = \sum_{i=1}^{M-1} \sum_{j=1}^{N(i)-1} (d(f(D_{i,j}), f(D_{i+1,j+1})) - 1)\beta i_0,$$

$$C_f^{\text{inter}} = \sum_{i=1}^{M-1} \sum_{j=1}^{N(i)-1} \sum_{k=1}^{f_i} (d(f(D_{i,j}), f(D_{i+1,j_k})) - 1)i_0,$$

with

M — number of levels of the structure,
$N(i)$ — number of subdomains at level i,
f_i — fan-out index at level i,
j_k — index of subdomains on level $i + 1$ which communicate with subdomain $D_{i,j}$ on level i.

Note that when the fanout $f_i = 1$, then the inter-layer traffic has an effective cost given by

$$C_f^{\text{inter}} = \sum_{i=1}^{M-1} \sum_{j=1}^{N(i)-1} (d(f(D_{i,j}), f(D_{i,j+1})) - 1)i_0.$$

The "obvious" sub-optimal mapping f_0 is obtained by mapping each individual layer as linear structure. In this case,

$$C_{f_0}^{\text{inter}} = 0,$$

and

$$C_{f_0}^{ef} = C_{f_0}^{\text{inter}}.$$

Proposition 3 (*Optimal mapping of a multi-level structure*): *Given a multi-level structure with the following properties*

- *it has 2^n subdomains,*

- *it has $M = 2^r$,*

- *the fanout indexes are all equal to one,*

- *the number of subdomains at all levels are equal*

$$N(1) = \cdots = N(M) = 2^{n-r}.$$

Then there is an optimal mapping of the multi-level structure into a hypercube of order n.

Proof: Let us denote the levels of the multi-level structure as

$$L_1, \ldots, L_{2^{r-1}}, L_{2^{r-1}+1}, \ldots, L_{2^r}$$

Consider now a path of length 2^n determined by the Gray codes of order n. Split this path into 2^r consecutive paths of length 2^{n-r} identified as

$$C_1, \ldots, C_{2^{r-1}}, C_{2^{r-1}+1} \ldots, C_{2^r}.$$

Map L_i to C_i for $1 \le i \le 2^r$. From Proposition 5, it follows immediately that adjacent levels of the structure are adjacent on the hypercube. Indeed Proposition 5 can be applied recursively and on any hypercube of order 2^{r+1} the two path of length 2^r are opposite hence they consist of 2^r pairs of opposite nodes.

6 Analysis of the Iteration and its Performance

The model in Sec. 4 with the embedding of Sec. 5 is used to investigate the performance of the multi-level iteration on a hypercube machine (an abstraction of the NCUBE). We study the speed up achieved as the problem size increases and as the parameters of the model vary.

In our earlier study of two level iteration, we determined some optimal values by exhaustively examing all cases, e.g., all ways to divide the processors between two levels. This approach is not practical for more than two levels. We supplement the earlier two level results by more information on the machine parameters and the characteristics of the numerical method.

Figures 12.3(a)–12.3(d) show the actual speed up obtained for two values of NSTART $= nx(0)^2$, 400 and 2800 using four models for the numerical properties of the iteration

- Gauss-Seidel and second order PDE discretization.

- Gauss-Seidel and fourth order PDE discretization.

- SOR and second order PDE discretization.

- SOR and fourth order PDE discretization.

Thus the total number of grid points, $(nx(0) + nx(2))^2$ is varied and we plot the speed up of the two level iteration over the one level iteration. All other parameters of the computation are kept the same (e.g., machine characteristics, convergence rates). All processors assigned to level 1 are idle in the second phase of the two level iteration after transmitting data to level 2. The machine model parameters are $t_s = 200$ and $t_C = 2$ and the synchronization factor uses $C_X = 0.04$. These are typical values for this computation on an NCUBE machine.

There are two curves in Figure 12.3, one for each of NSTART $= 400$ and 2800. These overlap in the plots although the speed ups are not identical. These three curves are given:

$$\begin{array}{lll} \textit{dashed: for NSTART} & = 400. \\ \textit{dotted: for NSTART} & = 2800. \\ \textit{solid: } \quad \text{overlap of NSTART} & = 400 \text{ and } 2800. \end{array}$$

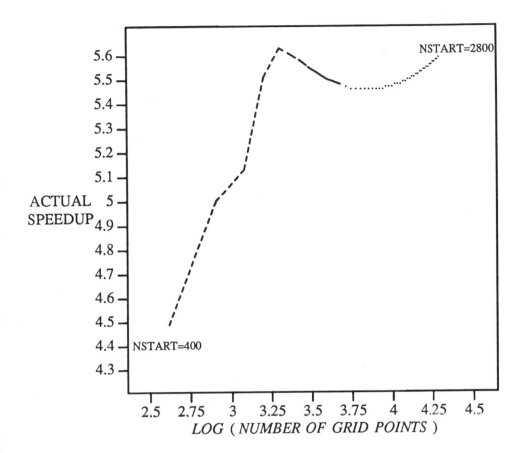

Figure 12.3(a) The *speed up* function of the logarithm of the number of grid points for two values of NSTART, Gauss-Seidel convergence rate, and second order discretization.

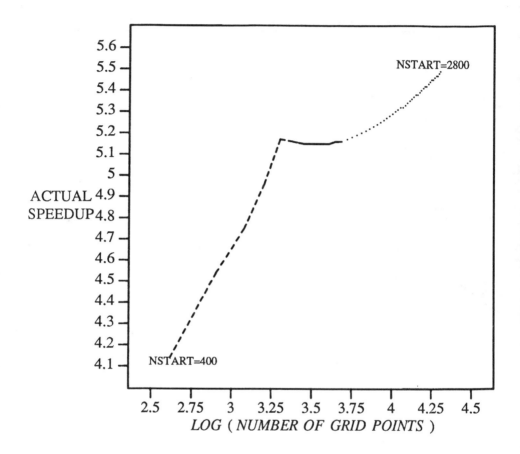

Figure 12.3(b) The *speed up* function of the logarithm of the number of grid points for two values of NSTART, Gauss-Seidel convergence rate, and fourth order PDE discretization error.

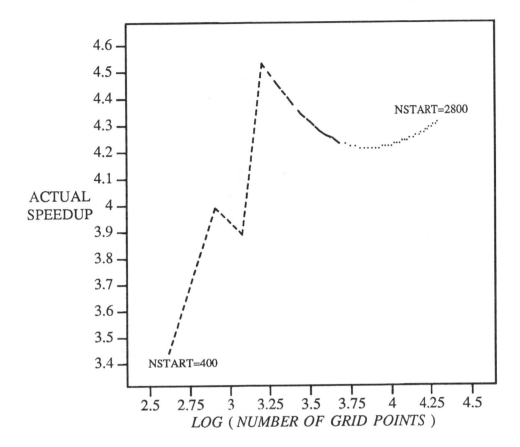

Figure 12.3(c) The *speed up* function of the logarithm of the number of grid points for two values of NSTART, SOR convergence rate, and second order PDE discretization.

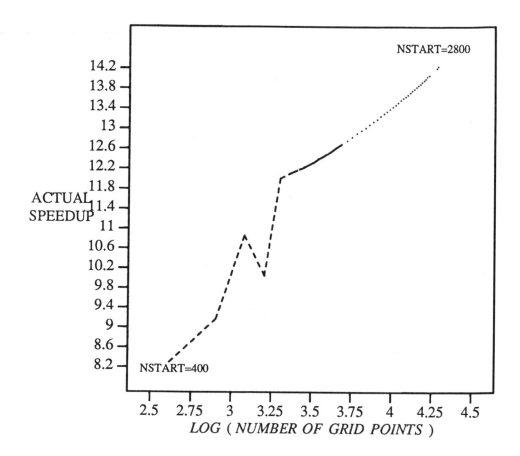

Figure 12.3(d) The *speed up* function of the logarithm of the number of grid points for two values of NSTART, SOR convergence rate, and fourth order PDE discretization.

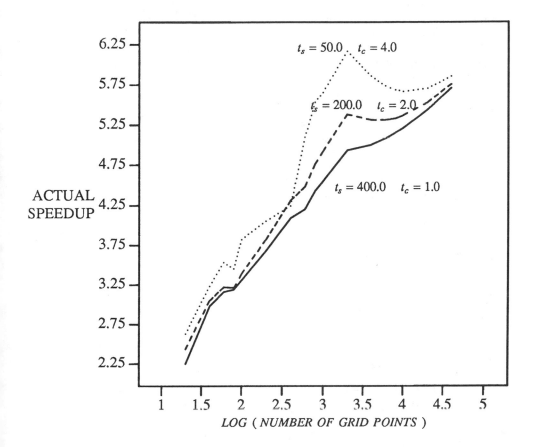

Figure 12.4 The *speed up* function of the logarithm of the number of grid points for three sets of machine parameters: t_s =communication start-up time, t_c =communication time per unit of message. The coefficient of variation of the execution time is $C_X = 0.2$.

There is considerable general similarity among the plots of Figure 12.3 which leads us to conclude that the effectiveness of the two level iteration is not heavily dependent on the properties of the numerical methods underlying the computation.

Our earlier study of the effect of the machine communication characteristics, t_s (start up time) and t_c (time per unit message) had these two parameters varying nearly proportionally. We used the (t_s, t_c) pairs (400,4), (200,2) and (50,1). Figure 12.4 shows the actual speed ups obtained with the (t_s, t_c) pairs (50,4), (200,2) and (400,1) to further explore the effect of these parameters. Comparing Figure 12.4 with Figure 8 of [4], we observe that

- Increasing the time per unit message, t_c, significantly increases the speed up obtained from the two level algorithm.

- The start up time parameter, t_s, has a minor effect on the speed up obtained from the two level algorithm.

Acknowledgements

This work was supported in part by the Strategic Defense Initiative under Army Research Office contract DAAL03-86-K-0106. This paper was originally typed by Georgia Conarroe in LATEX and reformatted in LATEXby Lisa Laguna.

References

[1] Marinescu, D.C., and J.R. Rice. "On the effects of synchronization in parallel computation," CSD-TR-750, Computer Science Department, Purdue University, March 1988.

[2] Marinescu, D.C., and J.R. Rice. "Synchronization of nonhomogeneous parallel computations," Parallel Processing for Scientific Computing, (Rodrigue, ed.), SIAM, 362–367, 1988.

[3] Marinescu, D.C., and J.R. Rice. "Domain oriented analysis of PDE splitting algorithms," Information Sciences 43 (1987) 3–24.

[4] Marinescu, D.C., and J.R. Rice. "Analysis of a two level asynchronous algorithm for PDEs," in Studies in Asynchronous Numerical Computations, (M. Wright, ed.), 23–33, North-Holland (1989).

[5] Rice, J.R., and D.C. Marinescu. "Analysis and modeling of Schwartz splitting algorithms for elliptic PDEs," in Advances in Computer Methods for Partial Differential Equations VI, (Steplemann and Vichnevetsky, eds.), IMACS, Rutgers University, 1987, 1–6.

[6] Saad, Y., and M. Schultz. "Topological Properties of Hypercubes," *IEEE Trans. on Computers* **37** (1988) 867–872. July 1988.

[7] Seitz, C.L. "The cosmic cube," *Commun.* ACM **28** (1985) 22–31.

[8] Wu, A.Y. "Embedding of tree networks into hypercubes," *J. Parallel Distributed Comput.* **2** (1985) 238-249.

Chapter 13

An Adaptive Algorithm for Richardson's Method

PAUL E. SAYLOR

University of Illinois at Urbana-Champaign

Dedicated to David M. Young, Jr., on the occasion of his sixty-fifth birthday.

Abstract

Complex linear algebraic equations arise in many scientific and engineering studies. Several well-known algorithms are applicable to complex matrices, such as SOR, the bi-conjugate gradient method, the conjugate gradient method, and Manteuffel's adaptive Chebyshev algorithm. For large problems, Richardson's method is attractive but requires a set of parameters. An early paper of David Young's presents optimum parameters as transformed roots of the Chebyshev polynomial in the symmetric positive definite case. An adaptive algorithm for optimum parameters will be described in the nonsymmetric case. The parameters are optimum in a least squares sense that includes Chebyshev parameters.

1 Introduction

In [32], David Young showed how to obtain optimal iteration parameters for Richardson's method for the solution of $Ax = b$ when A is a symmetric positive definite matrix. In several recent papers [6, 18, 4, 20, 29, 26], Young's result has been extended to general matrices A; also see [11, 15, 16, 8]. In this paper, an adaptive technique is added to one [26] of these extensions.

In Young's 1954 paper, *optimal* means that a bound on the l_2-norm of the error is minimized, and this feature is also characteristic of recent extensions, although in general only an approximation to the bound is minimized and that approximation is not necessarily the same from one algorithm to the next. More explanation will be given in Sec. 4. In each extension, however, optimal parameters depend on the eigenvalues of A. To be precise, let $\sigma(A)$ be the spectrum of A and let Ω be any region approximately containing $\sigma(A)$. There are various algorithms to compute optimal iteration parameters *if* Ω is given (and satisfies certain appropriate conditions). The adaptive algorithm given here dynamically computes Ω to be the convex hull of $\sigma(A)$.

1.1 Outline

Richardson's method is stated in Sec. 2. There are three distinct least squares problems associated with Richardson's method, and they are sketched herein. They determine an approximate convex hull, optimal parameters, and an initial approximation for Richardson's method. In Secs. 3–5, each is treated in detail. Finally, an algorithm is given in Sec. 6.

1.2 The Convex Hull of $\sigma(A)$

At this point in the introduction, a somewhat more careful discussion of optimality is called for. Suppose a residual polynomial of given degree is minimized with respect to a norm over a region Ω. (A residual polynomial is any polynomial constrained to be 1 at the origin.) Optimal parameters are defined to be the reciprocals of the roots of the residual polynomial of minimum norm. As a general rule, the more closely that Ω conforms to $\sigma(A)$, the faster Richardson's method converges. For example, when A is symmetric positive definite, Ω could be any interval containing the eigenvalues and not containing zero, but the *fastest*[1] convergence results when $\Omega = [\lambda_{\min}, \lambda_{\max}]$, where $\lambda_{\min}$ and $\lambda_{\max}$ are the smallest and largest eigenvalues of A, respectively. In this case, it is clear what is meant by Ω closely conforming to $\sigma(A)$, but in general the irregular geometric structure of $\sigma(A)$ in the complex plane means that the closest conforming Ω cannot be defined in a succinct geometric way. (Of course, it could be defined to be that region for which the rate of convergence is smallest, but this is generally impractical.) Here, even though it is not always satisfactory, the closest conforming region will be taken to be the convex hull of A.

1.3 Motivation

The special properties of an application problem naturally influence the method for its numerical solution, and many examples illustrate this. Thus, regular elliptic difference equations make the red-black ordering possible, important for the SOR method. In the case of this paper, which is motivated by the application to the linear systems arising from 3D simulations, the special property of the application is the brute size of the problem: Reasonable discretizations of a three dimensional domain

---

[1] *Fastest* means only that a residual polynomial is minimized. In many circumstances, a different approach would yield a better method. For example, given a Hermitian definite matrix with only two distinct eigenvalues, the fastest convergence is by the conjugate gradient method, which takes only two steps to obtain the exact solution.

by either the finite element or finite difference method yield leviathan systems of ten million or more unknowns. One consequence of such large vectors, taken in this paper, may impress the reader as dubious, and that is an emphasis on the normal equations to solve the least squares problems occurring in the adaptive method and in the method to determine an initial approximation. Ordinarily, one is advised to shun the normal equations, but the normal equations use the vectors of a Krylov sequence and these are computable in parallel by chaining the matrix multiplications, which is possible for example if the matrix is block tridiagonal. (See [3] and Sec. 3.1 of [19].) Another consequence of a large number of unknowns is the choice of an iterative method. If the correct conditions are satisfied [9, 1], the conjugate gradient method is applicable. This and the adaptive Chebyshev method of Manteuffel [13, 2] are discussed in the next two paragraphs. A method for the solution of non-Hermitian systems is often taken from the family of incomplete conjugate gradient-like methods. However, large vectors discourage use of these methods for non-Hermitian matrices since many vectors may be required for convergence [25]. Among various methods, Richardson's method requires the least amount of vector manipulation. Moreover, optional variants of Richardson's method exist that require fewer arithmetic operations and for which there are additional advantages in the manipulation of vectors [22].

The particular 3D simulation problem that motivated this paper is the numerical solution of Maxwell's equations, the matrices for which are complex, a feature also bearing on the choice of Richardson's method: general complex non-Hermitian matrices exclude the direct use of the adaptive Chebyshev algorithm of Manteuffel since it assumes that $\sigma(A)$ is symmetric about the real axis. Of course, by separating the real and imaginary parts, $Ax = b$ may be converted to the (usual) equivalent real system, but the convex hull of A is a subset of the convex hull of the spectrum of the equivalent real system, a condition that retards convergence. It should be noted that incomplete conjugate gradient-like methods do not require that the system matrix be real, but I/O transfers for these methods may be greater.

The appeal of the conjugate gradient method leads to further comments. A *true* conjugate gradient method may also be applicable, i.e., a method for which the error is minimized at each step with respect to an inner product induced norm and is such that at most four vectors need be stored (excluding the system matrix and, if any, preconditioning matrices) [9]. In particular, suppose that A is symmetric, but *not* Hermitian, a condition that often is satisfied in practice. If the real part of A is definite (it is symmetric if A is), then it may be used to precondition A, and the conjugate residual method applied to the result, a method stated in Elman [5]. If the imaginary part is definite, that could be used as a preconditioner for the conjugate residual method [1].

For a treatment of computational electromagnetics, the reader is referred to [23, 24, 27]. Computational examples are the determination of radar cross-sections, the design of power equipment, motors, and recording heads, and the engineering problems of induction logging [14, 30].

1.4   Conventions and Notation

It is generally recommended that a linear set $Ax = b$ be preconditioned before applying an iterative method. Preconditioning is equivalent to transforming $Ax = b$ into a system such as $CAx = Cb$ with more desirable properties such as positive definiteness or a smaller condition number. The methods to be described apply as well to the preconditioned system and it will therefore be assumed that A is the preconditioned matrix. (Preconditionings may yield equivalent systems of a different form, for example, $CAQQ^{-1}x = Cb$, but the same remarks hold for these cases as well.)

If A is singular, Richardson's method converges to a solution of $Ax = b$ provided that the system is consistent. Reference is often made to the spectrum of the matrix. If A is singular, replace "spectrum" by "the set of nonzero eigenvalues."

The number of unknowns of the linear set is denoted by N. Vectors and matrices are assumed to be general complex. In the complex case, inner products must be carefully stated to make it clear which vector is conjugated. The inner product of vectors $u = (\mu_1, \ldots, \mu_m)$ and $v = (\nu_1, \ldots, \nu_m)$ is defined to be

$$(u, v) = \sum_{i=1}^{m} \mu_i \bar{\nu}_i.$$

A norm with no subscript means the standard Euclidean norm of a complex vector and is defined by

$$\|u\|^2 = (u, u).$$

The Hermitian transpose of a matrix $M = (\mu_{ij})$ is defined to be

$$M^* = (\bar{\mu}_{ji}).$$

2   The Numerical Framework

In this section, the numerical problems associated with optimal parameters and the initial approximation are outlined.

2.1   Richardson's Method

Let $Ax = b$ be a linear algebraic system. Richardson's method [12], for $0 \leq i \leq k-1$, is defined by

$$\begin{aligned} x^{(0)} &= \text{given}, \\ r^{(i)} &= b - Ax^{(i)}, \\ x^{(i+1)} &= x^{(i)} + \tau_i r^{(i)}, \end{aligned}$$

where $\{\tau_i : i = 0, 1, \ldots, k-1\}$ is a *cycle* of *iteration parameters*. Integer k is called the *period*.

2.2 Eigenvalue Least Squares Problem

Richardson's method depends on a sequence of parameters that depend in turn on eigenvalues of the system matrix. The power method will be used to compute approximate eigenvalues, a computation requiring the solution of a least squares (LS) problem, which will be called the *eigenvalue LS problem*.

2.3 Optimal Residual Polynomial LS Problem

Iteration parameters are the reciprocals of the roots of certain polynomials, called residual polynomials. Optimal parameters are from an optimal residual polynomial, defined here to be that residual polynomial of a fixed degree that minimizes a weighted l_2-norm around a contour in the complex plane defined by an approximate convex hull of A. Minimizing a weighted l_2-norm is equivalent to solving another LS problem, the *optimal residual LS problem*.

2.4 The Minimum Residual LS Problem

At this point, a cycle of optimal parameters has been computed and Richardson's method can be executed. An initial guess for the solution is available, $x^{(0)}$, but this initial approximation was used in the computation of the eigenvalues and to generate a Krylov subspace (to be defined in Sec. 3.) The Krylov subspace and the inner products needed for the eigenvalue LS problem can next be applied to computing an approximation for which the residual is minimized over the Krylov subspace. Minimizing the residual yields a third LS problem, the *minimum residual LS problem*.

3 The Power Method for Eigenvalues

The power method for computing more than one eigenvalue is described in [31]. It is used in [13] together with Chebyshev polynomials to compute eigenvalues dynamically for the Chebyshev iteration; also see [2]. Before starting with the details, it is worth noting, in this conjugate gradient/Lanczos era, the equivalence of the familiar power method with the Arnoldi method, which is a generalization of the symmetric Lanczos method; see [13].

Let $\{v_i, \lambda_i\}$ be the eigensystem of matrix A, i.e, $Av_i = \lambda_i v_i$, and assume that the eigenvalues are arranged in the order $|\lambda_i| \geq |\lambda_{i+1}|$. Also assume that the eigenvectors span the space. Let s be some *seed* vector and let

$$s = \sum_{i=1}^{N} \alpha_i v_i$$

be the expansion of s in terms of the eigenvectors of A. The initial residual, $r^{(0)}$, is the seed vector used in a later section. The powers of A applied to s yield the

Krylov subspace

$$V_{m+1} = \text{span}\{s, As, \ldots, A^m s\}.$$

Vectors of the form $A^i s$ will be called *Krylov vectors*.

3.1   A Linear Combination of Krylov Vectors

Approximate eigenvalues of A are computed from a linear combination of the Krylov vectors. To begin, let

$$s_i = A^i s$$

and consider the expansion of s_m,

$$s_m = \alpha_1 \lambda_1^m v_1 + \alpha_2 \lambda_2^m v_2 + \cdots.$$

Let $j \leq m$. In order to compute an estimate of the first j eigenvalues $\lambda_1, \lambda_2, \ldots, \lambda_j$, assume that $\lambda_j > \lambda_{j+1}$. A linear combination of $s_{m-j}, \ldots, s_m$ will yield a polynomial the roots of which are estimates of these eigenvalues. To see this, let

$$\pi_j(\lambda) = \prod_{i=1}^{j} (\lambda - \lambda_i) = \tilde{c}_0 + \cdots + \tilde{c}_{j-1}\lambda^{j-1} + \lambda^j.$$

It follows that

$$\begin{aligned}
\tilde{c}_0 s_{m-j} &+ \cdots + \tilde{c}_{j-1} s_{m-1} + s_m \\
&= \alpha_1 \lambda_1^{m-j} \left(\pi_j(\lambda_1) \right) v_1 + \cdots \\
&\quad + \alpha_{j+1} \lambda_{j+1}^{m-j} \left(\pi_j(\lambda_{j+1}) \right) v_{j+1} + \cdots + \alpha_N \lambda_N^{m-j} \left(\pi_j(\lambda_N) \right) v_N \\
&= \alpha_{j+1} \lambda_{j+1}^{m-j} \left(\pi_j(\lambda_{j+1}) \right) v_{j+1} + \cdots + \alpha_N \lambda_N^{m-j} \left(\pi_j(\lambda_N) \right) v_N.
\end{aligned}$$

The Euclidean norm of the term on the right is negligible compared to λ_1^m, i.e.,

$$(3.1) \qquad \frac{\| \tilde{c}_0 s_{m-j} + \cdots + \tilde{c}_{j-1} s_{m-1} + s_m \|}{\lambda_1^m} \approx 0.$$

3.2   The Eigenvalue LS Problem

A set of approximate coefficients, $\{c_j\}$, is determined by the condition that, from (3.1),

$$(3.2) \qquad \| c_0 s_{m-j} + \cdots + c_{j-1} s_{m-1} + s_m \| = \text{minimum}.$$

In this subsection, a different version of (3.2) is described, and a matrix formulation is given.

Biased Least Squares Formulation

Problem (3.2) is a *biased* least squares problem in the sense that the coefficient of s_m is 1. An *unbiased* version of the least squares problem would be to determine a different set of coefficients $\{d_i : i = 0, 1, \ldots, j\}$ such that

$$\|d_0 s_{m-j} + \cdots + d_{j-1} s_{m-1} + d_j s_m\| = \text{minimum}$$

subject to the unbiased constraint

$$d_0^2 + \cdots + d_j^2 = 1.$$

The two versions yield distinct solutions. For the normal case, the unbiased eigenvalue approximation is superior, but in the non-normal case analysis [21] (in response to an observation communicated by Manteuffel) suggests the unbiased and biased versions are equivalent. (Manteuffel also pointed out that the estimates from the biased version lie in the field of values whereas those from the unbiased do not necessarily lie in the field of values.) Since the biased version is easier to solve, the unbiased version will not be treated further.

Matrix Vector Form

It is helpful to write the least squares problem in matrix notation. Let S be the matrix the columns of which are $s_{m-j}, \cdots, s_{m-1}$

$$S = [s_{m-j} \cdots s_{m-1}]$$

and let

$$c = [\bar{c}_0, \cdots, \bar{c}_{j-1}]^*.$$

The coefficient vector c is determined by the condition that

$$\|Sc + s_m\| = \text{minimum}.$$

This least squares problem will be called the *eigenvalue least squares (LS) problem.*

3.3   Solution of the Eigenvalue LS Problem

It is well known that the solution of the above least squares problem is mathematically equivalent to the solution of the *(Gaussian) normal equations,*

$$S^* S c = -S^* s_m.$$

For $j = m = 4$ the normal equations are

$$\begin{bmatrix} s_0^* \\ s_1^* \\ s_2^* \\ s_3^* \end{bmatrix} [s_0 s_1 s_2 s_3] \begin{bmatrix} c_0 \\ c_1 \\ c_2 \\ c_3 \end{bmatrix} =$$

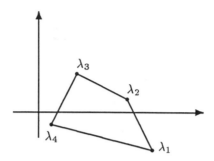

Figure 13.1 The convex hull of four eigenvalue estimates

$$
\begin{bmatrix}
\|s_0\|^2 & (s_1, s_0) & (s_2, s_0) & (s_3, s_0) \\
 & \|s_1\|^2 & (s_2, s_1) & (s_3, s_1) \\
 & & \|s_2\|^2 & (s_3, s_2) \\
 & & & \|s_3\|^2
\end{bmatrix}
\begin{bmatrix}
c_0 \\ c_1 \\ c_2 \\ c_3
\end{bmatrix}
= -
\begin{bmatrix}
(s_4, s_0) \\ (s_4, s_1) \\ (s_4, s_2) \\ (s_4, s_3)
\end{bmatrix}
$$

in which the lower half is omitted, since the matrix is symmetric.

Generally other methods for solving a least squares problem would be recommended in place of the normal equations. The obvious advantage of the normal equations approach is simplicity: vectors $s_i = A^i r^{(0)}$ can be computed in parallel if, for example, A is block tridiagonal [3, 19]. On the other hand, orthogonalizing the Krylov subspace such as by the Gram-Schmidt or modified Gram-Schmidt method breaks the matrix-vector rhythm. The normal equations are notoriously unstable, but in this application interest is in the dominant eigenvalues only and these are computed well; it is this version of the normal equations used in the original implementation of the Manteuffel algorithm, where results were satisfactory for $j = m = 4$. (A clarification is needed on dominant eigenvalues. In refining the convex hull, subdominant eigenvalues are computed. When computed, however, these will be dominant in s_0.)

4   Finding the Optimal Richardson Parameters

Discussion of numerical methods began with the power method to determine a set of j eigenvalues which will be taken to be an approximation of the convex hull of the spectrum of A.

The boundary of the convex hull is a polygon γ (which may reduce to an interval on the real axis). With basic information now available about the spectrum, the next step is to determine a set of optimal iteration parameters.

4.1   Residual Polynomials

Define $e^{(k)} = x - x^{(k)}$ and $r^{(k)} = b - Ax^{(k)}$ to be the *error* and the *residual error*, respectively. It is easy to show that

$$e^{(k)} = R_k(A)e^{(0)}$$

and

$$r^{(k)} = R_k(A)r^{(0)}$$

where R_k is the polynomial

$$R_k(\lambda) = \prod_{i=1}^{k}(1 - \tau_{i-1}\lambda).$$

Any polynomial with the defining property $R_k(0) = 1$, is called a *residual polynomial* [28]. Parameters are chosen to minimize $R_k(A)$, which will now be taken up.

4.2   Inner Products, Norms, and Optimal Residual Polynomials

Some notation is convenient. Define

$$(f,g)_w = \frac{1}{L}\int_\gamma f(\zeta)\overline{g(\zeta)}w(\zeta)d|\zeta|$$

where w is a positive weight function,

$$L = \int_\gamma w(\zeta)d|\zeta|,$$

and γ will be an interval or a union of intervals in the complex plane, usually the boundary of a polygon, for example, the boundary of the convex hull of the spectrum. (Of course, γ could be more general, but this assumption is adequate for computing.) Thus, if $w(\zeta) \equiv 1$, L is the length of the arc. In practice, rather than the continuous inner product, one would use a discrete inner product of the form

$$(f,g)_w = \frac{1}{M_w}\sum_{i=1}^{M} f(\zeta_i)\overline{g(\zeta_i)}w(\zeta_i)m(\zeta_i)$$

where $m(\zeta_i)$ is a positive measure, such as $m(\zeta_i) = |\zeta_{i-1/2} - \zeta_{i+1/2}|$, for example, (the midpoint rule for numerical integration) and where

$$M_w = \sum_{i=1}^{M} w(\zeta_i)m(\zeta_i).$$

A norm is defined by

$$\|f\|_w^2 = (f, f)_w.$$

An optimal residual polynomial of degree k is defined to be that residual polynomial, R_k, that solves the weighted least squares problem defined by

$$\|R_k\|_w^2 \leq \|P_k\|_w^2$$

where P_k is any residual polynomial of degree k.

4.3   Solving the Optimal Residual Polynomial LS Problem

Suppose that the weighted norm is the discrete norm,

$$\|R_k\|_w^2 = \frac{1}{M_w} \sum_{i=1}^{M} |R_k(\zeta_i)|^2 w(\zeta_i) m(\zeta_i).$$

Let

$$R_k(\zeta) = 1 + \sum_{i=1}^{k} \rho_i \zeta^i.$$

Note that $R_k(0) = 1$ (the definition of a residual polynomial). The discrete least squares problem is to minimize

$$M_w \|R_k\|_w^2 = \|F(\bar{\rho}_1, \ldots, \bar{\rho}_k)^* - y\|^2,$$

where $F = (f_{ij})$ is the $M \times k$ matrix defined by

$$f_{ij} = \zeta_i^j \, [w(\zeta_i) m(\zeta_i)]^{\frac{1}{2}}$$

and

$$(4.1) \qquad y = \left(-[w(\zeta_1) m(\zeta_1)]^{\frac{1}{2}}, \ldots, -[w(\zeta_M) m(\zeta_M)]^{\frac{1}{2}} \right)^*.$$

(Since the components are real, the conjugation symbol is omitted.) It is assumed that there are more node points, $\{\zeta_i\}$, than the number of coefficients, and so the LS problem is overdetermined. Thus there might be 100 node points defining the convex hull of the spectrum of A, and we might want only a fourth degree polynomial, in which case there would be four unknown coefficients, and 100 rows of F.

The problem may be solved by standard LS methods such as computing the QR-decomposition of F. The QR-decomposition is recommended because it yields the minimum value of

$$(4.2) \qquad M_w \|R_k\|_w^2 = \|F(\bar{\rho}_1, \ldots, \bar{\rho}_k)^* - y\|^2,$$

which is useful in deciding whether to update eigenvalue estimates. Thus, to explain, suppose that $F = QR$ is the QR-decomposition of F, where $Q = Q_{M \times M}$ is orthogonal, and $R = R_{M \times k}$ is upper triangular. Let $Q^H y = (\bar{\eta}_1, \ldots, \bar{\eta}_M)^*$ where y is defined by (4.1). The minimum value of (4.2) is then

$$(4.3) \qquad \text{minimum} = \sum_{j=k+1}^{M} |\eta_i^2|.$$

It should be stressed that the optimal residual polynomial LS problem is small compared to the eigenvalue LS problem. The number of unknowns in the linear system may be large, for example, $N = 10^7$, whereas the number of node points is on the order of $M = 100$. *The optimal polynomial LS problem may be assumed to be a cost-free computation.*

4.4   The Optimal Residual Polynomial

The coefficients of the optimal residual polynomial are now determined, and the roots of the polynomial may be computed from a standard polynomial rootfinder, for example, one based on computing the eigenvalues of the companion matrix. The reciprocals of the roots are the Richardson method parameters.

5   The Minimum Residual Method

Preparations for the iterative method are nearly complete. Richardson's method could now begin with the initial guess $x^{(0)}$, but a better choice is possible by using a linear combination of the Krylov vectors, chosen so as to minimize the residual. This yields a third LS problem, the *minimum residual LS problem*. Obtaining an approximate solution by minimizing the residual over a Krylov subspace generated by $s = r^{(m)}, m \neq 0$, is called the *generalized minimum residual method* (GMRES) [25], and if $m = 0$ the *minimum residual method* (MR). When MR is used to obtain an approximation for the Chebyshev iteration, it is the hybrid method presented in [7].

5.1   The Minimum Residual Krylov Subspace

The seed vector is $s = r^{(0)}$, where $r^{(0)} = b - Ax^{(0)}$ is the initial residual. The Krylov subspace for generating eigenvalue estimates is therefore

$$V_{m+1} = \{r^{(0)}, Ar^{(0)}, \ldots, A^m r^{(0)}\}.$$

Recall that $s_i = A^i r^{(0)}$.

5.2   The Minimum Residual LS Problem

The initial guess, $x^{(0)}$, will be used in combination with the first $m-1$ vectors of the Krylov subspace to yield an approximation

$$(5.1) \qquad x^{(m)} = x^{(0)} + \sum_{i=0}^{m-1} \alpha_i s_i$$

for which

$$(5.2) \qquad \|A(x - x^{(m)})\| = \text{minimum.}$$

Solving (5.2) will be called the minimum residual LS problem. Note that

$$
\begin{aligned}
A(x - x^{(m)}) &= A\left(x - (x^{(0)} + \alpha_0 s_0 + \cdots + \alpha_{m-1} s_{m-1})\right) \\
&= r^{(0)} - (\alpha_0 s_1 + \cdots + \alpha_{m-1} s_m) \\
&= s_0 - (\alpha_0 s_1 + \cdots + \alpha_{m-1} s_m).
\end{aligned}
$$

5.3   Complementary LS Problems

The problem now derived resembles the eigenvalue LS problem. The eigenvalue LS problem is (with $j = m$, where j is the number of eigenvalues to be computed)

$$(\text{EGVLLS}) \qquad \|c_0 s_0 + \cdots + c_{m-1} s_{m-1} + s_m\| = \text{minimum}$$

whereas the minimum residual LS problem is

$$(\text{MRLS}) \qquad \|s_0 - (\alpha_0 s_1 + \cdots + \alpha_{m-1} s_m)\| = \text{minimum.}$$

5.4   Matrix Form

For the eigenvalue LS problem,

$$(\text{EGVLLS}) \qquad S = [s_0 \ldots s_{m-1}]$$

(S was defined in Sec. 3.3) whereas for the minimum residual problem, define

$$(\text{MRLS}) \qquad S_{MR} = [s_1 \ldots s_m].$$

For the eigenvalue LS problem,

$$(\text{EGVLLS}) \qquad \|Sc + s_m\| = \text{minimum}$$

and for the minimum residual LS problem,

$$(\text{MRLS}) \qquad \|S_{MR}a - s_0\| = \text{minimum}$$

where a is defined to be

$$a = (\bar{\alpha}_0, \ \bar{\alpha}_1, \dots, \bar{\alpha}_{m-1})^*.$$

For the case $j = m = 4$ the eigenvalue normal equations are

$$\begin{bmatrix} s_0^* \\ s_1^* \\ s_2^* \\ s_3^* \end{bmatrix} [s_0 s_1 s_2 s_3] \begin{bmatrix} c_0 \\ c_1 \\ c_2 \\ c_3 \end{bmatrix} =$$

(EGVLLS)
$$\begin{bmatrix} \|s_0\|^2 & (s_1, s_0) & (s_2, s_0) & (s_3, s_0) \\ & \|s_1\|^2 & (s_2, s_1) & (s_3, s_1) \\ & & \|s_2\|^2 & (s_3, s_2) \\ & & & \|s_3\|^2 \end{bmatrix} \begin{bmatrix} c_0 \\ c_1 \\ c_2 \\ c_3 \end{bmatrix} = - \begin{bmatrix} (s_4, s_0) \\ (s_4, s_1) \\ (s_4, s_2) \\ (s_4, s_3) \end{bmatrix}$$

and the minimum residual normal equations are

$$\begin{bmatrix} s_1^* \\ s_2^* \\ s_3^* \\ s_4^* \end{bmatrix} [s_1 s_2 s_3 s_4] \begin{bmatrix} \alpha_0 \\ \alpha_1 \\ \alpha_2 \\ \alpha_3 \end{bmatrix} =$$

(MRLS)
$$\begin{bmatrix} \|s_1\|^2 & (s_2, s_1) & (s_3, s_1) & (s_4, s_1) \\ & \|s_2\|^2 & (s_3, s_2) & (s_4, s_2) \\ & & \|s_3\|^2 & (s_4, s_3) \\ & & & \|s_4\|^2 \end{bmatrix} \begin{bmatrix} \alpha_0 \\ \alpha_1 \\ \alpha_2 \\ \alpha_3 \end{bmatrix} = - \begin{bmatrix} (s_0, s_1) \\ (s_0, s_2) \\ (s_0, s_3) \\ (s_0, s_4) \end{bmatrix}.$$

Note that the minimum residual LS problem requires computing $\|s_4\|$. Otherwise the work for the minimum residual LS problem is the same as for the eigenvalue LS problem.

6 Algorithm

In this section, an algorithm is assembled. Before a statement can be given, however, there are certain features that require preliminary attention.

6.1   The Convex Hull

Computing the convex hull of a set of eigenvalue approximations is a negligible cost. For an algorithm, the reader is referred to [17]. The vertices of the convex hull are the eigenvalue approximations, and are connected by edges. Although it is visualized as a polygonal figure in the complex plane, the convex hull may be as simple as a line segment, determined by only two eigenvalue estimates. A difficulty arises if the convex hull contains the origin, which occurs for example with a Hermitian indefinite matrix: the convex hull of the spectrum is an interval including zero, whereas contour γ should be the union of positive and negative intervals determined by the positive and negative eigenvalues. The convex hull is not a satisfactory approximation of the spectrum near the origin. Also see Sec. 3.3 of [11] in which the authors observe that it may be more appropriate to solve $A^*Ax = A^*b$.

The reader is warned that the Hermitian indefinite case is *terra incognito*, and that the appearance of solving this case by the adaptive algorithm given below is only a shaky pretense.

6.2   Ordering the Parameters

In [32], Young stated that "... the growth of roundoff errors may be inhibited by using the $[\tau_i's]$ in a certain order in each cycle." and then suggests an ordering for the Hermitian positive definite case.

If the parameters are obtained by a conformal mapping, stable orderings have been suggested independently by [10] and S. Talezer (private communication, February, 1987). These techniques do not apply directly to least squares parameters. However, the so-called grand-leap formulation [22] of Richardson's method computes $x^{(k)}$ as $x^{(k)} = x^{(0)} + C_{k-1}(A)r^{(0)}$, where C_{k-1} is a polynomial. (Since the same polynomial approximates A^{-1}, it is important also in polynomial preconditioning.) If the roots of C_{k-1} are known then $C_{k-1}(A)$ may be evaluated without special concern for the order of the roots. In the least squares case, there is an algorithm in [22] for computing the roots of C_{k-1}.

6.3   Richardson's Method Variant

The so-called leapfrog form of Richardson's method is given in the algorithm below. In this variant of Richardson's method, alternate steps are omitted. The advantages are a slight reduction in the number of arithmetic operations, formulas with more opportunity for chaining on certain processors, and a reduction in the number of memory accesses under certain conditions. See [22] for the details.

6.4   An Algorithm

Algorithm.   *(Leapfrog Richardson's Method with Adaptive Parameter Computation.)*

Purpose. Compute an approximate convex hull of the spectrum of A, determine an MR approximation as the initial guess for Richardson's method, and execute Richardson's method omitting alternate steps.

Input. Matrix A, right side b, initial guess $x^{(0)}$, period k, m, where $m + 1$ is the dimension of the Krylov subspace, and j, the number of computed eigenvalues. In [26] k is taken to be 5 (but k must be even here), and in the Manteuffel algorithm, $j = 4$. Parameter m may be set equal to j. The user must also provide a maximum number of cycles of iterations and an error criterion to halt the algorithm. An initial estimate of the convex hull of the spectrum of A could be provided if available.

Output. Iterate $x^{(k)}$, the iterate reached after the last cycle of k parameters in the standard execution of Richardson's method; an approximate convex hull of the spectrum of A.

Restrictions. If the matrix is singular, the algorithm converges to a solution if the system is consistent. Period k is even. Dimension $m + 1$ of the Krylov subspace must be large enough that j eigenvalues can be computed: $j \leq m$.

Notes.

(1) A routine must be provided to evaluate the inner products $\sum_{i=1}^{M} f(\zeta_i) \, \overline{g(\zeta_i)}$ $w(\zeta_i) \, m(\zeta_i)/M_w$. The M points $\{\zeta_i\}$ lie on the boundary of a convex hull. It is reasonable to include the vertices among the ζ_i's then distribute the remaining points in some equitable fashion. For example, if ν is the number of vertices (and thus $\nu - 1$ is the number of edges) then one could allow a total of $5(\nu - 1)$ points lying in the interiors of edges and allocate these among the edges in a way proportional to the length of each edge.

(2) The algorithm is adaptive: eigenvalue estimates are revised if necessary for convergence by a criterion based on $\|r^{(k)}\|_2 \leq \|R_k\|_\infty \|r^{(0)}\|_2$ (valid if A is normal), where $\|R_k\|_\infty$ is the max-norm of R_k on γ. The heuristic approximation $\|R_k\|_\infty \approx \|R_k\|_w$ is used together with (4.3) to yield the criterion that if $\|r^{(k)}\|/\|r^{(0)}\| \geq 3 \left(\sum_{j=k+1}^{M} \eta_j^2 \right) / M_w^{1/2}$ then new eigenvalue estimates should be computed.

(3) Symbol C_H denotes an approximate convex hull of the spectrum of A.

Algorithm Statement.

(1) If there is an estimate of C_H, then continue; else set C_H equal to the null set and continue.

(2) Set $s_i = A^i r^{(0)}$, for $i = 0, 1, \ldots, m$.

(3) Solve the eigenvalue LS problem of Sec. 5.3 for coefficients $c_0, c_1, \ldots, c_{j-1}$ and compute the roots of $c_0 + c_1\lambda + \cdots + c_{j-1}\lambda^{j-1} + \lambda^j = 0$, for approximate eigenvalues $\lambda_1, \lambda_2, \ldots, \lambda_j$.

(4) Determine the convex hull of $C_H \cup \{\lambda_1, \lambda_2, \ldots, \lambda_j\}$; determine points $\{\zeta_i\}$ to define an inner product, Note (1) above;
compute $M_w = \sum_{i=1}^{M} w(\zeta_i) m(\zeta_i)$.

(5) Compute roots $1/\tau_0, 1/\tau_1, \ldots, 1/\tau_{k-1}$ of the optimal residual polynomial by solving the optimal residual polynomial LS problem in Sec. 4.3.

(6) Solve the minimum residual LS problem, Sec. 5.4, and set $x^{(0)}$ equal to the minimum residual LS approximation $x^{(m)}$ as given by (5.1).
Set $r^{(0)} := b - Ax^{(0)}$.

(7) For $i = 0, 1, \ldots, k$ by 2, DO:

    7.1) Set $\alpha := \tau_{i-2} + \tau_{i-1}$; set $\nu := \tau_{i-2}\tau_{i-1}$.

    7.2) Set $r^{(i-2)} := b - Ax^{(i-2)}$; set $t := Ar^{(i-2)}$.

    7.3) Set $x^{(i)} := x^{(i-2)} + \alpha r^{(i-2)} - \nu t$.

    7.4) Endfor

(8) If $\|r^{(k)}\|/\|r^{(0)}\|$ satisfies an accuracy criterion, exit;
else set $x^{(0)} := x^{(k)}$.

(9) If $\|r^{(k)}\| \geq 3\|R_k\|_\infty \|r^{(0)}\|_2/M_w^{1/2}$, set $r^{(0)} := r^{(k)}$ and go to 2);
else go to 7) .

Summary

Recent work with Richardson's method has yielded various methods for computing parameters if a region containing the spectrum is known. This generalizes pioneering work of David Young from 1953.

In this paper a dynamic algorithm is presented for computing an approximation to the convex hull of the spectrum of the system matrix. It is a straightforward modification of the adaptive method employed in the Manteuffel algorithm. The Manteuffel algorithm is a dynamic algorithm for the computation of the optimal ellipse parameters for the Chebyshev method. It is restricted to real matrices, whereas if, in altered form, it is applied to Richardson's method there is no restriction. Before starting the Richardson iteration, an initial approximation of the solution is obtained by solving a minimum residual LS problem, an idea used in the hybrid algorithm in [7]. There are three associated least squares problems, to determine: (i) the convex hull; (ii) optimal parameters; (iii) and an initial approximation. These are outlined in detail, and an adaptive algorithm described that implements the convex hull computation, the minimum residual approximation, and the execution of the leapfrog variant of Richardson's method.

Acknowledgments

I am indebted to Jerry Minerbo for many discussions about the difficulties of solving large problems. Gene Golub pointed out the distinction between biased and unbiased least squares problems. He also suggested, on a different occasion, that polynomial C_{k-1} avoids difficulties in ordering the parameters for Richardson's method. Partial support was provided by NSF DMS-8-7-03226. This paper was originally typed in troff by the author and converted to LaTeX by Mike Holst and reformatted by Lisa Laguna in LaTeX.

References

[1] Ashby, S. F., T. A. Manteuffel, and P. E. Saylor. *A Taxonomy for Conjugate Gradient Methods.* Research Report UIUCDCS-R-88-1414, Univ. of Illinois at Urbana-Champaign, Dept. of Computer Science, 1988.

[2] Ashby, S. F. *CHEBYCODE: A FORTRAN Implementation of Manteuffel's Adaptive Chebyshev Algorithm.* Research Report UIUCDCS-R-85-1203, Univ. of Illinois at Urbana-Champaign, Dept. of Computer Science, 1985.

[3] Chronopoulos, A. *A Class of Parallel Iterative Methods Implemented on Multiprocessors.* Research Report UIUCDCS-R-86-1267, Univ. of Illinois at Urbana-Champaign, Dept. of Computer Science, November 1986.

[4] Smolarski, D. C. *Optimum Semi-Iterative Methods of the Solution of Any Linear Algebraic System with a Square Matrix.* Research Report UIUCDCS-81-1077, Univ. of Illinois at Urbana-Champaign, Dept. of Computer Science, December 1981.

[5] Elman, H. C. *Iterative Methods for Large, Sparse Nonsymmetric Systems of Linear Equations.* Ph.D. thesis, Yale University, Computer Science Dept., 1982.

[6] Elman, H. C., and R. L. Streit. *Polynomial Iteration for Nonsymmetric Indefinite Linear Systems.* Research Report 380, Yale University, Dept. of Computer Science, 1985.

[7] Elman, H.C., Y. Saad, and P.E. Saylor. "A hybrid Chebyshev Krylov subspace algorithm for solving nonsymmetric systems of linear equations." *SIAM J. on Sci. Stat. Comp.*, 7(3):840–855, Oct., 1985.

[8] Eiermann, M., R. S. Varga, and W. Niethammer. "Iterationsverfahren fur nichtsymmetrische gleichungssysteme und approximationsmethoden im komplexen." *Jber. d. Dt. Math.-Verein*, 89:1–32, 1987.

[9] Faber, V., and T. Manteuffel. "Necessary and sufficient conditions for the existence of a conjugate gradient method." *SIAM J. Numer. Anal.*, 21(2):352–362, 1984.

[10] Fischer, B. and L. Reichel. "A stable Richardson iteration method for complex linear systems." *Numer. Math.*, to appear.

[11] Gragg, W. B., and L. Reichel. "On the application of orthogonal polynomials to the iterative solution of linear systems of equations with indefinite or non-hermitian matrices." *Linear Alg. Applic.*, 88-89:349–371, 1987.

[12] Hageman, L. A., and D. M. Young. *Applied Iterative Methods*. Academic Press, New York, 1981.

[13] Manteuffel, T. A. "Adaptive procedure for estimating parameters for the non-symmetric TChebyshev iteration." *Numer. Math.*, 31:183–208, 1978.

[14] Moran, J. H., and K. S. Kunz. "Basic theory of induction logging and application to study of two-coil sondes." *Geophysics*, XXVII:6:829–858, Dec.,1962.

[15] Opfer, G., and G. Schober. "Richardson's iteration for nonsymmetric matrices." *Linear Alg. Applic.*, 58:343–361, 1984.

[16] Parsons, B. N. "General k-part stationary iterative solutions to linear systems." *SIAM J. Numer. Anal.*, 24(1):188–198, 1984.

[17] Preparata, F. P., and M.I. Shamos. *Computational Geometry, An Introduction*. Springer-Verlag, New York, 1985.

[18] Reichel, L. "Polynomials by conformal mapping for the Richardson's iteration method for complex linear systems." *SIAM J. Numer. Anal.*, Dec., 1988.

[19] Saad, Y. "Practical use of polynomial preconditionings for the conjugate gradient method." *SIAM J. on Sci. Stat. Comp.*, 7(3):865–881, Oct., 1985.

[20] Saad, Y. "Least squares polynomials in the complex plane and their use for solving nonsymmetric linear systems." *SIAM J. Numer. Anal.*, 24(1):155–169, Feb.,1987.

[21] Saylor, P.E. "Adaptive parameter computations for the iterative solution of linear algebraic equations." In *Proceedings of the Fourth IMACS International Symposium on Advances in Computer Methods for Partial Differential Equations*, Lehigh, 1981.

[22] Saylor, P.E. "Leapfrog variants of iterative methods for linear algebraic equations." *Journal of Computational and Applied Mathematics*, to appear, 1988.

[23] Silvester, R. P., and R. L. Ferrari. *Finite Elements for Electrical Engineers*. Cambridge University Press, New York, 1983.

[24] Smith, C.F. *The Performance of Preconditioned Iterative Methods in Computational Electromagnetics*. Ph.D. thesis, Univ. of Illinois at Urbana-Champaign, Dept. of Electrical and Computer Engineering, 1987.

[25] Saad, Y., and M. H. Schultz. "GMRES: a generalized minimal residual algorithm for solving nonsymmetric linear systems." *SIAM J. Sci. Stat. Comput.*, 7:3:856–869, 1986.

[26] Smolarski, D.C., and P.E. Saylor. "An optimum iterative method for solving any linear system with a square matrix." *BIT*, to appear, 1988.

[27] Steele, C.W. *Numerical Computation of Electric and Magnetic Fields*. Van Nostrand Reinhold Co., New York, 1987.

[28] Stiefel, E. "Kernel polynomials in linear algebra and their numerical applications." *National Bureau of Standards Math. Series*, 49:1–22, 1958.

[29] Tal-Ezer, H. *Polynomial Approximation of Functions of Matrices and Applications*. NASA Contractor Report 178376, ICASE Report No. 87-63, Institute for Computer Applications in Science and Engineering, NASA Langley Research Center, Hampton, Va. 23665, 1987.

[30] Ursin, B. Review of elastic and electromagnetic wave propagation in horizontally layered media. *Geophysics*, 48:1063–1081, Aug.,1983.

[31] Wilkinson, J. H. *The Algebraic Eigenvalue Problem*. Clarendon Press, Oxford, 1965.

[32] Young, D.M. "On Richardson's method for solving linear systems with positive definite matrices." *J. Math. Phys.*, 32:243–255, 1954.

Chapter 14

A Note on the SSOR and USSOR Iterative Methods Applied to p-Cyclic Matrices

XIEZHANG LI and RICHARD S. VARGA
Kent State University

Dedicated to David M. Young, Jr., on the occasion of his sixty-fifth birthday.

Abstract

In this note, we determine a new functional equation

$$[\lambda - (1 - \omega)(1 - \hat{\omega})]^p =$$
$$\lambda^k \left[\lambda \omega + \hat{\omega} - \omega \hat{\omega}\right]^{|\varsigma_L| - k} \left[\lambda \hat{\omega} + \omega - \omega \hat{\omega}\right]^{|\varsigma_u| - k} \left(\omega + \hat{\omega} - \omega \hat{\omega}\right)^{2k} \mu^p,$$

which couples the nonzero eigenvalues of the USSOR (unsymmetric successive overrelaxation) iteration matrix $\mathbf{T}_{\omega,\hat{\omega}}$ with the eigenvalues of an associated block Jacobi matrix $\mathbf{B}$ in the p-cyclic case. Such function equations are of course direct descendants of the now famous functional equation

$$(\lambda + \omega - 1)^2 = \lambda \omega^2 \mu^2,$$

derived in 1950 by David M. Young, Jr. in his Harvard University thesis.

1 Introduction

There have been a number of recent research articles, all concerned with the symmetric successive overrelaxation (SSOR) iterative method and the unsymmetric successive overrelaxation (USSOR) iterative method, applied to p-cyclic matrices.

These research articles give generalizations of the following functional equation, derived by Varga, Niethammer and Cai [4]:

$$(1.1) \qquad [\lambda - (1-\omega)^2]^p = \lambda[\lambda + 1 - \omega]^{p-2}(2-\omega)^2\omega^p\mu^p,$$

which connects the eigenvalues λ of the associated SSOR matrix $\mathbf{S}_\omega$ to the eigenvalues μ of a particular weakly cyclic of index p Jacobi matrix $\mathbf{B}$ (where $p \geq 2$). Of course, the functional equation (1.1) strongly resembles in character the related well-known functional equations

$$(1.2) \qquad (\lambda + \omega - 1)^2 = \lambda\omega^2\mu^2$$

of Young [7,8], and
$$(1.2') \qquad (\lambda + \omega - 1)^p = \lambda^{p-1}\omega^p\mu^p$$

of Varga [5,6], which similarly connect the eigenvalues λ of an associated successive overrelaxation matrix $\mathcal{L}_\omega$ to the eigenvalues μ of a consistently ordered weakly cyclic of index p Jacobi matrix $\mathbf{B}$ (where $p \geq 2$).

The purpose of this note is threefold. First, we develop the following *new* functional equation (cf. also (2.1) of Theorem 1):

$$[\lambda - (1-\omega)(1-\hat{\omega})]^p =$$
$$\lambda^k [\lambda\omega + \hat{\omega} - \omega\hat{\omega}]^{|\varsigma_L|-k} [\lambda\hat{\omega} + \omega - \omega\hat{\omega}]^{|\varsigma_U|-k} (\omega + \hat{\omega} - \omega\hat{\omega})^{2k} \mu^p,$$

which serves to generalize and unify *all* the recent research articles on the SSOR and USSOR iterative methods applied to a block p-cyclic matrix. Second, we give a *graph-theoretic interpretation* of the exponent k in the equation above. As it turns out, a similar analysis applies to a graph-theoretic interpretation for the associated known SOR case. (This is remarked in Sec. 2.) Finally, (1.3) and Theorem 1 generalize the recent result of Gong and Cai [1] on the SSOR iterative method for p-cyclic matrices, which has been published only in Chinese. Our final purpose in this note is to connect our new Theorem 1 with known results in the literature, and to bring this result of Gong and Cai [1] to a larger audience.

For the remainder of this section, we give background and notation for our problem. For the iterative solution of the matrix equation

$$(1.4) \qquad\qquad \mathbf{Ax} = \mathbf{k},$$

where $\mathbf{A}$ is a given $n \times n$ complex matrix, assume that the matrix $\mathbf{A}$ can be written in block-partitioned form as

$$(1.5) \qquad \mathbf{A} = \begin{bmatrix} \mathbf{A}_{1,1} & \mathbf{A}_{1,2} & \cdots & \mathbf{A}_{1,p} \\ \mathbf{A}_{2,1} & \mathbf{A}_{2,2} & \cdots & \mathbf{A}_{2,p} \\ \vdots & \vdots & \ddots & \vdots \\ \mathbf{A}_{p,1} & \mathbf{A}_{p,2} & \cdots & \mathbf{A}_{p,p} \end{bmatrix},$$

where each diagonal submatrix $\mathbf{A}_{i,i}$ is square and nonsingular $(1 \leq i \leq p)$. (We assume throughout that $p \geq 2$.) With

$$\mathbf{D} := \operatorname{diag}[\mathbf{A}_{1,1},\ \mathbf{A}_{2,2},\ \cdots,\ \mathbf{A}_{p,p}],$$

the associated block-Jacobi matrix $\mathbf{B}$ is defined by

(1.6) $$\mathbf{B} := \mathbf{I} - \mathbf{D}^{-1}\mathbf{A},$$

which we can write, from the partitioning in (1.5), as

(1.7) $$\mathbf{B} = [\mathbf{B}_{i,j}] := \begin{bmatrix} \mathbf{0} & \mathbf{B}_{1,2} & \cdots & \mathbf{B}_{1,p} \\ \mathbf{B}_{2,1} & \mathbf{0} & \cdots & \mathbf{B}_{2,p} \\ \vdots & \vdots & \ddots & \vdots \\ \mathbf{B}_{p,1} & \mathbf{B}_{p,2} & \cdots & \mathbf{0} \end{bmatrix}.$$

As the block diagonal submatrices of $\mathbf{B}$ are by definition all null, we can also express $\mathbf{B}$ as the sum

(1.8) $$\mathbf{B} = \mathbf{L} + \mathbf{U},$$

where $\mathbf{L}$ and $\mathbf{U}$ are respectively strictly lower and strictly upper triangular matrices.

From (1.8), the associated unsymmetric successive overrelaxation (USSOR) iteration matrix $\mathbf{T}_{\omega,\hat{\omega}}$ is then defined by

(1.9) $$\mathbf{T}_{\omega,\hat{\omega}} := (\mathbf{I} - \hat{\omega}\mathbf{U})^{-1}[(1 - \hat{\omega})\mathbf{I} + \hat{\omega}\mathbf{L}](\mathbf{I} - \omega\mathbf{L})^{-1}[(1 - \omega)\mathbf{I} + \omega\mathbf{U}],$$

where ω and $\hat{\omega}$ are relaxation parameters. The associated symmetric successive overrelaxation (SSOR) iteration matrix $\mathbf{S}_\omega$ for (1.8) reduces to the case when $\omega = \hat{\omega}$ in (1.9), i.e.,

(1.10) $$\mathbf{S}_\omega := \mathbf{T}_{\omega,\omega}.$$

Our interest here is in the case where the block-Jacobi matrix $\mathbf{B}$ of (1.7) has the property that there is a cyclic permutation (a 1-1 onto mapping) of the integers $\{1, 2, \cdots, p\}$, expressed in cyclic form as $\sigma = (\sigma_1, \sigma_2, \cdots, \sigma_p)$, such that

(1.11) $$\mathbf{B}_{\sigma_j,k} \equiv \mathbf{0} \quad \text{for all } k \neq \sigma_{j+1} \quad (1 \leq j, k \leq p),$$

where $\sigma_{p+1} := \sigma_1$. It is easily seen that if the block-partitioned matrix $\mathbf{B}$ of (1.7) satisfies (1.11), then $\mathbf{B}$ is *weakly cyclic of index* p (cf. [6, p. 39]), and, conversely, if the partitioned matrix $\mathbf{B}$ is weakly cyclic of index p, then $\mathbf{B}$ satisfies (1.11) for a suitable cyclic permutation σ. Thus, we define the block-partitioned matrix $\mathbf{B}$ of (1.7) to be a *weakly cyclic matrix generated according to the cyclic permutation* $\sigma = (\sigma_1, \sigma_2, \cdots, \sigma_p)$ if (1.11) is satisfied. (We do remark that a block-partitioned matrix $\mathbf{B}$, which is weakly cyclic of index p, can, for a different partitioning of $\mathbf{B}$, be weakly cyclic of some index p' with $p' \neq p$.)

Assume that $\mathbf{B} = \mathbf{L} + \mathbf{U}$ of (1.7) is a weakly cyclic of index p matrix generated by a cyclic permutation $\sigma = (\sigma_1, \sigma_2 \ldots, \sigma_p)$, so that (1.11) is valid. Then, it follows from (1.11) that $\mathbf{B}^p$ is a block-diagonal matrix whose σ_j-th diagonal block is given by the product

$$(1.12) \qquad \mathbf{B}_{\sigma_j, \sigma_{j+1}} \cdot \mathbf{B}_{\sigma_{j+1}, \sigma_{j+2}} \cdots \mathbf{B}_{\sigma_j+p-1, \sigma_j} \qquad (1 \le j \le p),$$

where $\sigma_i := \sigma_{i-p}$ if $i > p$. To avoid trivial cases, we further assume that none of the square matrices in (1.12) is a null matrix. This implies that

$$(1.13) \qquad \mathbf{B}_{\sigma_j, \sigma_{j+1}} \not\equiv \mathbf{0} \qquad (1 \le j \le p).$$

Then, with the cyclic permutation $\sigma = (\sigma_1, \sigma_2, \cdots, \sigma_p)$, we define its associated disjoint subsets ζ_L and ζ_U of $\{1, 2, \cdots, p\}$ as

$$(1.14) \qquad \begin{cases} \zeta_L := \{\sigma_j : \sigma_j > \sigma_{j+1}\}, \\ \zeta_U := \{\sigma_j : \sigma_j < \sigma_{j+1}\}. \end{cases}$$

With $|R|$ denoting the *cardinality* of an arbitrary set R, then, by definition, $|\zeta_L|$ and $|\zeta_U|$ are precisely the number of nonzero block submatrices of $\mathbf{B}$ which are in $\mathbf{L}$ and in $\mathbf{U}$, respectively. Also, as $\zeta_L \bigcup \zeta_U = \{1, 2, \cdots, p\}$ and as $\zeta_L \bigcap \zeta_U = \emptyset$, then

$$(1.15) \qquad |\zeta_L| + |\zeta_U| = p.$$

To determine which entries of the product $\mathbf{LU}$, for the block-partitioning of (1.7), are nonzero, we define the disjoint (and possibly empty) subsets η_L and η_U of ζ_U as

$$\begin{cases} \eta_L := \{\sigma_j : \sigma_{j-1} > \sigma_j, \ \sigma_{j+1} > \sigma_j, \text{ and } \sigma_{j-1} > \sigma_{j+1}\} \\ \eta_U := \{\sigma_j : \sigma_{j-1} > \sigma_j, \ \sigma_{j+1} > \sigma_j, \text{ and } \sigma_{j-1} < \sigma_{j+1}\}, \quad (\text{where } \sigma_0 := \sigma_p). \end{cases}$$
(1.16)

Again by definition, $|\eta_L|$ and $|\eta_U|$ are precisely the *number* of nonzero block submatrices of $\mathbf{LU}$ which occur in the strictly block-lower and strictly block-upper triangular parts, respectively, of the partitioning for $\mathbf{LU}$. We further set

$$(1.17) \qquad k := \begin{cases} |\eta_L| + |\eta_U| & \text{if } p > 2, \\ 1 & \text{if } p = 2. \end{cases}$$

If ℓ is such that $\sigma_\ell = 1$, then evidently $\sigma_{\ell-1} > \sigma_\ell$ and $\sigma_{\ell+1} > \sigma_\ell$, so that (cf. (1.16)) σ_ℓ is necessarily either an element of η_L or of n_U for $p > 2$. Consequently, (cf. (1.17)), $k \ge 1$ if $p > 2$. Similarly, if σ_ℓ satisfies $\sigma_{\ell-1} > \sigma_\ell$ and $\sigma_{\ell+1} > \sigma_\ell$, then neither $\sigma_{\ell-1}$ nor $\sigma_{\ell+1}$ can be an element of η_L or η_U, so that $k \le [[p/2]]$, giving

$$(1.18) \qquad 1 \le k \le [[p/2]],$$

where $[[x]]$ denotes the integer part of a real number x. As can be verified, k is precisely the number of nonzero block submatrices of LU. It is further evident that $|\zeta_L| \geq k$ and $|\zeta_U| \geq k$.

We finally give in this section a *directed graph* interpretation of the positive integer k of (1.17). Specifically, let $G_\pi[\mathbf{B}]$ denote the *directed graph of type 2* for the block-partitioned matrix $\mathbf{B}$ of (1.7), i.e., (cf. [6, p. 121]), we associate with the matrix $\mathbf{B}$ of (1.7) a directed graph with p vertices, $V_1, V_2, \cdots, V_p$, where an arc from vertex V_i to the vertex V_j is drawn with a *double arrow* only if $\mathbf{B}_{i,j} \not\equiv \mathbf{0}$ and if $j > i$, while an arc from vertex V_i to the vertex V_j is drawn with a *single arrow* only if $\mathbf{B}_{i,j} \not\equiv \mathbf{0}$ and if $j < i$. Then, for any simple closed path of length p starting at any vertex V_i and ending at the same vertex V_i (this path consisting of consecutive single- and/or double-arrowed arcs), the positive integer k of (1.17) is *precisely* the number of times (in traveling this closed path) that a double-arrowed arc *follows* a single-arrowed arc. This will be illustrated in three examples in Sec. 2.

2   Statement of Main Result and Discussion

With the notations and definitions of Sec. 1, our main result is

Theorem 1.   *Assume that the block-partitioned matrix* $\mathbf{A}$ *of* (1.5) *is such that all diagonal submatrices* $\mathbf{A}_{i,i}$ *are square and nonsingular* $(1 \leq i \leq p)$, *and assume that its block-Jacobi matrix* $\mathbf{B}$ *of* (1.7) *is a weakly cyclic matrix of index* p, *generated by the cyclic permutation* $\sigma = (\sigma_1, \sigma_2, \cdots, \sigma_p)$. *If* $\omega + \hat{\omega} - \omega\hat{\omega} \neq 0$, *if* λ *is an nonzero eigenvalue of the USSOR matrix* $\mathbf{T}_{\omega,\hat{\omega}}$ *of* (1.9), *and if* μ *satisfies*

$$(2.1)\quad [\lambda - (1-\omega)(1-\hat{\omega})]^p = \lambda^k [\lambda\omega + \hat{\omega} - \omega\hat{\omega}]^{|\zeta_L|-k} [\lambda\hat{\omega} + \omega - \omega\hat{\omega}]^{|\zeta_U|-k} (\omega + \hat{\omega} - \omega\hat{\omega})^{2k} \mu^p$$

(where $k, |\zeta_L|,$ *and* $|\zeta_U|$ *are defined from* σ *in Sec. 1, and where the convention* $0^0 := 1$ *is used in* (2.1)), *then* μ *is an eigenvalue of* $\mathbf{B}$. *Conversely, if* μ *is an eigenvalue of* $\mathbf{B}$ *and if* $\hat{\lambda}$ *satisfies* (2.1), *then* $\hat{\lambda}$ *is an eigenvalue of* $\mathbf{T}_{\omega,\hat{\omega}}$.

The proof of this theorem will be given in Sec. 3. We remark that in the case $\omega = \hat{\omega}$, (2.1) reduces with (1.15) to

$$(2.1')\qquad [\lambda - (1-\omega)^2]^p = \lambda^k [\lambda + 1 - \omega]^{p-2k} (2-\omega)^{2k} \omega^p \mu^p,$$

which was given in Gong and Cai [1, Eq. (1.4)].

To complete this section, we show how this new functional equation (2.1) relates to recent results in this area.

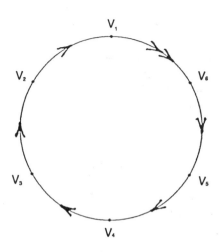

Figure 14.1 $G_\pi(\mathbf{B}_1)$

Example 1.   Consider the block-partitioned Jacobi matrix $\mathbf{B}_1$ given by

$$(2.2) \qquad \mathbf{B}_1 = \begin{bmatrix} 0 & 0 & \cdots & 0 & \mathbf{B}_{1,p} \\ \mathbf{B}_{2,1} & 0 & \cdots & 0 & 0 \\ 0 & \mathbf{B}_{3,2} & \ddots & & 0 \\ \vdots & \vdots & \ddots & \ddots & \vdots \\ 0 & 0 & \cdots & \mathbf{B}_{p,p-1} & 0 \end{bmatrix},$$

where $p \geq 2$. In this case, $\mathbf{B}_1$ is a weakly cyclic of index p matrix, generated by the cyclic permutation $(1, p, p-1, \cdots, 3, 2)$. From the definitions of Sec. 1, we have

$$\zeta_L = \{2, 3, \cdots, p\} \text{ and } |\zeta_L| = p - 1; \quad \zeta_U = \{1\} \text{ and } |\zeta_U| = 1,$$

$p > 2: \ \eta_L = \varnothing \text{ and } |\eta_L| = 0; \qquad\qquad \eta_U = \{1\} \text{ and } |\eta_U| = 1; \ k = 1,$

$p = 2: \ \eta_L = \varnothing \text{ and } |\eta_L| = 0; \qquad\qquad \eta_U = \varnothing \ \text{ and } |\eta_U| = 0; \ k = 1.$

For the case $p = 6$, the block-directed graph of type 2 for the matrix $\mathbf{B}_1$ of (2.2) is shown in Fig. 14.1.

In this case, the functional equation (2.1) reduces to

$$(2.3) \qquad [\lambda - (1 - \omega)(1 - \hat{\omega})]^p = \lambda[\lambda\omega + \hat{\omega} - \omega\hat{\omega}]^{p-2}(\omega + \hat{\omega} - \omega\hat{\omega})^2\mu^p,$$

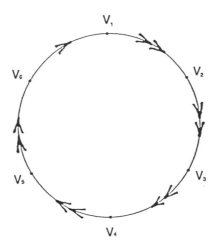

Figure 14.2 $G_\pi(\mathbf{B}_2)$

which is the functional equation for $\mathbf{T}_{\omega,\bar\omega}$, derived by Saridakis [3], for the block-Jacobi matrix of (2.2).

Example 2. Consider the block-partitioned Jacobi matrix $\mathbf{B}_2$ given by

$$(2.4) \qquad \mathbf{B}_2 = \begin{bmatrix} \mathbf{0} & \mathbf{B}_{1,2} & \mathbf{0} & \cdots & \mathbf{0} \\ \mathbf{0} & \mathbf{0} & \mathbf{B}_{2,3} & \cdots & \mathbf{0} \\ \vdots & \vdots & \ddots & \ddots & \vdots \\ \mathbf{0} & \mathbf{0} & \mathbf{0} & \ddots & \mathbf{B}_{p-1,p} \\ \mathbf{B}_{p,1} & \mathbf{0} & \mathbf{0} & \cdots & \mathbf{0} \end{bmatrix}.$$

In this case, $\mathbf{B}_2$ is a weakly cyclic of index p matrix generated by the cyclic permutation $(1, 2, \cdots, p)$, and we have

$$\zeta_L = \{p\} \text{ and } |\zeta_L| = 1; \quad \zeta_U = \{1, 2, \cdots, p-1\} \text{ and } |\zeta_U| = p-1,$$

$$p > 2: \ \eta_L = \{1\} \text{ and } |\eta_L| = 1; \quad \eta_U = \varnothing \text{ and } |\eta_U| = 0; \quad k = 1,$$

$$p = 2: \ \eta_L = \varnothing \text{ and } |\eta_L| = 0; \quad \eta_U = \varnothing \text{ and } |\eta_U| = 0; \quad k = 1.$$

For the case $p = 6$, the block-directed graph of type 2 for the matrix $\mathbf{B}_2$ of (2.4) is shown in Fig. 14.2.

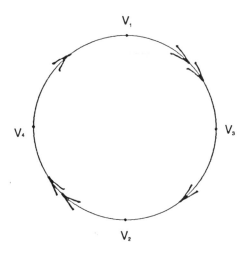

Figure 14.3 $G_\pi(\mathbf{B}_3)$

In this case, the functional equation (2.1) reduces to

$$(2.5) \qquad [\lambda - (1 - \omega)(1 - \hat\omega)]^p = \lambda[\lambda\hat\omega + \omega - \omega\hat\omega]^{p-2}(\omega + \hat\omega - \omega\hat\omega)^2\mu^p.$$

For the special case $\omega = \hat\omega$, the above functional equation (for $\mathbf{S}_\omega$) was obtained in Varga, Niethammer and Cai [4]. For general ω and $\hat\omega$, (2.5) was also obtained by Saridakis [3].

Example 3. Consider the block-partitioned Jacobi matrix $\mathbf{B}_3$ given by

$$(2.6) \qquad \mathbf{B}_3 = \begin{bmatrix} 0 & 0 & \mathbf{B}_{1,3} & 0 \\ 0 & 0 & 0 & \mathbf{B}_{2,4} \\ 0 & \mathbf{B}_{3,2} & 0 & 0 \\ \mathbf{B}_{4,1} & 0 & 0 & 0 \end{bmatrix}$$

In this case, $\mathbf{B}_3$ is a weakly cyclic of order 4 matrix generated by the cyclic permutation (1, 3, 2, 4). Thus,

$$\zeta_L = \{3,4\} \text{ and } |\zeta_L| = 2; \quad \zeta_U = \{1,2\} \text{ and } |\zeta_U| = 2,$$
$$\eta_L = \{1\} \text{ and } |\eta_L| = 1; \quad \eta_U = \{2\} \text{ and } |\eta_U| = 1; \quad k = 2,$$

and the block-directed graph of type 2 for the matrix $\mathbf{B}_3$ of (2.6) is given in Fig. 14.3.

In this case, the functional equation (2.1) reduces to

$$(2.7) \qquad [\lambda - (1-\omega)(1-\hat{\omega})]^4 = \lambda^2(\omega + \hat{\omega} - \omega\hat{\omega})^4\mu^4.$$

For the special case $\omega = \hat{\omega}$, the above functional equation was obtained in Varga, Niethammer and Cai [4, Eq. (2.36)], and, again for $\omega = \hat{\omega}$, was given as an example in Gong and Cai [1, Eq. (1.6)].

As mentioned in Sec. 1, we can also apply the above graph-theoretic ideas to the analysis of the SOR (successive overrelaxation) iterative method. Specifically, associated with the block-Jacobi matrix $\mathbf{B}$ of (1.6) (1.8) for the matrix problem (1.4), is the well-known SOR iteration matrix $\mathcal{L}_\omega$, defined by

$$(2.8) \qquad \mathcal{L}_\omega := (\mathbf{I} - \omega\mathbf{L})^{-1}[(1-\omega)\mathbf{I} + \omega\mathbf{U}].$$

If $\mathbf{B}$ is a weakly cyclic matrix of index p, generated by the cyclic permutation $(\sigma_1, \sigma_2, \cdots, \sigma_p)$, then the functional equation (analogous to (1.2), (1.2′), and (2.1)), which couples the eigenvalues μ of $\mathbf{B}$ to the eigenvalues λ of $\mathcal{L}_\omega$, is known (cf. Nickel and Fox [2] and [6, p. 109, Exercise 2]) to be

$$(2.9) \qquad (\lambda + \omega - 1)^p = \lambda^\tau \omega^p \mu^p.$$

It turns out (as is easily seen) that the exponent τ in (2.9) is *precisely* $|\zeta_L|$, and $|\zeta_L|$ is, from our discussions in Sec. 1, exactly the number of nonzero lower triangular block submatrices of $\mathbf{B}$. Equivalently, in terms of the associated directed graph $G_\pi(\mathbf{B})$ of type 2 described in Sec. 1, τ is *precisely* the number of single-arrowed arcs in any simple closed path of length p starting at any vertex V_i and ending at the same vertex V_i.

3   Proof of the Theorem

It can be verified from (1.9) that

$$(3.1) \qquad \lambda\mathbf{I} - \mathbf{T}_{\omega,\hat{\omega}} = (\mathbf{I} - \hat{\omega}\mathbf{U})^{-1}(\mathbf{I} - \omega\mathbf{L})^{-1}(\gamma\mathbf{I} - \alpha\mathbf{L} - \beta\mathbf{U} - \delta\mathbf{LU}),$$

where

$$(3.2) \qquad \begin{cases} \gamma := \lambda - (1-\omega)(1-\hat{\omega}), \\ \alpha := \lambda\omega + \hat{\omega} - \omega\hat{\omega}, \\ \beta := \lambda\hat{\omega} + \omega - \omega\hat{\omega}, \\ \delta := (1-\lambda)\omega\hat{\omega}. \end{cases}$$

Hence, λ is an eigenvalue of $\mathbf{T}_{\omega,\hat{\omega}}$ if and only if

$$(3.3) \qquad \det\{\gamma\mathbf{I} - \alpha\mathbf{L} - \beta\mathbf{U} - \delta\mathbf{LU}\} = 0.$$

Before we prove Theorem 1, we first establish Lemmas 2, 3, and 4. For notation, we introduce two $p \times p$ block-partitioned matrices $\mathbf{H}_L := [\mathbf{H}_{i,h}]$ and $\mathbf{H}_U := [\tilde{\mathbf{H}}_{h,j}]$, associated with the block-partitioned matrix $\mathbf{B}$ of (1.7), where

$$(3.4) \qquad \mathbf{H}_{i,h} := \begin{cases} \mathbf{B}_{i,h} & \text{if } h \in \eta_L, \\ 0 & \text{otherwise}, \end{cases}$$

and

$$(3.5) \qquad \tilde{\mathbf{H}}_{h,j} := \begin{cases} \mathbf{B}_{h,j} & \text{if } h \in \eta_U, \\ 0 & \text{otherwise}. \end{cases}$$

For example, in the case of $\mathbf{B}_3$ of (2.6), we have

$$\mathbf{H}_L = \begin{bmatrix} 0 & 0 & 0 & 0 \\ 0 & 0 & 0 & 0 \\ 0 & 0 & 0 & 0 \\ \mathbf{B}_{4,1} & 0 & 0 & 0 \end{bmatrix}, \quad \mathbf{H}_U = \begin{bmatrix} 0 & 0 & 0 & 0 \\ 0 & 0 & 0 & \mathbf{B}_{2,4} \\ 0 & 0 & 0 & 0 \\ 0 & 0 & 0 & 0 \end{bmatrix}.$$

Lemma 2. *Let $\mathbf{B} = \mathbf{L} + \mathbf{U}$ of (1.7) be a weakly cyclic of index p matrix, generated by a cyclic permutation $\sigma = (\sigma_1, \sigma_2, \cdots, \sigma_p)$. Then, for arbitrary complex numbers α, β, δ and δ with $\alpha \neq 0$ and $\beta \neq 0$,*

$$(3.6) \quad \det\{\gamma\mathbf{I} - \alpha\mathbf{L} - \beta\mathbf{U} - \delta\mathbf{L}\mathbf{U}\} = x \det\left\{\gamma\mathbf{I} - \left(\frac{\alpha\beta + \gamma\delta}{\beta}\right)\mathbf{H}_L - \alpha(\mathbf{L} - \mathbf{H}_L) \right.$$
$$\left. - \left(\frac{\alpha\beta + \gamma\delta}{\alpha}\right)\mathbf{H}_U - \beta(\mathbf{U} - \mathbf{H}_U)\right\},$$

where the matrices $\mathbf{H}_L$ and $\mathbf{H}_U$ are defined in (3.4) and (3.5).

Proof. With (1.8), set

$$(3.7) \qquad \mathbf{E} := \gamma\mathbf{I} - \alpha\mathbf{L} - \beta\mathbf{U} - \delta\mathbf{L}\mathbf{U}.$$

As we shall see, eliminating from the matrix $\mathbf{E}$ (by means of elementary block-row and block-column transformations applied to the matrix $\mathbf{E}$) those nonzero submatrices of $\mathbf{L}\mathbf{U} := [C_{i,j}]$, will directly give the desired result of (3.6).

It follows from the definition of η_L and η_U (cf. (1.16)) that for each $\mathbf{C}_{i,j} \neq 0$ with $i > j$ (in the lower triangular part of LU), there exists a unique h in η_L such that $\mathbf{C}_{i,j} = \mathbf{B}_{i,h}\mathbf{B}_{h,j}$. Focusing on the six associated submatrices in $\mathbf{E}$ (namely,

$\mathbf{E}_{h,h}, \mathbf{E}_{h,j}, \mathbf{E}_{h,i}$ and $\mathbf{E}_{i,h}, \mathbf{E}_{i,j}, \mathbf{E}_{i,i})$ we have from the form of $\mathbf{E}$ that

$$(3.8) \qquad \mathbf{E} = \begin{bmatrix} & \vdots & & \vdots & & \vdots & \\ \cdots & \gamma \mathbf{I}_{h,h} & \cdots & -\beta \mathbf{B}_{h,j} & \cdots & \mathbf{0} & \cdots \\ & \vdots & & & & \vdots & \\ & \vdots & & & & \vdots & \\ & \vdots & & & & \vdots & \\ & \vdots & & & & \vdots & \\ \cdots & -\alpha \mathbf{B}_{i,h} & \cdots & -\delta \mathbf{B}_{i,h} \mathbf{B}_{h,j} & \cdots & \gamma \mathbf{I}_{i,i} & \cdots \\ & \vdots & & \vdots & & \vdots & \end{bmatrix}.$$

Because $\beta \neq 0$ by assumption, consider the lower block-triangular matrix $\mathbf{Q}$, defined by

$$(3.9) \qquad \mathbf{Q} := \begin{bmatrix} \mathbf{I}_{1,1} & \mathbf{0} & \cdots & \mathbf{0} & \mathbf{0} \\ \mathbf{0} & I_{h,h} & \cdots & \mathbf{0} & \mathbf{0} \\ \vdots & & \ddots & \ddots & \vdots & \vdots \\ \mathbf{0} & -\frac{\delta}{\beta} \mathbf{B}_{i,h} & \ddots & \mathbf{I}_{i,i} & \mathbf{0} \\ \mathbf{0} & \mathbf{0} & \cdots & \mathbf{0} & \mathbf{I}_{p,p} \end{bmatrix},$$

where $\mathbf{Q}$ has a sole nonzero block, i.e., $-\delta \mathbf{B}_{i,h}/\beta$, in its strictly lower block-triangular part. Then, it is easily seen that the matrix product $\mathbf{QE}$ (corresponding to an elementary block-row transformation of $\mathbf{E}$) satisfies

$$\det(\mathbf{QE}) = \det \mathbf{E}; \quad (\mathbf{QE})_{i,j} \equiv \mathbf{0}; \quad (\mathbf{QE})_{i,h} = -\frac{\gamma \delta}{\beta} \mathbf{H}_{i,h} - \alpha \mathbf{L}_{i,h},$$

so that the submatrix $\mathbf{C}_{i,j}$ has been reduced to zero in this step. In this fashion, *all* nonzero submatrices $\mathbf{C}_{i,j}$ (with $i > j$) can be eliminated by such block-row elementary transformations, and the resulting lower triangular part of the transformed matrix $\mathbf{E}$ is $-(\gamma\delta/\beta)\mathbf{H}_L - \alpha \mathbf{L}$. Similarly, for all nonzero submatrices $\mathbf{C}_{i,j}$ in the upper triangular part $(i < j)$ of $\mathbf{LU}$, we apply corresponding block-column elementary transformations to $\mathbf{E}$. Then, the resulting upper triangular part in $\mathbf{E}$ becomes $-(\gamma\delta/\alpha)\mathbf{H}_U - \beta \mathbf{U}$. As such elementary transformations leave the associated determinants univariant, the lemma is proved. ∎

Lemma 3. *Let* $\mathbf{B} = \mathbf{L} + \mathbf{U}$ *of* (1.7) *be a weakly cyclic of index* p *matrix generated by a cyclic permutation* $\sigma = (\sigma_1, \sigma_2, \cdots, \sigma_p)$. *Then, for arbitrary complex numbers*

$\gamma, a, b, c,$ and d,

(3.10)
$$\det\{\gamma\mathbf{I} - a\mathbf{H}_L - b(\mathbf{L} - \mathbf{H}_L) - c\mathbf{H}_U - d(\mathbf{U} - \mathbf{H}_U)\}$$
$$= \det\{\gamma\mathbf{I} - t^{1/p}\mathbf{B}\},$$

where $t := a^{|\eta_L|}b^{|\zeta_L|-|\eta_L|}c^{|\eta_U|}d^{|\zeta_U|-|\eta_U|}$, where the matrices $\mathbf{H}_L$ and $\mathbf{H}_U$ are defined in (3.4) and (3.5), and where the convention $0^0 = 1$ is used in the definition of t.

Proof. Assume first that $abcd \neq 0$. We define the $p \times p$ block-partitioned matrix $\mathbf{M}(a, b, c, d)$ by

(3.11) $$\mathbf{M}(a, b, c, d) := t^{-1/p}\{a\mathbf{H}_L + b(\mathbf{L} - \mathbf{H}_L) + c\mathbf{H}_U + d(\mathbf{U} - \mathbf{H}_U)\}.$$

On comparing the matrix $\mathbf{M} := \mathbf{M}(a, b, c, d)$ with the matrix $\mathbf{B}$, it is easily seen that the matrix $\mathbf{M}$ has exactly the same partitioning structure as the matrix $\mathbf{B}$, except for scalar multipliers of its nonzero submatrices. Thus, $\mathbf{M}$ is a weakly cyclic of index p matrix, generated by the same permutation $\sigma = (\sigma_1, \sigma_2, \cdots, \sigma_p)$, and $\mathbf{M}^p$ and $\mathbf{B}^p$ are both block-diagonal matrices having the same diagonal submatrices, except for scalar multipliers. Since there are $|\eta_L|$ and $|\eta_U|$ nonzero submatrices in the matrices $\mathbf{H}_L$ and $\mathbf{H}_U$, respectively, then there are $|\zeta_L| - |\eta_L|$ and $|\zeta_U| - |\eta_U|$ nonzero submatrices in matrices $\mathbf{L} - \mathbf{H}_L$ and $\mathbf{U} - \mathbf{H}_U$, respectively. Recalling from (1.15) that $|\zeta_L| + |\zeta_U| = p$, it follows from the definition of t and by direct computation that the scalar multiplier of each diagonal submatrix in $\mathbf{M}^p$ is

$$t^{-1}a^{|\eta_L|}\, b^{|\zeta_L|-|\eta_L|}\, c^{|\eta_U|}\, d^{|\zeta_U|-|\eta_U|} = 1.$$

Thus,
(3.12) $$[\mathbf{M}(a, b, c, d)]^p = \mathbf{B}^p,$$

and the eigenvalues of matrix $\mathbf{M}(a, b, c, d)$ are *independent* of $a, b, c,$ and d. Note that as $\mathbf{M}(1, 1, 1, 1) = \mathbf{B}$, we have

(3.13) $$\det\{\gamma\mathbf{I} - t^{1/p}\mathbf{M}(a, b, c, d)\} = \det\{\gamma\mathbf{I} - t^{1/p}\mathbf{B}\},$$

which is the desired result of (3.10) when $abcd \neq 0$.

The remaining case, $abcd = 0$, similarly follows by continuity since both sides of (3.10) are *continuous* functions of the parameters $a, b, c,$ and d. For example, if, as in Example 1, $|\zeta_U| = 1 = |\eta_U|$, then $d^{|\zeta_U|-|\eta_U|} \equiv 1$ for all $d \neq 0$. Thus, on letting $d \to 0$, $d^{|\zeta_U|-|\eta_U|}$, arising as a factor of t in (3.10), has the value unity (which explains our use of the convention $0^0 := 1$).   ∎

By applying Lemma 2 and Lemma 3, we can establish the following result, Lemma 4, which gives a general determinantal invariance associated with weakly cyclic of index p matrices.

Lemma 4. *Let* $\mathbf{B} = \mathbf{L} + \mathbf{U}$ *of* (1.7) *be a weakly cyclic of index p matrix, generated by a cyclic permutation* $\sigma = (\sigma_1, \sigma_2, \cdots, \sigma_p)$. *Then, for arbitrary complex numbers* α, β, γ, *and* δ,

$$(3.14)\ \det\{\gamma\mathbf{I} - \alpha\mathbf{L} - \beta\mathbf{U} - \delta\mathbf{LU}\} = \det\{\gamma\mathbf{I} - [\alpha^{|\zeta_L|-k}\beta^{|\zeta_U|-k}(\alpha\beta + \gamma\delta)^k]^{1/p}\mathbf{B}\},$$

where $|\zeta_L|$, $|\zeta_U|$ *and* k *are as defined in Sec. 1, and where the convention* $0^0 := 1$ *is used in* (3.14).

Proof. For $\alpha \neq 0$ and $\beta \neq 0$, Lemma 4 is the straightforward consequence of (1.17) and Lemmas 2 and 3. As in the proof of Lemma 3, continuity considerations then allow us to extend (3.14) to cases when $\alpha = 0$ and $\beta = 0$, provided that the convention $0^0 := 1$ is used. ∎

This brings us to the
Proof of Theorem 1. If $\phi(\lambda) := \det(\lambda\mathbf{I} - \mathbf{T}_{\omega,\hat{\omega}})$, then from (3.1) it follows that $\phi(\lambda) = \det\{\gamma\mathbf{I} - \alpha\mathbf{L} - \beta\mathbf{U} - \delta\mathbf{LU}\}$ from (3.1). Thus, from (3.14) of Lemma 4, we further have

$$(3.15) \qquad \phi(\lambda) = \det\left\{\gamma\mathbf{I} - [\alpha^{|\zeta_L|-k}\beta^{|\zeta_U|-k}(\alpha\beta + \gamma\delta)^k]^{1/p}\mathbf{B}\right\}.$$

As remarked at the very beginning of this section, λ is an eigenvalue of $\mathbf{T}_{\omega,\hat{\omega}}$ if and only if $\phi(\lambda) = 0$, i.e. (cf. (3.15)), if and only if

$$(3.16) \qquad \det\left\{\gamma\mathbf{I} - [\alpha^{|\zeta_L|-k}\beta^{|\zeta_u|-k}(\alpha\beta + \gamma\delta)^k]^{1/p}\mathbf{B}\right\} = 0.$$

Now, the proof follows the procedure of the proof of (1.2′) (cf. [6, Th. 4.3]). First, from the definitions of (3.2), there follows

$$(3.17) \qquad \alpha\beta + \gamma\delta = \lambda(\omega + \hat{\omega} - \omega\hat{\omega})^2.$$

Since B is weakly cyclic of index p, it follows from (3.15),(3.2), and Romanovsky's Theorem (cf. [6, p. 40]) that

$$
\begin{aligned}
(3.18) \quad \phi(\lambda) &= \gamma^m \prod_{i=1}^{r}\{\gamma^p - \alpha^{|\zeta_L|-k}\beta^{|\zeta_U|-k}(\alpha\beta + \gamma\delta)^k\mu_i^p\} \\
&= [(\lambda - (1-\omega)(1-\hat{\omega})]^m \prod_{i=1}^{r}\{[\lambda - (1-\omega)(1-\hat{\omega})]^p - \\
&\quad \lambda^k(\omega + \hat{\omega} - \omega\hat{\omega})^{2k}(\lambda\omega + \hat{\omega} - \omega\hat{\omega})^{|\zeta_L|-k}(\lambda\hat{\omega} + \omega - \omega\hat{\omega})^{|\zeta_U|-k}\mu_i^p\},
\end{aligned}
$$

where the μ_i are nonzero eigenvalue of $\mathbf{B}$ if $r \geq 1$ and where m is a nonnegative integer. To establish the second part of this theorem, let μ be an eigenvalue of $\mathbf{B}$ and let $\hat{\lambda}$ satisfy (2.1). Then, one of the factors of $\phi(\hat{\lambda})$ of (3.18) vanishes, proving that $\hat{\lambda}$ is an eigenvalue of $\mathbf{T}_{\omega,\hat{\omega}}$, the desired second part of Theorem 1. To establish the first part of Theorem 1, let $\omega + \hat{\omega} - \omega\hat{\omega} \neq 0$ and let λ be an nonzero eigenvalue of $\mathbf{T}_{\omega,\hat{\omega}}$. It follows that at least one factor of (3.18) vanishes. It is convenient to note that (2.1), from (3.2) and (3.17), can be expressed as

$$(3.19) \qquad \gamma = \lambda^k \alpha^{|\zeta_L|-k} \beta^{|\zeta_U|-k} (\omega + \hat{\omega} - \omega\hat{\omega})^{2k} \mu^p.$$

If $\mu \neq 0$ and μ satisfies (3.19), then, assuming in addition that $\alpha\beta \neq 0$, we must have that $\gamma = \lambda - (1 - \omega)(1 - \hat{\omega}) \neq 0$. Thus, (2.1) is valid for some nonzero μ_i where $1 \leq i \leq r$. Combining this with (2.1), we have that $\mu^p = \mu_i^p$. Taking pth roots, then

$$(3.20) \qquad \mu = \mu_i e^{2\pi i s/p},$$

where s is a nonnegative integer satisfying $0 \leq s < p$. But, from the weakly cyclic of index p nature of the matrix $\mathbf{B}$, it is evident that μ is also an eigenvalue of $\mathbf{B}$, which is the desired first part of Theorem 1. To conclude the proof, if $\omega + \hat{\omega} - \omega\hat{\omega} \neq 0$, if λ is a nonzero eigenvalue of $\mathbf{T}_{\omega,\hat{\omega}}$, and if $\mu = 0$ satisfies (3.19), then we must show that $\mu = 0$ is an eigenvalue of $\mathbf{B}$. But with these hypotheses and $\alpha\beta \neq 0$, it is evident from (3.19) that $\gamma = 0$. In this case, (3.16) reduces to

$$(3.21) \qquad \det\left\{ -\left[\alpha^{|\zeta_L|-k} \beta^{|\zeta_U|-k} \lambda^k (\omega + \hat{\omega} - \omega\hat{\omega})^{2k} \right]^{1/p} \mathbf{B} \right\} = 0.$$

But, as the multiplicative factor of $\mathbf{B}$ in (3.21) is nonzero, then $\det \mathbf{B} = 0$. Hence, $\mu = 0$ is an eigenvalue of $\mathbf{B}$ which is again the desired first part of Theorem 1, under the added assumption that $\alpha\beta \neq 0$. To establish the first part of Theorem 1 when $\alpha\beta = 0$ is similar but tedious, and this is omitted. ∎

Acknowledgements

The research of the authors was supported in part by the Air Force Office of Scientific Research under AFOSR-85-0245. This paper was originally typed by Gail deBostis in LATEX and reformatted by Lisa Laguna.

References

[1] Gong, L., and D.Y. Cai. "Relationship between eigenvalues of Jacobi and SSOR iterative matrix with p-weak cyclic matrix," *J. Computational Mathematics of Colleges and Universities* **1** (1985) 79–84 (in Chinese).

[2] Nickols, N.K., and L. Fox. "Generalized consistent ordering and the optimum successive over-relaxation factor," *Numer. Math.* **13** (1969) 425–433.

[3] Saridakis, Y.G. "On the analysis of the unsymmetric successive overrelaxation method when applied to p-cyclic matrices," *Numer. Math.* **49** (1986) 461–473.

[4] Varga, R.S., W. Niethammer, and D.Y. Cai. "p-cyclic matrices and the symmetric successive overrelaxation method," *Linear Algebra Appl.* **58** (1984) 425–439.

[5] Varga, R.S. "p-cyclic matrices: a generalization of the Young-Frankel successive overrelaxation scheme," *Pacific J. Math.* **9** (1959) 617–628.

[6] Varga, R.S. *Matrix Iterative Analysis.* Englewood Cliffs, N.J.: Prentice Hall, 1962.

[7] Young, D.M., Jr. "Iterative methods for solving partial difference equations of elliptic type," *Trans. Amer. Math. Soc.* **76** (1954) 92–111.

[8] Young, D.M. *Iterative Solution of Large Linear Systems.* New York: Academic Press, 1971.

Chapter 15

The ADI Minimax Problem for Complex Spectra

EUGENE L. WACHSPRESS
University of Tennessee at Knoxville

This paper is dedicated to two mathematicians, one still in his prime and the other approaching middle age. The prime is David Young, in whose honor this conference has been organized on the occasion of his 65th birthday. Dave's contributions to iterative solution of large systems is universally recognized. I have had the privilege of working and playing (tennis, that is) with Dave during the past thirty years and am honored with the opportunity to share in this momentous occasion.

The other mathematician to whom I dedicate this paper is William B. Jordan of Scotia, NY. *SIAM Review* problem-solvers will recognize Bill as one of the more frequent submitters of elegant published solutions. Bill retired from GE fifteen years ago and celebrated his eightieth birthday a few months ago. It was Bill who introduced me to elliptic functions and the beautiful theory underlying solution of the ADI minimax problem. When I uncovered the generalization to complex spectra reported here, I asked Bill if he had anything on complex moduli on the unit circle. His response was that "back in '56 I derived a general formula for the transformation of parameter in Theta functions ... one of which is applicable ...". After consulting my calculator, I decided that Bill was indeed referring to 1956.

Thank you Dave and Bill for your friendship and guidance over the years.

Abstract

The ADI minimax problem has been solved for a range of complex spectral domains. This analysis was motivated by new application to solution of Sylvester's equation. The commutation property required for application to the Dirichlet problem is not restrictive in this new application. However, complex spectra to which conventional ADI theory does not apply are often encountered. Theory of elliptic functions plays a prominent role in the analysis. Optimum parameters are derived only for a class of "elliptic-function domains" which are not apt to occur in practice. However, this theory is then applied to obtain nearly-optimum parameters for realistic spectral regions. Nonsymmetric systems create other problems. Convergence of ADI iteration is retarded by deficient eigenvector spaces. A Lanczos-type generation of a basis for deficient spaces is described. This basis is used to develop a low-order system which may be solved by a direct method to correct the result of the ADI iteration.

1 Introduction and Review of Results for Real Spectra

Initial applications of the Alternating Direction Implicit (ADI) method were for finding the solution of elliptic difference equations with real spectra. The theory for the choice of optimum iteration parameters [1,2] drew heavily on modular transformations of elliptic functions. Recent application of ADI iteration to numerical solution of the Sylvester and Lyapunov matrix equations [3] required generalization to complex spectra. Just as "backward error analysis" proved fruitful in studies of numerical stability, "backward spectrum analysis" has proved to be useful for ADI iteration. The ADI parameters determined for real spectra are also optimal for a class of complex spectra which we denote as "elliptic-function regions." One may imbed a given spectral region in an elliptic-function region for which optimum parameters and the resulting error reduction are known. Alternatively, one may draw upon the known spacing of optimum parameters for the elliptic-function regions to guide selection of effective parameters for actual regions.

The ADI equations for iterative solution of the linear system $\mathbf{Au} = \mathbf{b}$ with given n-vector $\mathbf{b}$, $n \times n$ SPD (symmetric positive definite) matrix $\mathbf{A}$ that splits into the sum of SPD matrices $\mathbf{H}$ and $\mathbf{V}$, and initial estimate $\mathbf{u}(0)$ are

$$[\mathbf{H} + p(j)\mathbf{I}]\mathbf{u}(j - 1/2) = -[\mathbf{V} - p(j)\mathbf{I}]\mathbf{u}(j - 1) + \mathbf{b}$$
$$[\mathbf{V} + q(j)\mathbf{I}]\mathbf{u}(j) = -[\mathbf{H} - q(j)\mathbf{I}]\mathbf{u}(j - 1/2) + \mathbf{b}$$
$$j = 1, 2, ..., J.$$

This is a "model" ADI problem when $\mathbf{H}$ and $\mathbf{V}$ commute. Theory for selection of the parameters $p(j)$ and $q(j)$ to yield the desired solution to a model problem in the least number of iterations is the ADI minimax problem.

Although $\mathbf{HV} - \mathbf{VH}$ is zero for only a small class of separable problems in the application to elliptic difference equations for which the method was initially developed, the Sylvester and Lyapunov equations always satisfy the "commutation condition." However, in this recently discovered application the spectra are often not real. Effective implementation of ADI iteration to this new problem area required the generalization of the theory to complex spectra. Initial research on this problem [4] was restricted to small perturbations from the real line. This theory has now been generalized to encompass a wide range of spectral domains, including spectra anticipated in practice.

Complex iteration parameters enter in conjugate pairs. By combining the two iterations with a conjugate pair, one can perform the iteration in real arithmetic with essentially no increase in computation time over that required for two iterations with real parameters.

This theory has been verified with the numerical solution of the Lyapunov matrix equation by ADI iteration. Some of these studies were reported by Saltzman [4].

The ADI minimax problem is to find the parameter J-tuple $w(j)$ which minimizes the maximum absolute value over a specified domain D of the function:

(1.1)
$$g(z) = \prod_{j=1}^{J} \frac{[w(j) - z]}{[w(j) + z]}.$$

When $D = \{z \mid 0 < a \leq z \leq b\}$, the analysis of W.B. Jordan [1] consistent with earlier work of Chebyshev and Solotareff [2] gives the definitive solution:

$$w(j) = b \quad \mathrm{dn}\left[\frac{(2j-1)K}{2J}, k\right]$$

(1.2)
$$j = 1, 2, ..., J$$

where dn is the elliptic dn function of modulus $k = \sqrt{1 - k'^2}$ and $k' = a/b$. When $k' << 1$, as is often the case, the elliptic function is approximated quite well (p. 191 in [1]) when $r(j) \geq 1/2$ by

(1.3)
$$\mathrm{dn}[r(j)K, \quad k] = \frac{2q'^{\frac{r}{2}}(1 + q'^{1-r} + q'^{1+r})}{(1 + 2q')(1 + q'^{r})},$$

where $q' = k'^2(1 + k'^2/2)/16$ and $r(j) = (2j-1)/2J$. The remaining values are computed from $\mathrm{dn}[(1-r)K, \quad k] = k'/\mathrm{dn}(rK, \quad k)$. The maximum value of $\mid g(z) \mid$ over $[a, b]$ may be approximated in this case by

(1.4)
$$R = \max_{a \leq z \leq b} \mid g(z) \mid = 2\exp\left[-\frac{\pi^2 J}{2\ln\frac{4}{k'}}\right].$$

The error reduction for J ADI iterations is bounded by R^2.

2   Early Analysis of Complex Spectra

The earliest reported analysis of complex spectra was in the Saltzman thesis [4] which applied primarily to spectra with relatively small imaginary components. A review of this analysis provides a springboard for study of more general spectra.

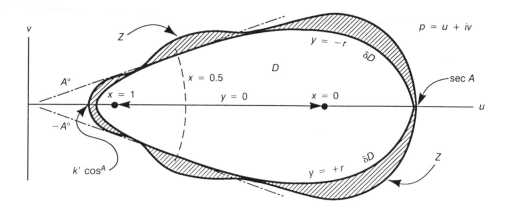

Figure 15.1 The eigenvalue domain.

Let the eigenvalues p over which the error reduction is to be minimized for J iterations be enclosed by the elliptic function region D:

$$(2.1) \qquad D = \{p = \mathrm{dn}(zK, \; k) \mid z = x + iy, \; 0 \le x \le 1 \text{ and } \mid y \mid \le r\}.$$

When $1 - k \ll 1$, D is is an egg-shaped region with parameters displayed in Fig. 15.1. The complementary modulus $k' = (1 - k^2)^{1/2}$ is $\ll 1$. The complete elliptic integral for modulus k is approximated well by $K = \ln(4/k')$, and for k' by $K' = \pi/2$. Region D is tangent at $\sqrt{k'} \exp[iKr]$ to the ray from the origin at angle $A = rK = r\ln(4/k')$. This tangent point in D corresponds to $x = 0.5$ and $y = -r$. It should be noted that the inversion of D in the circle of radius $\sqrt{k'}$ is D.

If the ADI parameters are computed from Eq. (1.3), then

$$(2.2) \qquad\qquad g(z) = k(J)^{1/2}\mathrm{sn}[(1 + 2Jz)K(J), \; k(J)],$$

where $k(J)$ may be computed from k and J. As J increases, $k(J)$ goes to zero, and a desired error reduction is attained by suitable choice of J. We assume henceforth that J is sufficiently large that $k(J) \ll 1$. It can then be shown by approximating elliptic functions of small modulus with trigonometric functions and those of modulus close to unity with hyperbolic functions that, for all z on a closed curve Z (shown in Fig. 15.1) which are close to the boundary of D, $\mid g(z) \mid$ has the constant value of

$$(2.3) \qquad\qquad R(Z) = 2\exp\left[-\frac{\pi^2 J}{2\ln(\frac{4}{k'})}\right]\cosh\left[\frac{J\pi A}{\ln(\frac{4}{k'})}\right].$$

Rouche's Theorem may then be used to prove [4] that this is the least possible value for R attainable with J parameters for the domain bounded by curve Z. We note that $\mid g(z) \mid \leq R$ of Eq. (2.3) for all p in D. As J increases, Z approaches the boundary of D. Note that if we define the x-intercepts as a and b, then the modulus of the elliptic function satisfies

$$(2.4) \qquad k' = \frac{a}{b}\sec^2 A.$$

When J is large enough that $\cosh[\cdot]$ can be approximated well by $0.5\exp[\cdot]$, the value of R may be approximated by

$$(2.5) \qquad R(Z) = \exp\left[-\frac{\pi J(\pi - 2A)}{2\ln(\frac{4}{k'})}\right].$$

Thus, it is seen that as A approaches $\pi/2, R$ approaches unity. This is the correct limit for eigenvalues on the imaginary axis. This form allows one to compute the loss in convergence as a function of A. For example, when $A = \pi/4$, twice as many iterations are required as when A is close to zero. However, before one can use these results on problems where A is not small, one must review some of the assumptions leading to Eq. (2.5). In particular, the value of k' in Eq. (2.4) is valid only for a particular range of a/b and A.

It is easily proved [4] that when the spectrum is bounded by a circle the optimum parameters are repeated use of the single value $w(j) = \sqrt{ab}$. This corresponds to $k' = 1$ and $A = \arccos\frac{2\sqrt{a/b}}{1+(a/b)}$. Clearly, Eq. (2.4) is not valid in this case. In the next section, theory will be developed for the entire range of elliptic-function domains varying from the real line to a disk and from the disk to a circle arc.

3 The Family of Elliptic Function Domains

We retain the domain of Eq. (2.1) but drop the approximations based on $k' \ll 1$. The ratio of the real intercepts is obtained from formulae 16.8.3 and 16.20.3 in Abramowitz and Stegun [5] as

$$(3.1) \qquad \frac{a}{b} = \frac{\mathrm{dn}[K(1 + ri), k]}{\mathrm{dn}[Kri, k]} = \frac{k'\mathrm{cn}^2(Kr, k')}{\mathrm{dn}^2(Kr, k')}.$$

By 16.9.1 in [5], $\mathrm{cn}(\mathrm{mod}\ k') = (\mathrm{dn}^2 - k^2)/k'^2$, and 3.1 may be solved for dn^2:

$$(3.2) \qquad \mathrm{dn}^2(Kr, k') = \frac{1 - k'^2}{1 - \frac{ak'}{b}}$$

We then obtain

$$(3.3) \qquad \mathrm{cn}^2(Kr, k') = \frac{\frac{a}{b}(1 - k'^2)}{k'[1 - \frac{ak'}{b}]},$$

and since $sn^2 = 1 - cn^2$, we have

(3.4)
$$sn^2(Kr, k') = \frac{1 - \frac{a}{bk'}}{1 - \frac{ak'}{b}}.$$

Note that when $r = 0, k' = a/b$ is the appropriate value for this real domain and this yields $dn = 1, cn = 1$, and $sn = 0$ in the above equations.

The maximum angle is attained when $x = 1/2$ and $|y| = r$ as in the previous analysis. However, it is crucial that we not assume $k' \ll 1$ in evaluating this angle. As the boundary becomes more circular, k' approaches unity. Formula 16.21.4 and Table 16.5 in [5] yield

(3.5)
$$\tan^2 A = (1 - k')^2 \frac{sn^2(Kr, k')}{cn^2(Kr, k')dn^2(Kr, k')}.$$

Substitution of Eqs. (3.1)–(3.4) into Eq. (3.5) results in

(3.6)
$$\tan^2 A = \frac{(k' - \frac{a}{b})(1 - \frac{ak'}{b})}{\frac{a}{b}(1 + k')^2}.$$

Given a domain with angle A and real intercept ratio a/b, we may solve Eq. (3.6) for k'. We define

(3.7)
$$\cos^2 B = \frac{2}{1 + \frac{1}{2}(\frac{a}{b} + \frac{b}{a})}$$

and

(3.8)
$$m = \frac{2\cos^2 A}{\cos^2 B} - 1.$$

If $A < B$, then $m > 1$ and we obtain from Eq. (3.6):

(3.9.1)
$$k' = \frac{1}{m + \sqrt{m^2 - 1}} \quad \text{in } (0, 1].$$

and

(3.9.2)
$$w(j) = \sqrt{\frac{ab}{k'}} dn \left[\frac{(2j - 1)K}{2J}, \quad k \right]$$
$$j = 1, 2, \ldots, J$$

Two limiting cases are of interest. When $m \gg 1$, we have $k' = p(a/b) \ll 1$ and from Eq. (3.6), $\tan^2 A = p - 1$, or $p = \sec^2 A$ as in Eq. (2.4) and in the early analysis.

Next let $m = 1$, the smallest value for which k' remains real. In this limit $k' = 1$ and Eq. (3.6) yields $\tan A = \frac{(1 - \frac{a}{b})}{2\sqrt{a/b}}$. Hence, $\cos A = \frac{2\sqrt{a/b}}{(1 + \frac{a}{b})}$ as was previously derived for the disk. It is thus established that these new relationships provide a transition between the real line and the disk. It should be noted that in this limit $Kr \to K'$ which becomes infinite at $k' = 1$ and that the elliptic- function region does approach a disk rather than a point.

To illustrate a spectrum in the transition region, let $a/b = 0.1$ and $A = 45°$. Then $m = 2.025$ and $k' = 0.264$. Note that $(a/b) \sec^2 A = 0.2$. The larger value of k' here reflects the contraction of the parameters toward $\sqrt{ab}$ as the domain moves from the real line to the disk. When $m \geq 1$ the optimum parameters are real. If $m < 1$ all the parameters are complex, except for one real value at $\sqrt{ab}$ when J is odd. We circumvent analysis with elliptic functions of complex moduli by defining a dual spectrum. To motivate the dual spectrum technique, we consider optimum parameters for the spectrum consisting of the arc of the unit circle between $-A$ and $+A$. The folding $z' = \frac{(z+1/z)}{2}$ transforms the arc into the real interval $[\cos A, 1]$. The dual interval $[a, 1/a]$ folds into $[1, \sec A]$ when $a = \tan(\pi/4 - A/2)$. Hence, if we first compute the optimum parameters over $[a, 1/a]$ for J iterations, these parameters will fold into parameters over the interval $[1, \sec A]$ which gives the proper Chebyshev alternating extremes property. The reciprocal of these parameters will retain this property over the interval $[\cos A, 1]$. The inverse transformation back to the arc will then yield the optimum parameters over the arc! When J is odd, the real parameter at angle $A(j) = 0$ is used only once and all the other parameters on $[\cos A, 1]$ transform back into the $+/-$ angles on the arc.

The recipe for computing the optimum $A(j)$ for $j = 1, 2, ..., J$ derived on this basis is:

(3.10.1)
$$k' = \tan^2(\pi/4 - A/2)$$

(3.10.2)
$$z(j) = \frac{1}{\sqrt{k'}} \mathrm{dn}\left[\frac{(2j-1)K}{2J}, k\right]$$

(3.10.3)
$$w(j) = \frac{1}{2}\left[z(j) + \frac{1}{z(j)}\right]$$

$$j = 1, 2, ..., \text{ integer-part-of } \left(\frac{1+J}{2}\right)$$

(3.10.4)
$$A(2j-1) = \arccos \frac{1}{w(j)},$$

(3.10.5)
$$A(2j) = -A(2j-1)$$

(3.10.6)
$$w(j) = \exp[iA(j)].$$

When J is odd, the value $A[(1+J)/2] = 0$ is not repeated. This technique generalizes to elliptic-function spectra. The actual and dual spectra fold into reciprocal spectra with m' for the dual spectrum > 1 when m for the actual spectrum is < 1. The algebra is not trivial, but the resulting equations are easily verified. The duality relationships are remarkable. They highlight an elegant application of classical analysis to a crucial problem of numerical analysis.

The elliptic spectrum is defined by the triplet $\{a, b, A\}$. The dual elliptic spectrum is defined by the triplet $\{a', 1/a', A'\}$, with $A' = B$ of Eq. (3.7) and

(3.11)
$$a' = \tan\left(\frac{\pi}{4} - \frac{A}{2}\right).$$

Substituting this value for a' into Eq. (3.7), we find that

(3.12)
$$B' = A$$

and, therefore,

(3.13)
$$m' = \frac{2\cos^2 B}{\cos^2 A} - 1$$

must be greater than 1 when $m < 1$. We use m' in place of m in Eq. (3.9) and compute the optimum real parameters $\{w'(j)\}$ for the dual problem. The corresponding parameters for the actual elliptic spectrum may then be computed from:

(3.14)
$$\cos A(j) = \frac{2}{w'(j) + \frac{1}{w'(j)}}$$

for $j = 1, 2, ...,$ integer-part-of $\left(\frac{1+J}{2}\right)$

(3.15)
$$w(2j - 1) = \sqrt{ab} \ \exp[iA(j)],$$

and
(3.16)
$$w(2j) = \sqrt{ab} \ \exp[-iA(j)].$$

When J is odd, Eq. (3.16) is dropped for $2j = J + 1$ and the last value computed by Eq. (3.15) is $w(J) = \sqrt{ab}$.

This procedure is illustrated with $\{a, b, A\} = \{0.1, 1.0, 60°\}$. We compute $\cos^2 B = 2/(1.0 + 10.1/2) = 0.3306$ and since this is greater than $\cos^2 60° = 0.25, m = 0.5125 < 1$. For the dual problem, $m' = 2(.3306)/.25 - 1. = 1.645$ and $k' = 0.3389$. For $J = 2$, we compute $w'(1) = 0.685114$, and $w'(2) = 1.45961$ so $2/[w'(1) + 1/w'(1)] = 0.93252$ and $A(1) = \arccos(0.93252) = 21.17°$. The optimum parameters are, therefore, $w(1) = 0.316 \exp(i21.17°)$ and $w(2) = 0.316 \exp(-i21.17°)$.

The range of elliptic-function domains for which we have now developed a theory for computing nearly optimum parameters is displayed in Fig. 15.2.

We have yet to estimate R when k' is not $\ll 1$. Dropping this assumption but retaining the assumption that J is large enough so that $R \ll 1$, we obtain for a real spectrum the replacement for Eq. (1.4) of

(3.17)
$$R = 2\exp\left[-\frac{\pi J K'}{K}\right].$$

Note that when $k' \ll 1, K' = \pi/2$ and $K = \ln(4/k')$ so that Eq. (3.17) reduces to Eq. (1.4) when k' is small.

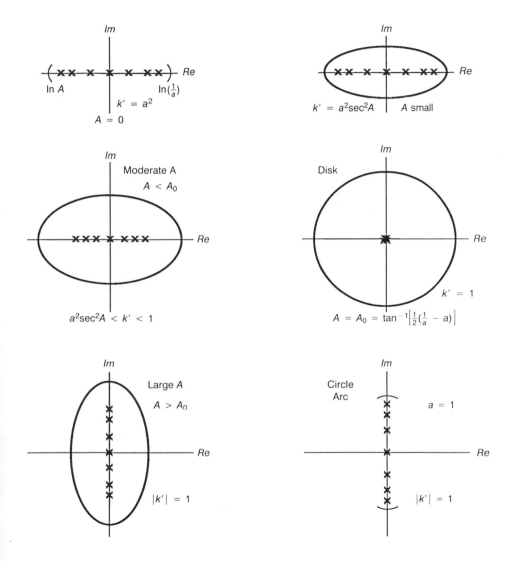

Figure 15.2 Logarithm of elliptic-function spectra for $0 \leq A < \pi/2$. $\times$ = optimum iteration parameters for $J = 7$

For the more general complex spectrum, this must be multiplied by the factor $\cosh(\pi J r)$, but r can no longer be approximated by $A/\ln(4/k')$. We first note that

$$(3.18) \qquad R(Z) = 2\exp\left[-\frac{\pi J K'}{K}\right]\cosh(\pi J r).$$

When the argument of the cosh is large enough for the exponential approximation, we obtain the replacement for Eq. (2.5) with v defined as $v = K' - rK$:

$$(3.19) \qquad R(Z) = \exp(-\frac{\pi v J}{K}).$$

When the spectrum is real and $a/b \ll 1$, the value for K' is close to $\pi/2$ and $r = 0$. As the spectrum varies from the real line segment to the disk, $K' - rK$ decreases toward $\sqrt{a/b}$ even though K' grows without bound. Thus, $v = K' - rK$ is the appropriate parameter to evaluate.

From Eq. 16.8.1 in [4] and Eq. (3.1), we determine that

$$(3.20) \qquad \operatorname{sn}(v, k') = \operatorname{cd}(rK, k') = \sqrt{\frac{a}{bk'}}.$$

When $k' \ll 1$, the sn function may be replaced by the sin and

$$(3.21) \qquad v = \arcsin\sqrt{\frac{a}{bk'}}\ , k' \text{ small}.$$

For small values of the angle A, $\quad k' = \frac{a}{b}\sec^2 A$, and

$$(3.22) \qquad v = \arcsin(\cos A) \quad = \quad (\frac{\pi}{2} - A)$$

and when $k' \ll 1, K$ is approximated by $\ln(4/k')$. This is in agreement with the approximation in Eq. (2.5), which is the small-k' and small-A approximation.

When $1 - k' \ll 1$, the sn function may be replaced by tanh, and

$$(3.23) \qquad v = \operatorname{arctanh}\sqrt{\frac{a}{bk'}}, \quad k' \text{ close to unity .}$$

The log form of the arctanh in 4.6.22 of [5] yields

$$(3.24) \qquad \exp(-v) \quad = \quad \left[\frac{1 - \sqrt{\frac{a}{bk'}}}{1 + \sqrt{\frac{a}{bk'}}}\right]^{1/2}$$

In this limit, $K = \pi/2$ and $R(Z) = \exp(-2vJ)$ approaches the correct value of

$$(3.25) \qquad R \text{ (disk)} \quad = \quad \left[\frac{1 - \sqrt{\frac{a}{b}}}{1 + \sqrt{\frac{a}{b}}}\right]^{J}$$

In general,

(3.26)
$$v = F\left[\arcsin\left(\sqrt{\frac{a}{bk'}}\right),\ k'\right],$$

where F is the incomplete elliptic integral of the first kind defined by 17.2.6 in [5]. To evaluate R, we must compute K and v. This is easily accomplished with the arithmetic geometric mean algorithm described in Sec. 17.6 of reference [5].

To compute K, we set $a(0) = 1$ and $b(0) = k'$. Then $a(s)$ is the arithmetic mean of $a(s-1)$ and $b(s-1)$ while $b(s)$ is the geometric mean of $a(s-1)$ and $b(s-1)$. The common limit is the arithmetic-geometric mean of 1 and k'. We denote this limit by $p(k')$. Then $K = \pi/2p(k')$. To compute v, we replace k' by $k = \sqrt{1 - k'^2}$ and denote the AGM sequences by $a'(s)$ and $b'(s)$ to obtain the AGM of 1 and k as $p(k)$. We then apply Algorithm 17.6.8 in [5], noting that the primed sequence is used since the modulus in Eq. (3.26) is k' rather than k. The initial angle for this algorithm is the amplitude of F in Eq. (3.26): $C(0) = \arcsin\sqrt{a/bk'}$. Successive angles are defined by

(3.27)
$$\tan[C(s) - C(s-1)] = \frac{b'(s)}{a'(s)} \tan C(s-1),$$

and when $b'(n)/a'(n)$ is sufficiently close to unity:

(3.28)
$$v = \frac{C(n)}{2^n a'(n)}.$$

The value of $a'(n)$ should be quite close to $p(k)$. Convergence is characterized by C doubling each iteration.

The procedure will now be illustrated for the spectrum $\{a, b, A\} = \{0.1, 1.0, 45°\}$, for which we have already determined that $k' = 0.26414$. We compute $k = 0.96448$ and $C(0) = 37.973°$ or 0.66276 radians. The AGM algorithms give $K = 2.74864$ and $v = 0.67475$. Substituting these values into Eq. (3.19), we obtain

(3.29)
$$R = (0.4624)^J.$$

We recall that our approximation is intended for J large enough that $R \ll 1$. A comparison of the true values with the values obtained by Eq. (3.29) reveals that Eq. (3.29) is within a few percent of truth when $J \geq 4$:

J	1	2	4
R(Truth)	0.5195	0.2365	0.0472
R(Eq. (3.29))	0.4719	0.2230	0.0495

The approximation to the dn function in Eq. (1.3) deteriorates as k' increases toward unity. A simple and accurate alternative will now be described. Both K and K' are computed from the arithmetic-geometric means for estimation of the error reduction and are thus available for determining the nomes q and q'. If $k' \geq 1/\sqrt{2}$, then $K \leq K'$ and $q = \exp(-\pi K'/K) \leq 0.0432$. The nome expansion 16.23.3 in [5] converges rapidly, and we need only retain the first few terms:

$$(3.30) \qquad \operatorname{dn}(zK, k) = \frac{\pi}{2K} + \frac{2\pi}{K}\left[\frac{q\cos\pi z}{1+q^2} + \frac{q^2\cos 2\pi z}{1+q^4}\right].$$

We use this approximation when $z = (2j-1)/2J \geq 1/2$, and the remaining parameters are computed from: $\operatorname{dn}[(1-z)K, k] = k'/\operatorname{dn}(zK, k)$. If $k' < 1/\sqrt{2}$, then $q' = \exp(-\pi K/K') < 0.0432$ and the approximation in Eq. (1.3) is adequate.

When $J = 2^n$, the quadratic recursion relationships on p. 199 in [1] yield the exact dn functions in the absence of roundoff error, and we may compare the approximations with these values. For example, when $k' = 0.8$ and $J = 4$, we have Table 15.1.

n	$a(n)$	$b(n)$	quadratic recursions	
0	1.00000	0.80000	$\{w\} = 0.9915$	0.99345
	$\downarrow$		0.80685	0.85703
			$\uparrow$	$\uparrow$
1	0.90000	0.89443	$\{w'\} = 0.89918$,	0.89524
	$\downarrow$		$\uparrow$	
2	0.8972136	0.8972093	$\rightarrow$ $w'' = 0.89721$	
	$p(0.8) = 0.8972114$			

Table 15.1

Note that if $J = 8$ and only six significant digits are carried, this algorithm gives the same values, each of multiplicity two. This would be quite acceptable for the ADI iteration with this value for k'. To use Eq. (3.30) instead, we first note that $p(0.8) = 0.89721 = \pi/2K$. The AGM algorithm with $k = 0.6$ replacing k' yields $0.78725 = \pi/2K'$. Thus, $q = \exp(-\pi 0.89721/0.78725) = 0.02787$. Eq. (3.30) with $z = 5/8$ and $z = 7/8$ yields $\operatorname{dn}(5K/8, k) = 0.85698$ and $\operatorname{dn}(7K/8) = 0.99153$. These values are correct to four significant digits. A full cycle of eight distinct parameters would be computed if J were equal to 8.

4 Spectral Boundary

The eigenvalue spectrum of a real matrix $\mathbf{B}$ may be bounded by various methods. A particularly useful bound is that of Bendixson's Theorem (p. 69 in [6]). Let

B have an SPD symmetric component $\mathbf{M} = (\mathbf{B} + \mathbf{B}^T)/2$ and an antisymmetric component of $\mathbf{N} = (\mathbf{B} - \mathbf{B}^T)/2$. The condition that **M** be SPD is consistent with applicability of ADI iteration. This is a stability property in control theory application. If the eigenvalue interval of **M** is $[a, b]$ and the spectral radius of **N** is g, then the eigenvalues of **B** are bounded by the rectangle $a \leq x \leq b$, $|y| \leq g$. This leads to a spectrum within a cone with apex half-angle $A = \arctan(g/a)$. The ADI convergence depends strongly on this angle, especially when $a \ll 1$. A sharper bound on A is obtained as follows:

Theorem 4.1 $A \leq \arctan$ *[spectral radius of* $\mathbf{M}^{-1}\mathbf{N}$*]*.

Proof: Let $\mathbf{By} = (p + iq)\mathbf{y}$ with $\mathbf{y} = \mathbf{u} + i\mathbf{v}$ be the eigensolution for which angle A is attained. Then

$$(4.1) \qquad \mathbf{y^*By} = (p + iq)\mathbf{y^*y} \quad \text{and} \quad A = \arctan \frac{2\mathbf{u}^T\mathbf{Nv}}{\mathbf{u}^T\mathbf{Mu} + \mathbf{v}^T\mathbf{Mv}}$$

since $\mathbf{u}^T\mathbf{Nu} = \mathbf{v}^T\mathbf{Nv} = 0$. Matrix $\mathbf{M}^{-1}\mathbf{N}$ is similar to matrix $\mathbf{M}^{-1/2}\mathbf{N}\mathbf{M}^{-1/2}$, which is a normal matrix. Hence, the spectral radius r of $\mathbf{M}^{-1}\mathbf{N}$ satisfies

$$(4.2) \qquad r(\mathbf{M}^{-1}\mathbf{N}) = \max_{\mathbf{w}} \left| \frac{\mathbf{w^*Nw}}{\mathbf{w^*Mw}} \right| \geq \left| \frac{\mathbf{y^*Ny}}{\mathbf{y^*My}} \right| = \tan A.$$

A spectrum may be designated by $\{a, b, g, A\}$. Typical problems for which a significant number of ADI iterations may be required are problems for which $a/b \ll 1$ and $g/b \ll 1$. An upper bound on A is $\arctan(g/a)$, which is attained when **M** and **N** commute.

The maximum eigenvalue of **M** may be bounded by the maximum absolute row sum:

$$(4.3) \qquad b = \max_i \sum_j |m(i, j)|.$$

The minimum eigenvalue may be estimated by several steps of inverse power iteration, starting with $\mathbf{u}(0)^T = [1, 1, \dots, 1]$:

$$(4.4) \qquad \mathbf{Mu^*} = \mathbf{u}(i - 1), \quad \mathbf{u}(i) = \frac{n\mathbf{u^*}}{\| \mathbf{u^*} \|_1}, \quad i = 1, 2,$$

until $\| \mathbf{u^*} \|_1$ remains constant within a prescribed tolerance. The minimum eigenvalue of **M** is then approximated by

$$(4.5) \qquad a = \frac{n}{\| \mathbf{u^*} \text{ last } \|_1}$$

Now let $\mathbf{v}(0)^T = [1, 1, -1, -1, 1, 1, -1, -1, \dots]$ and compute

$$(4.6) \qquad \mathbf{Mv^*} = \mathbf{Nv}(i - 1), \quad \mathbf{v}(i) = \frac{n\mathbf{v^*}}{\| \mathbf{v^*} \|_1}, \quad i = 1, 2,$$

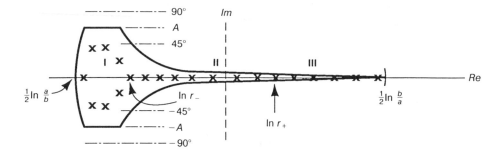

Figure 15.3 Logarithm of a Representative Spectrum.

until $\| \mathbf{v}^* \|_1$ remains constant within prescribed tolerances. Then $\frac{1}{n} \| \mathbf{v}^* \|_1$ is an estimate for the spectral radius of $\mathbf{M}^{-1}\mathbf{N}$ which according to our theorem is a rigorous bound on the tangent of the maximum angle A of the eigenvalues of $\mathbf{B}$:

$$(4.7) \qquad A = \arctan\left[\frac{1}{n} \| \mathbf{v}^* \text{ (last) } \|_1 \right].$$

The magnitude of the imaginary components is bounded by the spectral radius of matrix $\mathbf{N}$, which in turn is bounded by

$$(4.8) \qquad g = \max_i \sum_j | n(i,j) | \; = \; \| N \|_\infty$$

These spectral bounds are all obtained in $O(n)$ flops when matrix $\mathbf{B}$ is $n \times n$. The plot of the logarithm of a representative spectrum in Fig. 15.3 is revealing:

A crucial property of this spectrum is its lack of logarithmic symmetry. An elliptic-function region has logarithmic symmetry. It is apparent that the theory for computing good parameters based on elliptic-function spectra is inadequate for anticipated spectra.

An estimate of location of appropriate parameters for the nonsymmetric spectrum is given by the x's in Fig. 15.3. We now address the computation of this more reasonable parameter selection.

5 Spectrum Partitioning

A qualitative analysis of ADI error reduction for a real spectrum provides a basis for parameter selection for complex spectra. We normalize the spectrum to the interval

$[a, 1/a]$ so that the values of $\log w(j)$ are symmetric about 0. They are spaced almost uniformly near 0 and are closer together toward the ends. The contribution to error reduction is predominantly from neighboring parameters. These fall away on both sides near the center but only on one side near the ends. Hence, they must be more closely spaced near the ends of the interval.

When the spectrum is complex, the spacing must be reduced to yield the appropriate reduction off the real axis. This reduction is a function of the angle subtended by the spectrum. For a rectangular spectrum with maximum imaginary value g and $g/b \ll 1$, the angle varies from its value of $\arctan(g/a)$ at $x = a$ to $\arctan(g/b)$ at $x = b$. The logarithmic spacing of optimum parameters, aside from the previously noted end effects, should therefore increase as x increases. Let $\mathrm{del}(R, A)$ denote the spacing of the parameter logarithms in a region where $k' \ll 1$. Then Eq. (1.3) applies, and the ratio of adjacent parameters for $j = (J-1)/2$ and $j = (J+1)/2$ is approximately equal to $q'^{-\frac{1}{2J}}$. The logarithm of this ratio with $K' = \pi/2$ is $\mathrm{del}(q', J) = -(\ln q')/2J = K/J$. From Eq. (3.19):

$$(5.1) \qquad \mathrm{del}(R, A) = \frac{K}{J} = -\frac{\pi v}{\ln R}.$$

The value of v varies as a function of the angle subtended by the spectrum, increasing as x increases. This suggests a strategy for choosing effective parameters. In the most general case, this strategy will require consideration of three distinct regions. Region I is the left end in which closely spaced (on a logarithmic scale) parameters, resulting from both the relatively large initial angle and the left end effect, gradually become more widely spaced. This spacing approaches the asymptotic spacing of Eq. (5.1) in the absence of the right end-effect.

Region III is the right end with decreasing logarithmic increments as x approaches b. Region II is the transition region in which the spacing gradually increases as x increases from the right end of region I toward the left end of region III as a result of the decreasing spectrum angle. We now make this more precise.

The analysis applies primarily to problems for which $g/b \ll 1$. When g/b is larger, an alternative strategy should be considered.

A. Region II: The Asymptotic Region

Given $\{a, b, g, A\}$ and a desired error reduction of R, we compute a set of parameters to be used in the central or asymptotic region.

We assume that

$$(5.2) \qquad k' = \frac{a}{b} \ll 1 \quad \text{and} \quad g' = \frac{g}{b} \ll 1$$

If the angle A bounding the spectral region D is less than $45°$, we choose the asymptotic region to include eigenvalues with real part between

$$(5.3) \qquad r(-) = b^{1/4}a^{3/4} \quad \text{and} \quad r(+) = a^{1/4}b^{3/4}$$

We then define

(5.4) $A(-) = \arctan \dfrac{g}{r(-)}$ and $A(+) = \arctan \dfrac{g}{r(+)}$.

If $A > 45°$, we choose

(5.5) $r(-) = g$ and $r(+) = g^{1/3}b^{2/3}$

so that

(5.6) $A(-) = 45°$ and $A(+) = \arctan(\dfrac{g}{b})^{2/3}$

For prescribed R and these angles, we compute del($-$) and del($+$) from Eq. (5.1). The log spacing just to the left of $r(-)$ is del($-$) and just to the right of $r(+)$ is del($+$). We assume spacing between $r(-)$ and $r(+)$ which varies geometrically:

$$\text{del}(m) = h^m\text{del}(-) \text{ for } m = 1, 2, ..., M - 1, \text{ and } \text{del}(M) = \text{del}(+).$$

We determine that

(5.7) $h = 1 + \dfrac{\text{del}(+) - \text{del}(-)}{\text{del}(-) + \ln \frac{r(+)}{r(-)}}$,

and that

(5.8) $M = \dfrac{1}{\ln h} \ln \dfrac{\text{del}(+)}{\text{del}(-)}$.

The iteration parameters over the asymptotic region are

$$w(1) = r(-), \quad w(m+1) = w(m)\exp[h^m\text{del}(-)], \quad w(M) = r(+)$$
$$m = 1, 2, ..., M - 2.$$

B. Region III. The Right End

The angle subtended at the right end of the spectrum is small: $g/b \ll 1$. Having already determined a value for $r(+)$, we set $k'(+) = [r(+)/b]^2$ and compute the optimum parameters by Eq. (1.3) since the spectrum is essentially real. The number of parameters, $t(+)$, required between $k'(+)b$ and b to attain the prescribed error reduction R is then computed.

We choose $J(+)$ as the smallest odd integer greater than $t(+)$ and determine the $[1 + J(+)]/2$ parameters between $r(+)$ and b:

$$w(+, j) = b \ \text{dn}\left[\frac{(2j-1)K(+)}{2J(+)}, k(+)\right]$$
$$j = 1, 2, ..., \frac{J(+) + 1}{2}.$$

C. Region I: The Left End

If $A < 45°$, then the left region is treated in much the same manner as the right end. Now $k'(-)$ is computed from Eqs. (3.7)–(3.9) with $r(-)$ replacing b in Eq. (3.7).

We compute

$$(5.9) \qquad t(-) = -\frac{K \ln R}{v\pi}$$

and choose $J(-)$ as the smallest odd integer greater than $t(-)$ and determine the $[1 + J(-)]/2$ parameters between a and $r(-)$:

$$
(5.10) \qquad
\begin{aligned}
w(-,j) &= \frac{\sqrt{ar(-)k'}}{[\mathrm{dn}\frac{(2j-1)K(-)}{2J(-)}, k(-)]}, \\
j &= 1, 2, ..., \frac{1 + J(-)}{2}.
\end{aligned}
$$

If $A > 45°$, we introduce additional parameters. The parameter w yields a factor in $g(z)$ of $\tan 22.5°$ at $z = we^{\pm i\frac{\pi}{4}}$. The prescribed error reduction can be attained only when $|\frac{w-z}{w+z}| \le \tan 22.5°$ for all z on the arc of radius w within the spectral region. When $A > 45°$ parameters are chosen on the arc of radius $w(-,j)$ to assure a factor product $\le \tan 22.5°$ along the arc. The parameters are computed as in Eqs. (3.10). The factor product along the arc is approximated well by Eq. (1.4) with $k' = \tan^2(\pi/4 - A/2)$. The number of parameters M on the arc is the smallest integer greater than

$$(5.11) \qquad M = -0.6381 \ln Y,$$

where $Y = \frac{1}{2}[\tan(\frac{\pi}{4} - \frac{A}{2})]$. The maximum angle for any value of M is readily computed from Eq. (5.11). A few values are:

$$
\begin{array}{ccccccc}
M = & 1 & 2 & 3 & 4 & 5 & 6 \\
A(\max) = & 45° & 80° & 87.9° & 89.6° & 89.91° & 89.98°
\end{array}
$$

This partitioning of the spectrum leads to parameters placed as indicated in Fig. 15.3.

6 Subspace Refinement

A. Excluded Subspaces

When the spectrum includes multiple eigenvalues and the Jordan form of the matrix is not diagonal, convergence may be seriously hampered. If matrix **B** has a generalized eigenvector $\mathbf{v}(e)$ such that $\mathbf{Bv}(e) = p\mathbf{v}(e) + q\mathbf{v}$ with $\mathbf{Bv} = p\mathbf{v}$, then iteration with parameter w acting on the vector $\mathbf{v}(e)$ yields the vector

$[(p - w)/(p + w)]\mathbf{v}(e) + [2qw/(p + w)^2]\mathbf{v}$ and n iterations with this same parameter w yields

(6.1) $$(\frac{p - w}{p + w})^n \mathbf{v}(e) + n \left(\frac{2qw}{(p + w)^2}\right) \left(\frac{p - w}{p + w}\right)^{n-1} \mathbf{v}.$$

If w is close to $p : w = (1 - 2s)p$ with $s \ll 1$, then the last term above goes to zero as ns^{n-1}. For an r-fold degeneracy, the error reduction varies as $\binom{n}{r-1}s^{n-r+1}$. Since ADI iteration is not performed with repeated use of a fixed parameter, the error in deficient subspaces will dominate eventually. One cannot apply more effective parameters without more detailed information on the Jordan form. A direct method seems preferable to an iterative method demanding such knowledge.

There is a theorem that the spectral radius of a matrix is a lower bound on all norms for the matrix and that one can always define a norm as close as desired to the spectral radius. One can therefore find a norm such that the error reduction described in Secs. 1–5 applies in that norm. When the matrix is symmetric, the spectral radius is equal to its Euclidean norm. When it is not symmetric, the Euclidean norm may be much larger than the spectral radius, and it may grow appreciably as the eigenvector space approaches degeneracy.

A supplementary procedure will now be described for handling this situation. This procedure has more general application. When the left end of the spectrum subtends an angle at the origin greater than $80°$, we may choose to ignore eigenvalues outside the $45°$ bound, thereby not reducing the error by the prescribed amount over the excluded space. We include this space with that of the deficient eigenvector space in the supplementary procedure.

B. Direct Solution Over a Subspace

When solving a linear system, one may follow the ADI iteration by an Arnoldi iteration. The ADI iteration filters out the bulk of the error and results in an error component in a relatively low-order subspace. This situation is ideally suited for Arnoldi iteration and is similar to the "generalized ADI iteration" [1] proposed even for symmetric problems. Then the system $\mathbf{Au} = \mathbf{s}$ with $\mathbf{A} = \mathbf{H} + \mathbf{V}$, in which $\mathbf{H}$ and $\mathbf{V}$ do not commute, is solved with ADI iteration for a close commuting problem as a preconditioner for a Lanczos iteration on the actual system.

The interest in applying ADI to nonsymmetric systems arose in connection with the Sylvester equation [3]. Application of subspace refinement to solution of the Sylvester equation will now be exposed. Let $\mathbf{X}(J)$ be the result of J ADI iterations on the system

(6.2) $$\mathbf{AX} + \mathbf{XB} = \mathbf{C},$$

where $\mathbf{A}$ is $n \times n$ real nonsymmetric, $\mathbf{B}$ is $m \times m$ real nonsymmetric, and $\mathbf{C}$ is $n \times m$ real. For the special case of the Lyapunov equation, $\mathbf{B} = \mathbf{A}^T$ and $\mathbf{C}$ is symmetric. One scheme for solving the Sylvester equation involves preliminary reduction of $\mathbf{A}$ and $\mathbf{B}$ to tridiagonal or near tridiagonal form so that even for the general case we are especially interested in sparse matrices $\mathbf{A}$ and $\mathbf{B}$.

Let the error after the ADI iteration be

(6.3) $$\mathbf{Y} = \mathbf{X} - \mathbf{X}(J).$$

Matrix $\mathbf{Y}$ satisfies the Sylvester equation

(6.4) $$\mathbf{AY} + \mathbf{YB} = \mathbf{D}$$

where

(6.5) $$\mathbf{D} = \mathbf{C} - \mathbf{AX}(J) - \mathbf{X}(J)\mathbf{B}.$$

This error should have dominant components in the eigenvector-deficient or subspace excluded from the ADI error reduction.

Matrix $\mathbf{Y}$ admits an expansion of the form:

(6.6)
$$\mathbf{Y} = \sum z(s, s')\mathbf{u}(s)\mathbf{v}(s')^T,$$
$$s = 1, 2, ..., S; \qquad s' = 1, 2, ..., S'.$$

Here, the vectors $\mathbf{u}$ and $\mathbf{v}$ are in the excluded spaces of $\mathbf{A}^T$ and $\mathbf{B}$, respectively. It is assumed that $S \ll n$ and that $S' \ll m$. If this is not anticipated for a class of problems, then a direct solution of the initial Sylvester equation would probably be more efficient for that class of problems.

The matrix $\mathbf{D}$ in Eq. (6.5) may also be expanded in terms of the excluded subspaces. We may generate the two sequences of orthonormal vectors $\mathbf{u}$ and $\mathbf{v}$ with the Lanczos algorithm on pp. 346–347 of [7] applied to the given matrix $\mathbf{D}$. The termination criteria may be relaxed to permit error on the same order as that prescribed for the ADI iteration. A drastic reduction in the error measure should occur after a number of steps equal to the dimension of the excluded subspace, this being much less than either m or n. The result of this factorization of $\mathbf{D}$ is

(6.7.1) $$\mathbf{U}^T\mathbf{DV} = \mathbf{E} \text{ where } \mathbf{D} = \mathbf{UEV}^T$$

and

(6.7.2) $$\mathbf{U}^T\mathbf{U} = \mathbf{V}^T\mathbf{V} = \mathbf{I}.$$

The orthogonal matrices $\mathbf{U}$ and $\mathbf{V}$ are rich in the excluded subspaces of $\mathbf{A}^T$ and $\mathbf{B}$, respectively. They are in general not invariant subspaces of these matrices. One must therefore introduce additional vectors until a prescribed tolerance is satisfied. We seek a matrix $\mathbf{U}$ such that $\mathbf{U}^T$ is a left invariant of $\mathbf{A}$, and a matrix $\mathbf{V}$ which is a right invariant of $\mathbf{B}$. The matrices $\mathbf{U}$ and $\mathbf{V}$ in Eqs. (6.7) are chosen as initial values $\mathbf{U}(0)$ and $\mathbf{V}(0)$ for generation of an extended subspace. Successive extension of $\mathbf{U}$ is accomplished by a block Arnoldi algorithm, modified by discarding of small components. We first compute $\mathbf{W}(1)$ orthogonal to $\mathbf{U}(0)$:

(6.8.1) $$\mathbf{U}(0)^T\mathbf{A} - [\mathbf{U}(0)^T\mathbf{AU}(0)]\mathbf{U}(0)^T = \mathbf{W}(1)^T$$

We discard rows of $\mathbf{W}(1)^T$ of norm less than a prescribed bound such as the ADI error reduction. We orthonormalize the remaining rows to obtain

$$(6.8.2) \qquad\qquad \mathbf{U}(1)^T = \mathbf{Q}(1)\mathbf{W}(1)^T.$$

Thereafter,

$$(6.8.3) \qquad\qquad \mathbf{U}(n-1)^T \mathbf{A}\mathbf{U}(j) = \mathbf{G}(j),$$

$$(6.8.4) \qquad \mathbf{U}(n-1)^T \mathbf{A} - \sum_{j=0}^{n-1} \mathbf{G}(j)\mathbf{U}(j)^T = \mathbf{W}(n)^T,$$

$$(6.8.5) \qquad \mathbf{U}(n)^T = \mathbf{Q}(n)\mathbf{W}(n)^T = \text{orthonormalized rows of } \mathbf{W}(n)^T$$
$$\text{excluding rows of small norm.}$$

The algorithm is terminated when all rows of $\mathbf{W}(n)^T$ have norm less than a prescribed value or when n reaches some maximum allowed value. We then define

$$(6.8.6) \qquad\qquad \mathbf{U} = [\mathbf{U}(0), \mathbf{U}(1), ..., \mathbf{U}(n)]$$

It is easy to show that $\mathbf{U}^T\mathbf{U} = \mathbf{I}$, and in the absence of roundoff error the space of $\mathbf{U}(0)$ is extended to the left- invariant subspace of $\mathbf{A}$ of lowest dimension containing $\mathbf{U}(0)$. We have

$$(6.9) \qquad \mathbf{G} = \mathbf{U}^T\mathbf{A}\mathbf{U}, \quad \mathbf{U}^T\mathbf{U} = I, \quad \mathbf{U}^T\mathbf{A} = \mathbf{G}\mathbf{U}^T + \mathbf{Q}, \quad \|\,\mathbf{Q}\,\| \ll 1.$$

The same procedure is applied with $\mathbf{V}$ and matrix $\mathbf{B}$, except that one now seeks a right invariant subspace. One defines $\mathbf{H} = \mathbf{V}^T\mathbf{B}\mathbf{V}$ and $\mathbf{F} = \mathbf{U}^T\mathbf{D}\mathbf{V}$ and solves for $\mathbf{Z}$ in

$$(6.10) \qquad\qquad \mathbf{G}\mathbf{Z} + \mathbf{Z}\mathbf{H} = \mathbf{F}.$$

One then computes as the correction to X:

$$(6.11) \qquad\qquad \mathbf{Y} = \mathbf{U}\mathbf{Z}\mathbf{V}^T.$$

For the Lyapunov equation, the symmetric Lanczos algorithm yields $\mathbf{D} = \mathbf{U}\mathbf{E}\mathbf{U}^T$, and one need only generate the left invariant subspace of $\mathbf{A}$ with Eqs. (6.8.1)–(6.8.2). Then $\mathbf{H} = \mathbf{G}^T$ in Eq. (6.10) and $\mathbf{Y} = \mathbf{U}\mathbf{Z}\mathbf{U}^T$ in Eq. (6.11).

If the updated $\mathbf{X}$ still does not satisfy the Sylvester equation within prescribed tolerances, one may compute another correction $\mathbf{Y}$ by solving the "iterative refinement" equation

$$(6.12) \qquad\qquad \mathbf{A}\mathbf{Y} + \mathbf{Y}\mathbf{B} = \mathbf{C} - (\mathbf{A}\mathbf{X} + \mathbf{X}\mathbf{B})$$

with ADI iteration.

Numerical studies with ADI solution of the Lyapunov equations have demonstrated the efficacy of this approach. However, problems for which subspace refinement is needed have not yet been examined. This general strategy must be qualified with extensive numerical experimentation.

Acknowledgements

This paper was originally typed by Christine Marks in AMST$_E$X and reformatted in L^{A}T$_E$X by Lisa Laguna.

References

[1] Wachspress, E.L. *Iterative Solution of Elliptic Systems*, Prentice Hall, 1966.

[2] Todd, J. "Applications of Transformation Theory: A Legacy from Zolotarev (1847–1978)," *Approximation Theory and Spline Functions*, S.P. Singh *et al.*, eds., 207–45, D. Reidel Publishing Co.

[3] Wachspress, E.L. "Iterative Solution of the Lyapunov Matrix Equation," *Applied Mathematics Letters* **1** (1988) 87–90.

[4] Saltzman, N. "ADI Parameters for Some Complex Spectra," Univ. of Tennessee Master's thesis, 1987. (Excerpts have been submitted for publication in the *SIAM Jour. of Numer. Analysis*.)

[5] Abramowitz, M., and I.A. Stegun. *Handbook of Mathematical Functions*, NBS AMS–55.

[6] Householder, A.S. *The Theory of Matrices in Numerical Analysis*, Blaisdell, 1964 (Dover–1975).

[7] Golub, G.H., and C. van Loan. *Matrix Computations*, John Hopkins University Press, 1983.

Chapter 16

Some Domain Decomposition Algorithms for Elliptic Problems

MAKSYMILIAN DRYJA
Warsaw University

and

OLOF B. WIDLUND
New York University

Dedicated to David M. Young, Jr., on the occasion of his sixty-fifth birthday.

Abstract

We discuss domain decomposition methods with which the often very large linear systems of algebraic equations, arising when elliptic problems are discretized by finite differences or finite elements, can be solved with the aid of exact or approximate solvers for the same equations restricted to subregions. The interaction between the subregions, to enforce appropriate continuity requirements, is handled by an iterative method, often a preconditioned conjugate gradient method. Much of the work is local and can be carried out in parallel. We first explore how ideas from structural engineering computations naturally lead to certain matrix splittings. In preparation for the detailed design and analysis of related domain decomposition methods, we then consider the Schwarz alternating algorithm, discovered in 1869. That algorithm can conveniently be expressed in terms of certain projections. We develop these ideas further and discuss an interesting additive variant of the Schwarz method. This also leads to the development of a general framework, which already has proven quite useful in the study of a variety of domain decomposition methods and certain related algorithms. We demonstrate this by developing several algorithms and by showing how their rates of convergence can be estimated. One of them is a Schwarz-type method, for which the subregions overlap, while the others are so called iterative substructuring methods, where the subregions do not overlap. Compared to previous studies

273

of iterative substructuring methods, our proof is simpler and in one case it can be completed without using a finite element extension theorem. Such a theorem has, to our knowledge, always been used in the previous analysis in all but the very simplest cases.

1 Introduction

Domain decomposition methods have recently become an important focus of research on numerical methods for partial differential equations. They appear to offer the best promise for the parallel solution of the often very large systems of linear or nonlinear algebraic systems of equations which arise when the elliptic problems of elasticity, fluid dynamics and many other important applications are discretized by finite elements or finite differences. The rapidly growing interest in this field is reflected in a series of SIAM sponsored symposia and mini-symposia exclusively devoted to research in this area; see [9, 15].

The domain decomposition methods considered here can be regarded as *divide and conquer* algorithms. In each step of an iteration, the original discrete elliptic problem is solved on the subregions into which the original region has been divided. (Under favorable circumstances, an already existing code can be used for this purpose.) The interaction between the different parts of the region is handled by transferring suitable data across the interfaces, which are created by the subdivision. These data are generated in each iteration from the residual of the original or a derived system of equations. A conjugate gradient method is often used to accelerate the convergence. In addition to the local problems, some global, coarse problem must be incorporated into the preconditioner in order to obtain a fast rate of convergence in the case of many subregions. That is the case of primary interest in parallel computing research. We note that the need for a coarse, global part of the preconditioner is well known in multi-grid work. More formally, we have shown in our previous work that if in each iteration step of a domain decomposition algorithm information is only exchanged between neighboring substructures, the rate of convergence can be no better than if the conjugate gradient method is used, without any preconditioning, for the coarse model obtained by using the subregions as elements; cf. Widlund [29].

There has been considerable progress in this research area and a number of fast algorithms have been designed and studied for which the condition number of the iteration matrix is uniformly bounded or grows only in proportion to the expression $(1 + \log(H/h))^q, q = 2$ or 3, where H is the diameter of a typical subregion and h the diameter of a typical element into which the subregions are divided; see, e.g., Bramble, Pasciak and Schatz [5, 6, 7, 8] , Dryja [11], Dryja, Proskurowski and Widlund [13], Dryja and Widlund [14] and Widlund [29]. Most of the important results in this field have been developed in a finite element framework and we also adopt such an approach in this paper.

To assess the complexity of domain decomposition algorithms, traditional tools must also be used to measure the cost of solving the different subproblems. If a parallel computing system is used, there are of course many additional aspects to estimating the overall cost of the computation. We provide no detailed discussion of such matters here. Instead, we focus on the design of preconditioners that give rapid rates of convergence while decreasing the amount of work per step, and on the development of a general framework inside which a variety of algorithms can be designed and analyzed. We note that several sets of careful numerical experiments with various domain decomposition algorithms have been carried out on parallel computers; cf. Keyes and Gropp [18, 19], who used a hypercube, and Greenbaum et al. [17], who used an ultra computer prototype built at the Courant Institute.

Domain decomposition methods can be classified according to whether the subregions (substructures) overlap or not. The algorithms which do not use any overlap are often called *iterative substructuring* methods while the others are called *Schwarz-type* methods. This is in recognition of the importance of ideas of structural engineering computing in the development of the former class of methods and of the fundamental contribution by H. A. Schwarz, who introduced his alternating method in 1869; cf. Schwarz [24]. The two classes of methods have much in common and there is a real promise that a unified theory can be developed. This paper is an effort in that direction. We are able to show that some interesting iterative substructuring methods can also be derived from a so called *additive variant* of the Schwarz algorithm. Technically this work also offers something new. One of the algorithms can be analyzed without using a so called *finite element extension theorem*. Such theorems have previously been central to the development of the theory; see Widlund [28] for a proof of such a theorem for general conforming finite element methods and a general discussion.

We note that in a recent paper, Bjørstad and Widlund [3] have returned to the two subregion case to show that a method, introduced by Chan and Resasco [10], is in fact the classical Schwarz's method accelerated by the conjugate gradient method. That study provides additional evidence that we need not strictly distinguish between methods which use overlapping subregions and those who do not.

In our development of the theory, a prominent role is played by subspaces and projections. Already fifty years ago Sobolev [25] showed that the classical Schwarz algorithm can be analyzed using a variational framework and more recently this approach has been further developed by Pierre-Louis Lions [20]. An additive Schwarz algorithms was first studied in Dryja and Widlund [12]. In this paper, we strengthen and simplify that result. We note that similar ideas have also been discussed by Matsokin and Nepomnyaschikh [22]. The general analytic framework that is now available for the study of these additive methods, has also proven very useful in our study of so called iterative refinement methods; cf. Widlund [32, 33].

A few years ago, much of the work on iterative substructuring methods was focused on elliptic problems defined on regions divided into two or a few subregions with separating curves (surfaces) which do not intersect; see e.g., Bjørstad and Widlund [2] and Widlund [28]. In this paper, we show that some of those algorithms and results can be recycled and combined with an iterative substructuring algorithm derived by using subspaces and projections to obtain previously known algorithms and results. This approach highlights how preconditioners can be built from parts which are strictly local to an individual substructure, parts which involve interaction between pairs of neighboring substructures and a coarse global model with relatively few degrees of freedom.

This paper is organized as follows. In Sec. 2, we review some of the ideas of substructuring that are very important in the development of computational methods of structural engineering. This discussion naturally leads to matrix splittings, which provide preconditioners for the large linear systems of algebraic equations, which arises in finite element work. In Sec. 3, we discuss different Schwarz methods and some general tools for estimating their rates of convergence. In the concluding sections, we show how two types of domain decomposition algorithms can be analyzed by using relatively simple tools of mathematical and finite element analysis. While we can do a lot with linear algebra, we ultimately have to resort to tools of analysis in order to complete the proofs of our main results.

2   Substructures, Subspaces and Projections

The domain decomposition methods, considered in this paper, provide preconditioners of systems of equations representing discrete elliptic problems. Preconditioners can be associated with matrix splittings but we note that those considered here differ in certain respects from most of those encountered the standard literature on iterative methods; cf Varga [27]. Domain decomposition methods can best be understood in terms of the substructuring ideas, which have provided a very successful framework for the development of very large programming systems for structural mechanics computing; cf. Bell, Hatlestad, Hansteen and Araldsen [1] or Przemieniecki [23].

We consider a linear, self adjoint, elliptic problem, which is discretized by a finite element method on a bounded Lipschitz region. The region Ω is a subset of $I\!\!R^n$, $n=2$ or 3, the differential operator is the Laplacian, and zero Dirichlet conditions and continuous, piecewise linear finite elements are used. The theory could equally well be developed for much more general linear elliptic problems, which can be formulated as minimization problems. Arbitrary conforming finite elements could also be considered without further major complications. Nonconforming finite elements, non-self adjoint problems and problems that give rise to indefinite symmetric systems of equations are also quite important. Some progress has already been made in such cases.

In the model case considered here, the continuous and discrete problems take the form

$$a(u,v) = f(v),\ \forall\ v \in\ V\ ,$$

and

(2.1) $$a(u_h, v_h) = f(v_h), \; \forall \; v_h \in V^h ,$$

respectively. The bilinear form is defined by

$$a(u, v) = \int_\Omega \nabla u \cdot \nabla v \, dx .$$

This form defines a semi-norm $|u|_{H^1(\Omega)} = (a(u,u))^{1/2}$ in the Sobolev space $H^1(\Omega)$. It is a norm for $V = H_0^1(\Omega)$. Here $H_0^1(\Omega)$ is the subspace of $H^1(\Omega)$ functions with zero trace; all elements of V and its subspace V^h vanish on $\partial\Omega$, the boundary of Ω. The triangulation of Ω is introduced in the following way. The region is divided into non-overlapping substructures Ω_i, $i = 1, 2, \ldots, N$. To simplify the description, our study is confined to triangular (simplicial) substructures. In such a case, the original region must of course be a polygon (polyhedron). All the substructures Ω_i are further divided into elements. The common assumption in finite element theory that all elements are shape regular is adopted and the same assumption is made concerning the substructures. On the element level this means that there is a bound on h_K/ρ_K, which is independent of the number of degrees of freedom and of K. Here h_K is the diameter of the element K and ρ_K the diameter of the largest sphere that can be inscribed in K.

Since $a(u_h, v_h) = a(u, v_h)$, $\forall \; v_h \in V^h$, the finite element solution is the projection of the exact solution onto the finite element space with respect to the inner product defined by the bilinear form. We will see that the problems defined on the subregions, from which preconditioners for the entire problem can be assembled, can similarly be viewed in terms of orthogonal projections onto subspaces directly associated with the subregion in question.

By using ideas of structural engineering, the stiffness matrix K can be constructed in the following way. The elements of K are given by

$$k_{i,j} = a(\varphi_i, \varphi_j) ,$$

where φ_i and φ_j are standard finite element basis functions. Since an integral over Ω can be written as a sum of integrals over the substructures, the stiffness matrix can be assembled from the stiffness matrices $K^{(k)}$ the elements of which are defined by

$$k_{i,j}^{(k)} = a_{\Omega_k}(\varphi_i, \varphi_j) ,$$

with i and j corresponding to the degrees of freedom of the substructure Ω_k. The form $a_{\Omega_k}(u_h, v_h)$ represents the contribution to the integral from that substructure.

This so called subassembly process can be summarized in the formula

$$x^T K y = \sum_i x^{(i)^T} K^{(i)} y^{(i)} ,$$

where $x^{(i)}$ is the subvector of parameter values associated with the substructure Ω_i and its boundary $\partial\Omega_i$.

In engineering computing practice, the large linear system with the coefficient matrix K is solved using programs, which are elaborate implementations of block Gaussian elimination; cf., e.g., [1]. Particular attention is placed on the use of data structures, etc., which allows for efficient I/O, and on the need to integrate the factorization and solution steps as much as possible with the generation of the blocks which together define the stiffness matrix. By the so-called *subassembly process*, described above in a special case, the contributions from the individual elements are first computed and these element stiffness matrices are then merged with their neighbors, creating so-called *super elements*. This process is often used recursively a substantial number of times. On each level, the variables associated with the formation of a particular super element can be divided into two sets; those which are common to other super elements and the interior variables which are not. In our description of domain decomposition algorithms only three levels are considered, the elements, with a characteristic diameter h, the substructures with a diameter H and the entire region Ω, which, without loss of generality, is assumed to have unit diameter.

The last phases of a standard engineering calculation of this type involve a substantial fraction of the arithmetical work and require more global communication than during the previous phases, where the work can proceed in a distributed fashion, without synchronization between the substructures. The iterative substructuring methods require synchronization once every iteration, but the best of them converge quite fast. They also demand much less communication of data between the substructures than methods which use direct methods throughout. The use of inexact solvers for the subproblems can also be considered when an iterative substructuring method is used. If a substructure has a simple geometry, special fast solvers may also be used.

If we divide the subvectors $x^{(i)}$ associated with the i-th substructure into two, $x_B^{(i)}$ and $x_I^{(i)}$, corresponding to the variables shared with other substructures and those which are interior to the substructure, then the matrix $K^{(i)}$ can be written as

$$\begin{pmatrix} K_{II}^{(i)} & K_{IB}^{(i)} \\ K_{IB}^{(i)T} & K_{BB}^{(i)} \end{pmatrix}.$$

Since the interior variables are associated with only one of the substructures, they can be eliminated locally and in parallel. The reduced matrix is a so-called *Schur complement* and has the form

$$S^{(i)} = K_{BB}^{(i)} - K_{IB}^{(i)T} K_{II}^{(i)-1} K_{IB}^{(i)}.$$

It is now easy to show that if the corresponding Schur complement of the global stiffness matrix K is denoted by S, then

$$(2.2) \qquad x^T S y = \sum_i x_B^{(i)T} S^{(i)} y_B^{(i)}.$$

The elimination of the interior variables from the substructures can be viewed in terms of orthogonal projections, with respect to the bilinear form, of the solution u_h of Eq. (2.1) onto the subspaces $H_0^1(\Omega_i) \bigcap V^h$, $i = 1, 2, \ldots, N$. These subspaces can easily be shown to be orthogonal, in the sense of the bilinear form, to the so-called *piecewise discrete harmonic functions* which satisfy

$$K_{II}^{(i)} x_I^{(i)} + K_{IB}^{(i)} x_B^{(i)} = 0, \; \forall i.$$

If the local problems are solved exactly, what remains is to find a sufficiently accurate approximation of the part of the solution which is piecewise discrete harmonic. This is done by approximately solving the reduced linear system with the matrix S. An iterative substructuring method is obtained by selecting a preconditioner for the the matrix S. Once an approximation of the solution has been found on the boundaries of the substructures, the solution can be found everywhere by solving local Dirichlet problems on each substructure separately.

The matrices $S^{(i)}$ are full. Complete information on the sparsity and block structure of S can be obtained by using Eq. (2.2). It is also natural to partition the vector $x_B^{(i)}$ and the Schur complement $S^{(i)}$ further. We group the variables associated with the interior of each edge (face) of the substructure into separate subvectors. Each of them is associated with exactly two substructures which are neighbors of each other. The remaining degrees of freedom of $x_B^{(i)}$ form a separate subvector. In two dimensions, the components of this vector, $x_V^{(i)}$, are the values at the vertices of Ω_i, while in three dimensions they are associated with all the nodes of the edges of the simplex, including its vertices. We can say that these are the nodes of the wire basket outlining the substructure. The Schur complement $S^{(i)}$ is thus represented by a four by four (five by five) block matrix. The blocks correspond to individual edges (faces) and the set of vertices the *wire basket*. If one or several edges of Ω_i is part of $\partial\Omega$, and thus not associated with any degrees of freedom in the Dirichlet case, then the numbers of blocks of $S^{(i)}$ is reduced. In what follows, we consider splittings where the off-diagonal blocks of $S^{(i)}$ are dropped, and others where all the contributions from certain substructures are left out in the construction of the preconditioner.

In the case of two substructures, there is only one set of interface variables and S is the sum of $S^{(1)}$ and $S^{(2)}$, the Schur complements originating from the two subregions respectively. The basic *Neumann-Dirichlet algorithm* amounts to using $S^{(1)}$ or $S^{(2)}$ as a preconditioner for S; cf. Bjørstad and Widlund [2] and Widlund [28]. This algorithm can be extended directly to the case of many substructures if a red-black ordering of the substructures exists; cf. Dryja, Proskurowski and Widlund [13] and Widlund [29]. This splitting corresponds to dropping all contributions from the red substructures in the sum given in Eq. (2.2). The resulting preconditioner corresponds to a global problem since the remaining black substructures are connected at their vertices (wire baskets). If we group all edge (face) variables and all the vertex (wire basket) variables into two subvectors $x_E^{(i)}$ and $x_V^{(i)}$, then we can write $S^{(i)}$ as

$$
\begin{pmatrix}
S_{EE}^{(i)} & S_{EV}^{(i)} \\
S_{EV}^{(i)T} & S_{VV}^{(i)}
\end{pmatrix}.
$$

In this quadratic form, which represents the sum over the black substructures only, the variables of $x_E^{(i)}$ appear only once. They are therefore coupled only to variables which are associated with the same substructure. The entire system of linear equations, which correspond to the Neumann-Dirichlet preconditioner, can, after a permutation, therefore be written in terms of a two-by-two block matrix where the leading block is the direct sum of the $S_{EE}^{(i)}$ which correspond to the black substructures. An analysis of the structure of this large matrix reveals that it is quite economical to use standard block Gaussian elimination, in the case of two dimensions. The Schur complement that remains after the elimination of the edge variables is sparse and its dimension is equal to the number of variables associated with substructure vertices which do not fall on $\partial\Omega$; see further Dryja, Proskurowski and Widlund [13]. It is shown in Widlund [29], that the condition number of the resulting preconditioner grows in proportion to $(1 + \log(H/h))^2$. The algorithm is less attractive in the three dimensional case, since the number of variables, which have to be treated in a special way, are relatively more numerous than for a problem in the plane. However, the same kind of bound can be established on the rate of convergence. There is an interesting variant of the algorithm, where the wire basket variables are handled differently and the preconditioner is cheaper to use; see Dryja [11]. We will not discuss it further in this paper.

Other splittings of S are known, which give as good (or better) results as the Neumann-Dirichlet method. We note that when we considered that algorithm, the values at the vertices (on the wire basket) were treated differently from those of the interior of the edges (faces). The methods considered by Bramble, Pasciak and Schatz [5, 8] also treat these sets of variables differently. As was shown in Widlund [29, 30], the part of the preconditioner introduced in [5] that relates to the vertex variables of plane problems can be viewed in terms of a projection onto a finite element subspace, where the substructures play the role of elements. The dimension of this subspace, V^H, is equal to the number of substructure vertices which belong to the open set Ω. In the splitting which corresponds to this preconditioner, the couplings between the different edges (faces) are also ignored. Thus in the four-by-four blocks representation of $S^{(i)}$ previously discussed, the off-diagonal blocks are set to zero. The diagonal blocks may also be changed; cf. Sec. 5. In Sec. 5, we will show that the variables associated with an individual edge can be naturally related to a subspace of functions which vanish outside the two substructures, which have this edge in common. The few preconditioners which have been studied in detail for the three dimensional case, cf. Bramble, Pasciak and Schatz [8] and Dryja [11], can also be described in similar general terms.

3 Schwarz Methods

We begin by briefly discussing the classical formulation of Schwarz's method in the continuous case. There are two fractional steps corresponding to two overlapping subregions, Ω'_1 and Ω'_2, the union of which is the region Ω. The initial guess $u^0 \in V$. The iterate u^{n+1} is determined from u^n by sequentially updating the approximate solution on the two subregions.

$$-\Delta u^{n+1/2} = f \qquad \text{in } \Omega'_1,$$
$$u^{n+1/2} = u^n \qquad \text{on } \partial\Omega'_1$$

and

$$-\Delta u^{n+1} = f \qquad \text{in } \Omega'_2,$$
$$u^{n+1} = u^{n+1/2} \qquad \text{on } \partial\Omega'_2.$$

We could equally well have written down the finite element version of the algorithm. From now on, we only consider that case. It is easy and convenient to describe this classical method in terms of two projections P_i, $i = 1, 2$, onto $V_i^h = H_0^1(\Omega'_i) \bigcap V^h$; cf. Lions [20]. They are defined by

$$(3.1) \qquad a(P_i v_h, \phi_h) = a(v_h, \phi_h), \ \forall \phi_h \in V_i^h.$$

It is also easy to show that the error propagation operator of this multiplicative Schwarz method is

$$(I - P_2)(I - P_1),$$

This algorithm can therefore be viewed as a simple iterative method for solving

$$(P_1 + P_2 - P_2 P_1)u_h = g_h,$$

with an appropriate right-hand side g_h.

This operator is a polynomial of degree two and thus not ideal for parallel computing, since two sequential steps are involved. If more than two subspaces are used, this effect is further pronounced, even if the degree of the polynomial representing the multiplicative algorithm often is lower than maximal. This is so because a product of two projections associated with subregions which do not overlap, vanishes; cf. the discussion in Widlund [31]. The basic idea behind the additive form of the algorithm is to work with the simplest possible polynomial in the projections. Therefore, the equation

$$(3.2) \qquad Pu_h = (P_1 + P_2 + \cdots + P_N)u_h = g'_h,$$

is solved by an iterative method. Since the operator P is symmetric, with respect to the bilinear form, and positive definite, the method of choice is the conjugate gradient method. Equation (3.2) must have the same solution as Eq. (2.1), i.e., the correct right-hand side must be found. Since by Eq. (2.1), $a(u_h, \phi_h) = f(\phi_h)$, the right-hand side g'_h can be constructed by solving Eq. (3.1) for all values of i and adding the results. It is similarly possible to apply the operator P of Eq. (3.2) to any given element of V^h by applying each projection P_i to the element and adding the results. Most of the work, in particular that which involves the individual projections, can be carried out in parallel.

We now describe the additive Schwarz method introduced in Dryja and Widlund [14]; cf. also Dryja [12]. We start with the same triangular (simplicial) nonover-lapping substructures Ω_i, that have been considered before. Since Schwarz-type domain decomposition algorithms use overlapping subregions, we extend each sub-structure to a larger region Ω'_i. We assume that the overlap is generous assuming that the distance between the boundaries $\partial\Omega_i$ and $\partial\Omega'_i$ is bounded from below by a fixed fraction of H_i, the diameter of Ω_i. We also assume that $\partial\Omega'_i$ does not cut through any element. We make the same construction for the substructures that meet the boundary except that we cut-off the part of Ω'_i that is outside of Ω.

The analysis of Schwarz methods is more complicated when the boundaries of the different subdomains Ω'_i intersect at one or several points; cf. the discussion in Lions [20]. Such a situation occurs if the region is L-shaped and is partitioned into two overlapping rectangles. We discuss such a more complicated situation in Sec. 5.

Our finite element space is represented as the sum of N+1 subspaces

$$V^h = V_0^h + V_1^h + \cdots + V_N^h.$$

The first subspace V_0^h is equal to V^H, the same coarse global space of continuous, piecewise linear functions on the coarse mesh defined by the substructures Ω_i that we introduced in Sec. 2. The other subspaces are related to the subdomains, in the same way as in a traditional Schwarz algorithm, i.e., $V_i^h = V^h \bigcap H_0^1(\Omega'_i)$. The computation of the projection of an arbitrary function onto the subspace V^H involves the solution of a standard finite element linear system of algebraic equations which is on the order of N. This coarse, global approximation of the elliptic equation is of the same type as the local problems associated with subdomains. The only real difference between the problem related to the first subspace and the others lies in the way that the right-hand-side of the linear system is generated as weighted averages with weights determined by the basis functions associated with the coarse mesh. We note that if we make the dimension of all the subspaces approximately equal, then we will have $N + 1$ linear systems, each with about than N unknowns, to solve in each step of the iterative solution of a linear system with about N^2 unknowns.

It is well-known that the number of steps required to decrease an appropriate norm of the error of a conjugate gradient iteration by a fixed factor is proportional to $\sqrt{\kappa}$, where κ is the condition number of P; see, e.g., Golub and Van Loan [16]. We therefore need to establish that the operator P of Eq. (3.2) is not only invertible but that satisfactory upper and lower bounds on its eigenvalues can be obtained. A constant upper bound can easily be obtained for P; cf. Sec. 4. For certain other additive algorithms a useful technique is based on strengthened Cauchy inequalities;

cf. Mandel and McCormick [21], Widlund [33] and Yserentant [34].

A lower bound can often conveniently be obtained by using a lemma, given by Lions [20] for the case of $N = 2$; a proof is also given in Widlund [33].

Lemma 3.1 *Let* $u_h = \sum_{i=1}^N u_{h,i}$, *where* $u_{h,i} \in V_i$, *be a representation of an element of* $V^h = V_1 + \cdots + V_N$. *If the representation can be chosen so that it follows that* $\sum_{i=1}^N a(u_{h,i}, u_{h,i}) \leq C_0^2 a(u_h, u_h), \forall\ u_h \in V^h$, *then* $\lambda_{min}(P) \geq C_0^{-2}$.

We remark that it is clear that C_0 decreases if we expand the subspaces. This follows from the fact that there is a larger choice in selecting $u_{h,i} \in V_i$. If we can expand the subspaces without worsing the upper bound, and that is often possible, our estimate of $\kappa(P)$ improves. On the other hand a larger subspace also means that the subproblems have more variables and that they are worse conditioned. For the special case of the classical Schwarz method on two regions, this tradeoff is well understood; for a discussion of precise estimates of the rate of convergence cf. Björstad and Widlund [3].

As was previously pointed out, the framework with subspaces and projections has already proven quite useful not only for the study of Schwarz-type methods, but also for iterative refinement methods. Similarly, it can be shown, that the bound for the condition number of the iteration operator of the hierarchical basis multigrid method, introduced by Yserentant [34] can be derived by using these techniques. Yserentant's algorithm involves the use of a direct sum of cleverly chosen subspaces, which are quite different from those considered here. All except one of the subproblems that result are very well conditioned. However, since a direct sum of subspaces is used, there is no flexibility in the representation of the elements of V^h. This fact adds to the understanding why Yserentant's method is much less attractive in the case of three dimensions since the best possible C_0 and the condition number grows rather rapidly in that case, when the mesh is refined.

4 Analysis of an Additive Schwarz Method

In this section, we will study the method introduced in the previous section and give a proof of the following result.

Theorem 4.1 *The operator* P *of the additive algorithm defined by the spaces* V^H *and* V_i^h *satisfies the estimate* $\kappa(P) \leq$ *const.*

Here as elsewhere in this paper, the constants in our estimates, are independent of h and H.

An upper bound for the spectrum of P is quite easy to obtain. We only have to note that for $i \geq 1$,

$$a(P_i u_h, u_h) = a(P_i u_h, P_i u_h) = a_{\Omega_i'}(P_i u_h, P_i u_h) \leq a_{\Omega_i'}(u_h, u_h).$$

We recall that the subscript indicates the domain of integration of the bilinear form. The basic observation here is that $P_i u_h$ can be regarded as a projection of

$H^1(\Omega_i') \bigcap V^h$ onto $H_0^1(\Omega_i') \bigcap V^h$. In the partion of the region considered here, each point is covered by subregions Ω_i' a finite number of times. A constant upper bound of the eigenvalues of P is therefore obtained by noting that, additionally, the norm of P_0 is equal to one.

The lower bound is obtained by using Lemma 3.1. We partition the finite element function u_h as follows. We first choose $u_{h,0} \in V^H$. By using smoothing and interpolation, cf., e.g., Strang [26], we can find a linear map $\hat{I}_H$ into V^H, which is bounded in $H_0^1(\Omega)$ and which satisfies

$$(4.1) \qquad \|u_h - \hat{I}_H u_h\|_{L_2(\Omega)} \le (\text{const.})H|u_h|_{H^1(\Omega)}.$$

Let $w_h = u_h - \hat{I}_H u_h$. The other terms in the representation of u_h are defined by $u_{h,i} = I_h(\theta_i w_h)$, $i = 1, 2, \ldots, N$. Here I_h is the interpolation operator into the space V^h and the θ_i define a partition of unity with $\theta_i \in C_0^\infty(\Omega_i')$ and $\sum \theta_i(x) = 1$. Because of the relative generous overlap of the subregions, introduced in Sec. 3, these functions can be chosen so that $\nabla \theta_i$ is bounded by const. $/H_i$. By using the linearity of I_h, we can easily show that we obtain a correct partitioning of u_h. In order to estimate the semi-norm of $u_{h,i}$, we work on one element K at a time. We obtain

$$|u_{h,i}|_{H^1(K)}^2 \le 2|\overline{\theta_i} w_h|_{H^1(K)}^2 + 2|I_h((\theta_i - \overline{\theta_i})w_h)|_{H^1(K)}^2.$$

Here $\overline{\theta_i}$ is the average value of θ_i over K. It is easy to see, by using an inverse inequality, that

$$|I_h((\theta_i - \overline{\theta_i})w_h)|_{H^1(K)} \le (\text{const.}) \, h^{-1}\|I_h((\theta_i - \overline{\theta_i})w_h)\|_{L_2(K)}.$$

We can now use the fact that on K, θ_i differs from its average by at most const.h/H_i. After summing over all elements of Ω_i', we arrive at the inequality

$$|u_{h,i}|_{H^1(\Omega_i')}^2 \le (\text{const.}) \, (|w_h|_{H^1(\Omega_i')}^2 + H_i^{-2}\|w_h\|_{L_2(\Omega_i')}^2).$$

We now sum over all i and use that each point in Ω is covered only a fixed number of times. We then obtain a uniform bound on C_0^2, and conclude the proof of Theorem 4.1, by estimating the two terms of

$$|w_h|_{H^1(\Omega)}^2 + H_i^{-2}\|w_h\|_{L_2(\Omega)}^2$$

by $|u_h|_{H_0^1(\Omega)}^2$. The bounds follow by using the boundedness of $\hat{I}_H$ in H^1 and inequality 4.1, respectively.

5   Iterative Substructuring Methods

In this section, we show how we can obtain an iterative substructuring method by using the framework with subspaces and projections developed in Sec. 3. We primarily consider problems in the plane. We give an estimate of the condition number of the operator P, which corresponds to a specific choice of subspaces. This proof is carried out without using an extension theorem. We then indicate how results for the the special case of two substructures can be used to derive different algorithms, among them one due to Bramble, Pasciak and Schatz [5]. We also briefly discuss the three dimensional case.

We assume that the region is divided into substructures as in Sec. 2. An additive Schwarz method is introduced by using the coarse space V^H and subspaces V_{ij}^h, which are related to an edge of a substructure. Thus, let Γ_{ij} be the edge which is common to two adjacent substructures Ω_i and Ω_j. Then, $V_{ij}^h = H_0^1(\Omega_{ij}) \bigcap V^h$, where $\Omega_{ij} = \Omega_i \bigcup \Gamma_{ij} \bigcup \Omega_j$. The region Ω, with the exception of the vertices of the substructures, is therefore covered by the subregions Ω_{ij} constructed from all pairs of adjacent substructures.

Compared with the subspaces used in the previous section, we use less overlap in the sense that only the elements of V^H differ from zero at the vertices of the substructures. This is reflected in a poorer bound on the condition number.

Theorem 5.1 *The operator P of the additive algorithm defined by the spaces V^H and V_{ij}^h satisfies the estimate $\kappa(P) \leq$ (const.)$(1 + log(H/h))^2$.*

By using the same method as in Sec. 4, we can easily show that the eigenvalues of $P \leq 4$. In the proof of the lower bound of the spectrum of P, we use Lemma 3.1 and the following lemma, which plays an important role in the more traditional theory for iterative substructuring algorithms. Variations of this result, which dates back at least to 1966, are given in a number of papers; see e.g., Bramble [4] , Bramble, Pasciak and Schatz [5] or Yserentant [34].

Lemma 5.1 *Let α be any value of $u_h(x)$, with $x \in \Omega_i$. Then*

$$\|u_h - \alpha\|_{L^\infty(\Omega_i)}^2 \leq (\text{const.}) \, (1 + \log(H/h))|u_h|_{H^1(\Omega_i)}^2 \, .$$

We note that this result holds only for regions in two dimensions and that it resembles a Sobolev inequality. Since all the elements of V_{ij}^h vanish at the vertices of the substructures, and we must pick the interpolant $I_H u_h$ as the element in V^H in the representation of u_h . It is easy to show that $|I_H u_h|_{H^1(\Omega_i)}^2$ can be estimated by

$$\sum_{k,l=1}^{3} (u_h(\tilde{V}_k) - u_h(\tilde{V}_l))^2 \, ,$$

where the $\tilde{V}_k's$ are the vertices of the substructure Ω_i . It then follows from Lemma 5.1 that

(5.1) $$|I_H u_h|_{H^1(\Omega_i)}^2 \leq (\text{const.})(1 + \log(H/h))|u_h|_{H^1(\Omega_i)}^2 \, .$$

Let $w_h = u_h - I_H u_h$. As in the previous section, we use a partion of unity. The argument is now more complicated since there is less overlap. The elements of V_{ij}^h, used in the representation of u_h, are given by the formula

$$u_{h,ij} = I_h(\theta_{ij} w_h) \, .$$

Since w_h vanishes at the vertices of the substructures and we only use values of θ_{ij} at nodal points, we only need a partion of unity at the nodal points which are not substructure vertices. The function θ_{ij} must be equal to 1 at all nodal points in the interior of Γ_{ij}, since all the other cut-off functions vanish there, and it must vanish at the corresponding nodes on the other edges of Ω_i and Ω_j. It is easy to see that $\nabla\theta_{ij}$ therefore must grow as fast as const. $/r$, where r is the distance to the closest endpoint of Γ_{ij}. Cut-off functions, for which this is also an upper bound, can indeed be constructed.

The necessary bound is obtained, one substructure at a time. In order to estimate $|u_{h,ij}|_{H^1(\Omega_i)}$ in terms of $|w_h|_{H^1(\Omega_i)}$, we first consider the few elements of Ω_i for which an end point of Γ_{ij} is a vertex. $u_{h,ij}$ and w_h vanish at such a vertex, and at the other nodes of these special elements the absolute value of $u_{h,ij}$ is no larger than that of w_h. A straightforward calculation shows that the contributions to $|u_{h,ij}|^2_{H^1(\Omega_i)}$ from these triangles are smaller than those to $|w_h|^2_{H^1(\Omega_i)}$. We now use arguments similar to those of the previous section to obtain an estimate of the integral over the rest of the substructure. Working with one element at a time, we obtain

$$\begin{aligned}
|u_{h,ij}|^2_{H^1(K)} &\leq 2|w_h|^2_{H^1(K)} + (\text{const.})r^{-2}\|w_h\|^2_{L_2(K)} \\
&\leq 2|w_h|^2_{H^1(K)} + (\text{const.})(h/r)^2\|w_h\|^2_{L_\infty(K)} \, .
\end{aligned}$$

By using the fact that w_h does not change if a constant is added to u_h, the formula for $u_{h,ij}$ and that $\|w_h\|_{L_\infty(K)} \leq 2\|u_h\|_{L_\infty(\Omega_i)}$, we obtain

$$|u_{h,ij}|^2_{H^1(K)} \leq 2|w_h|^2_{H^1(K)} + (\text{const.})(h/r)^2\|u_h - \alpha\|^2_{L_\infty(\Omega_i)}, \forall \, \alpha \, .$$

The sum of the first term over the elements can be estimated by using inequality (5.1). Since the number of elements decreases to zero linearly with the distance r, the sum of the second expression over all the elements of the substructure, except those at the endpoints of Γ_{ij}, can be estimated by

$$(\text{const.})\|u_h - \alpha\|^2_{L_\infty(\Omega_i)} \int_h^H r^{-1} dr$$

Here H represents the diameter of Ω_i and h the minimum distance of any other nodal point to the end points of Γ_{ij}. The estimate on C_0^2 and the proof of the whole theorem is now concluded by using Lemma 5.1.

This algorithm, introduced and analyzed as a method based on the subspaces V^H and V_{ij}^h and the related projections, can equally well be understood in terms of a splitting. Let us consider the case where the components of the right hand side of the system of equations corresponding to the interiors have been set to zero in a preliminary step. The right hand side of a linear system related to Ω_{ij} and V_{ij}^h then differs from zero only on Γ_{ij}. The Schur complement associated with this edge can be shown, straightforwardly, to be the sum of the corresponding blocks in the four by four block representation of the Schur complements $S^{(i)}$ and $S^{(j)}$, which were introduced in Sec. 2.

One of the attractive features of the framework first introduced in Sec. 3 is the ease by which the subproblems can be replaced by preconditioners. Let us consider the additive Schwarz method for the problem discussed in the beginning of Sec. 3. We can write the projection P_1 in matrix terms. After a suitable permutation of the variables, it is seen to correspond to

$$
y = P_1 x = \begin{pmatrix} K^{(1)-1} & 0 \\ 0 & 0 \end{pmatrix} Kx.
$$

It is easy to see that this matrix is symmetric in the $K-$ inner product which corresponds to the bilinear form. If $K^{(1)-1}$, and the other matrices which play similar roles, are replaced by inverses of preconditioners for the subproblems, then it is easy to see that the resulting algorithm converges and that its condition number can be estimated immediately in terms of $\kappa(P)$ and bounds for the local preconditioners. Using this method, a number of algorithms can be derived from a basic method based on projections and subspaces. Thus, if the Schur complement corresponding to the problem defined on Ω_{ij} is replaced by the square root of the discrete, one-dimensional Laplacian, denoted by $l_0^{1/2}$ in Bramble, Pasciak and Schatz [5], we obtain the main algorithm of their paper. The estimate of the condition number of the resulting method can be obtained, by using the argument just given, combining Theorem 5.1 and a bound for the condition number of problems defined on the union of two subregions.

We conclude this paper by a remark on the part of preconditioner which corresponds to the coarse global problem. In two dimensions, we have used the subspace V^H for this purpose. The related quadratic form has, in a special case, the form

$$
\sum_i \sum_{k,l=1}^{3} (u_h(\tilde{V}_k^{(i)}) - u_h(\tilde{V}_l^{(i)}))^2 .
$$

It is easy to show that this form equally well can be written as a double sum over $(u_h(\tilde{V}_k^{(i)}) - \overline{u}_h^{(i)})^2$, where $\overline{u}_h^{(i)}$ is the average value of the three values associated with the vertices of Ω_i. In three dimensions, the corresponding quadratic form, with the sums and averages calculated with respect to the variables associated with the wire baskets of the individual substructures, provide an important, global part of the powerful iterative substructuring methods, which have been developed by Bramble, Pasciak and Schatz [8] and Dryja [11].

Acknowledgements

This work was supported in part by the National Science Foundation under Grant NSF-CCR-8703768 and, in part, by the U.S. Department of Energy under contract

DE-AC02-76ER03077-V at the Courant Mathematics and Computing Laboratory. This paper was originally typed by the second author in LaTeX and reformatted by Lisa Laguna.

References

[1] Bell, K., B. Hatlestad, O. E. Hansteen, and Per O. Araldsen. NORSAM, *a programming system for the finite element method. Users manual, Part 1, General description.* NTH, Trondheim, 1973.

[2] Bjørstad, Petter E., and Olof B. Widlund. "Iterative methods for the solution of elliptic problems on regions partitioned into substructures," *SIAM J. Numer. Anal.* **23** (1986) 1093–1120.

[3] Bjørstad, Petter E., and Olof B. Widlund. "To overlap or not to overlap: a note on a domain decomposition method for elliptic problems," *SIAM J. Sci. Stat. Comput.* **10** (1989) 1053–1061.

[4] Bramble, James H. "A second order finite difference analogue of the first biharmonic boundary value problem," *Numer. Math.* **9** (1966) 236–249.

[5] Bramble, James H., Joseph E. Pasciak, and Alfred H. Schatz. "The construction of preconditioners for elliptic problems by substructuring, I," *Math. Comp.* **47** (1986) 103–134.

[6] Bramble, James H., Joseph E. Pasciak, and Alfred H. Schatz. "The construction of preconditioners for elliptic problems by substructuring, II," *Math. Comp.* **49** (1987) 1–16.

[7] Bramble, James H., Joseph E. Pasciak, and Alfred H. Schatz. *The Construction of Preconditioners for Elliptic Problems by Substructuring, III.* Technical report, Cornell University, 1987.

[8] Bramble, James H., Joseph E. Pasciak, and Alfred H. Schatz. *The Construction of Preconditioners for Elliptic Problems by Substructuring, IV.* Technical report, Cornell University, 1988. To appear in Math. Comp.

[9] Chan, Tony F., Roland Glowinski, Gérard A. Meurant, Jacques Périaux, and Olof Widlund, eds. *Domain Decomposition Methods*, SIAM, Philadelphia, 1989. Proceedings of the Second International Symposium on Domain Decomposition Methods , Los Angeles, California , January 14–16, 1988.

[10] Chan, Tony F., and Diana C. Resasco. "Analysis of domain decomposition preconditioners on irregular regions." In R. Vichnevetsky and R. Stepleman, eds., *Advances in Computer Methods for Partial Differential Equations.* IMACS, 1987.

[11] Maksymilian Dryja. "A method of domain decomposition for 3-D finite element problems." In Roland Glowinski, Gene H. Golub, Gérard A. Meurant, and Jacques Périaux, eds., *Domain Decomposition Methods for Partial Differential Equations*, SIAM, Philadelphia, 1988.

[12] Maksymilian Dryja. "An additive Schwarz algorithm for two- and three-dimensional finite element elliptic problems." In Tony Chan, Roland Glowinski, Gérard A. Meurant, Jacques Périaux, and Olof Widlund, eds., *Domain Decomposition Methods*, SIAM, Philadelphia, 1989.

[13] Dryja, Maksymilian, Wlodek Proskurowski, and Olof Widlund. "A method of domain decomposition with crosspoints for elliptic finite element problems." In Bl. Sendov, ed., *Optimal Algorithms*, 97–111, Sofia, Bulgaria, 1986. Bulgarian Academy of Sciences.

[14] Dryja, Maksymilian, and Olof B. Widlund. *An Additive Variant of the Schwarz Alternating Method for the Case of Many Subregions*. Technical Report 339, also Ultracomputer Note 131, Department of Computer Science, Courant Institute, 1987.

[15] Glowinski, Roland, Gene H. Golub, Gérard A. Meurant, and Jacques Périaux, eds. *Domain Decomposition Methods for Partial Differential Equations*, SIAM, Philadelphia, 1988. Proceedings of the First International Symposium on Domain Decomposition Methods for Partial Differential Equations, Paris, France, January 1987.

[16] Golub, Gene H., and Charles F. Van Loan. *Matrix Computations*. Johns Hopkins Univ. Press, 1983.

[17] Greenbaum, Anne, Congming Li, and Han Zheng Chao. *Parallelizing Preconditioned Conjugate Gradient Algorithms*. Technical report, Courant Institute, 1988. To appear in *Computer Physics Communications*.

[18] Keyes, David E., and William D. Gropp. "A comparison of domain decomposition techniques for elliptic partial differential equations and their parallel implementation," *SIAM J. Sci. Stat. Comput.* **8** (1987) 166–202.

[19] Keyes, David E., and William D. Gropp. "Domain decomposition techniques for the parallel solution of nonsymmetric systems of elliptic bvps." In Tony Chan, Roland Glowinski, Gérard A. Meurant, Jacques Périaux, and Olof Widlund, eds., *Domain Decomposition Methods*, SIAM, Philadelphia, 1989.

[20] Lions, Pierre Louis. "On the Schwarz alternating method. I." In Roland Glowinski, Gene H. Golub, Gérard A. Meurant, and Jacques Périaux, eds., *First International Symposium on Domain Decomposition Methods for Partial Differential Equations*, SIAM, Philadelphia, 1988.

[21] Mandel, Jan, and Steve McCormick. "Iterative solution of elliptic equations with refinement: The two-level case." In Tony Chan, Roland Glowinski, Gérard A. Meurant, Jacques Périaux, and Olof Widlund, eds., *Domain Decomposition Methods*, SIAM, Philadelphia, 1989.

[22] A. M. Matsokin and S. V. Nepomnyaschikh. "A Schwarz alternating method in a subspace," *Soviet Mathematics* **29** (1985) 78–84.

[23] Przemieniecki, J.S. "Matrix structural analysis of substructures," *Am. Inst. Aero. Astro. J.* **1** (1963) 138–147.

[24] Schwarz, H. A. *Gesammelete Mathematische Abhandlungen*, Vol. 2, 133–143. Springer, Berlin, 1890. First published in Vierteljahrsschrift der Naturforschenden Gesellschaft in Zürich **15** (1870) 272–286.

[25] Sobolev, S. L. "The Schwarz algorithm in the theory of elasticity," *Dokl. Acad. N. USSR*, **IV(XIII)** (1936) 236–238. (in Russian).

[26] Strang, Gilbert. "Approximation in the finite element method," *Numer. Math.* **19** (1972) 81–98.

[27] Varga, Richard S. *Matrix Iterative Analysis*. Prentice-Hall, 1962.

[28] Widlund, Olof B. "An extension theorem for finite element spaces with three applications." In Wolfgang Hackbusch and Kristian Witsch, eds., *Numerical Techniques in Continuum Mechanics*, 110–122, Braunschweig/Wiesbaden, 1987. Notes on Numerical Fluid Mechanics, Vol. 16, Friedr. Vieweg und Sohn. Proceedings of the Second GAMM-Seminar, Kiel, January , 1986.

[29] Widlund, Olof B. "Iterative substructuring methods: Algorithms and theory for problems in the plane." In Roland Glowinski, Gene H. Golub, Gérard A. Meurant, and Jacques Périaux, eds., *First International Symposium on Domain Decomposition Methods for Partial Differential Equations*, SIAM, Philadelphia, 1988.

[30] Widlund, Olof B. "Iterative substructuring methods:The general elliptic case." In *Computational Processes and Systems*, *6*, Moscow, 1988. Nauka. Proceedings of Modern Problems in Numerical Analysis, a conference held in Moscow, USSR, September , 1986. (In Russian, also available from the author, in English, as a technical report.)

[31] Widlund, Olof B. *On the Rate of Convergence of the Classical Schwarz Alternating Method in the Case of More Than Two Subregions*. Technical report, Department of Computer Science, Courant Institute, 1988.

[32] Widlund, Olof B. *Some Domain Decomposition and Iterative Refinement Algorithms for Elliptic Finite Element Problems*. Technical Report 386, Department of Computer Science, Courant Institute, 1988. To appear in *J. Comp. Math.*

as the proceedings of the China–U.S. Seminar on Boundary Integral Equations and Boundary Element Methods in Physics and Engineering, held at the Xi'an Jiatong University, Xi'an, The People's Republic of China, December 27, 1987–January 1, 1988.

[33] Widlund, Olof B. "Optimal iterative refinement methods." In Tony Chan, Roland Glowinski, Gérard A. Meurant, Jacques Périaux, and Olof Widlund, eds., *Domain Decomposition Methods*, SIAM, Philadelphia, 1989.

[34] Yserentant, Harry. "On the multi-level splitting of finite element spaces," *Numer. Math.* **49** (1986) 379–412.

Chapter 17

The Search for Omega

DAVID M. YOUNG
University of Texas at Austin

and

TSUN-ZEE MAI
University of Alabama–Tuscaloosa

Abstract

For the effective use of iterative algorithms for solving large sparse linear systems it is often necessary to select certain iteration parameters. Examples of iteration parameters are the relaxation factor omega for the SOR and SSOR methods, and the largest and smallest eigenvalues of the matrix for a basic iterative method when Chebyshev acceleration is used to speed up the convergence. For many iterative algorithms the performance is extremely sensitive to the choice of iteration parameters. Moreover, uncertainty as to how to choose iteration parameters has often, in the past, tended to discourage the use of iterative methods, as opposed to direct methods, for certain classes of problems.

The purpose of this paper is to review the development of procedures for choosing iteration parameters, with special emphasis on methods applicable to linear systems arising from the numerical solution of partial differential equations. The discussion will include a priori procedures including analytic techniques, spectral methods, and methods based on related differential equations. Automatic, or "adaptive," procedures, wherein the iteration parameters are improved as the computation proceeds, will also be discussed. Some of these procedures have been incorporated into the `ITPACK` software packages for solving large sparse linear systems. Numerical experiments indicate that the amount of overhead needed to determine satisfactory parameters is usually not excessive. Methods for choosing iteration parameters for nonsymmetric systems will also be considered.

293

1   Introduction

For many iterative algorithms for solving large sparse linear systems the rate of convergence is very sensitive to the choice of the iteration parameters. It seems fair to say that uncertainty as to how to choose iteration parameters has been an important stumbling block that has tended to discourage the use of iterative methods, as opposed to direct methods, for solving certain classes of problems. In this paper we review some of the procedures which have been developed over the years for choosing iteration parameters.

A prime example of an iteration parameter is the relaxation factor, omega, which is used for the SOR method and for the SSOR method. Much of our discussion will be devoted to the "search for omega" for these methods. Another example of an iterative algorithm involving iteration parameters arises when Chebyshev acceleration is used to speed up the convergence of a basic iterative method with an iteration matrix $\mathbf{G}$. Here the iteration parameters needed are the smallest and largest eigenvalues of $\mathbf{G}$.

Other examples of iteration parameters include: the shift parameter for the shifted incomplete Cholesky method proposed by Manteuffel [32]; the parameters for the strongly implicit method, see, e.g., Stone [39] and Dupont et al. [7]; and the parameters for the alternating direction implicit method of Peaceman and Rachford [34].

Before discussing procedures for choosing iteration parameters we give a review, in Sec. 2, of various types of iterative algorithms where iteration parameters arise. In Sec. 3 we describe a priori procedures for choosing iteration parameters. Here one attempts to estimate the parameters in advance. The techniques used include analytic techniques, spectral techniques, and techniques involving the use of related partial differential equations. While such techniques are often useful, in many cases they do not yield sufficiently accurate parameter estimates. In Sec. 4 we describe adaptive techniques where improved estimates are determined automatically as the computational process proceeds. In many cases these procedures yield satisfactory parameters without an excessive amount of additional computational effort. Some of the adaptive procedures have been incorporated into the ITPACK software packages. The discussion includes methods used in searching for omega for the SOR method as well as adaptive methods for finding acceleration parameters for Chebyshev acceleration. Adaptive SSOR with Chebyshev and conjugate gradient acceleration and variational-based adaptive methods are also considered.

For most of our discussion it is assumed that the matrix $\mathbf{A}$ of the linear system is symmetric and positive definite (SPD). The case where the matrix $\mathbf{A}$ is non-symmetric is much more complicated. Various approaches for choosing iteration parameters for the SOR method and for Chebyshev acceleration are considered in Sec. 5. A composite adaptive procedure, based on the use of a generalized conjugate gradient method, is described for finding parameters associated with certain basic iterative methods.

We wish to emphasize that our review is by no means complete. There are no doubt many excellent schemes of which we are not aware. We would be grateful to have these called to our attention.

2 Iterative Algorithms and Iteration Parameters

In this section we describe some iterative algorithms for solving the linear system

$$(2.1) \qquad\qquad \mathbf{Au} = \mathbf{b}$$

where $\mathbf{A}$ is given square matrix and $\mathbf{b}$ is a given column vector. Usually, as in the case of a linear system derived from the discretization of a partial differential equation, the matrix $\mathbf{A}$ is very large and very sparse.

A typical iterative algorithm for solving the linear system (2.1) consists of a basic iterative method together with an acceleration procedure. A basic iterative method is a one-step procedure of the form

$$(2.2) \qquad\qquad \mathbf{u}^{(n+1)} = \mathbf{G}\mathbf{u}^{(n)} + \mathbf{k}$$

where

$$(2.3) \qquad \begin{cases} \mathbf{G} = \mathbf{I} - \mathbf{Q}^{-1}\mathbf{A} \\ \mathbf{k} = \mathbf{Q}^{-1}\mathbf{b} \end{cases}$$

Here $\mathbf{Q}$ is the "splitting" matrix corresponding to the basic iterative method and the matrix $\mathbf{G}$ is the iteration matrix. The matrix $\mathbf{Q}$ is usually chosen to be a simple easily-inverted matrix such as a diagonal, tridiagonal, upper triangular or lower triangular matrix or as a product of such matrices.

Examples of frequently-used basic iterative methods include:

Richardson's method : $\mathbf{Q} = \mathbf{I}$

Jacobi method   : $\mathbf{Q} = \mathbf{D}$

SOR method   : $\mathbf{Q} = \frac{1}{\omega}\mathbf{D} - \mathbf{C}_L$

SSOR method   : $\mathbf{Q} = \frac{\omega}{2-\omega}(\frac{1}{\omega}\mathbf{D} - \mathbf{C}_L)\mathbf{D}^{-1}(\frac{1}{\omega}\mathbf{D} - \mathbf{C}_U)$

Here $\mathbf{A} = \mathbf{D} - \mathbf{C}_L - \mathbf{C}_U$ where $\mathbf{D}$ is the diagonal matrix with the same diagonal elements as $\mathbf{A}$, and where $\mathbf{C}_L$ and $\mathbf{C}_U$ are strictly lower and strictly upper triangular matrices, respectively. The parameter ω is known as the "relaxation factor."

For some basic iterative methods, often referred to as "approximate factorization methods," $\mathbf{Q}$ has the form $\mathbf{LU}$ where $\mathbf{L}$ and $\mathbf{U}$ are lower and upper triangular matrices, respectively. (Often $\mathbf{L}$ and $\mathbf{U}$ have the same "sparsity" as $\mathbf{A}$). Examples of such methods are the incomplete Cholesky (IC) method considered by Varga [42] and by Meijerink and Van der Vorst [33], the shifted incomplete Cholesky (SIC) method of Manteuffel [32], and the modified incomplete Cholesky (MIC) method of Gustafsson [17]. Another example is the strongly implicit (SIP) method of Stone

[39]. See also Dupont et al. [7]. Except for the IC method, each of the above methods involves a parameter. We refer to a parameter associated with a basic iterative method as a "splitting parameter."

In many problems arising from the numerical solution of partial differential equations by discretization methods the rate of convergence of the SOR method with the optimum value of ω is faster by an order-of-magnitude than that of the Jacobi method. However, the rate of convergence is very sensitive to ω. Similarly, the rates of convergence of many other basic iterative methods are sensitive to the choice of the splitting parameter.

A basic iterative method (2.2) is "symmetrizable" if the matrix $\mathbf{I} - \mathbf{G}$ is similar to an SPD matrix or, equivalently, if $\mathbf{Z}(\mathbf{I} - \mathbf{G})$ is SPD for some SPD matrix $\mathbf{Z}$. Most of the standard iterative methods, including Richardson's method, the Jacobi method, the SSOR method, and the IC method are symmetrizable if $\mathbf{A}$ is SPD. More generally, a basic iterative method is symmetrizable if $\mathbf{A}$ and $\mathbf{Q}$ are SPD. It should be noted, however, that the SOR method is not symmetrizable.

If a basic iterative method (2.2) is symmetrizable, then the eigenvalues of $\mathbf{G}$ are real and less than unity. The convergence of a symmetrizable basic iterative method can be speeded up by the use of a polynominal acceleration procedure. Examples of polynominal acceleration procedures are Chebyshev acceleration, see, e.g., Varga [41,43], Golub and Varga [15], and Hageman and Young [20], and conjugate gradient (CG) acceleration.

The formulas for Chebyshev acceleration are given by

$$(2.4) \quad \mathbf{u}^{(n+1)} = \rho_{n+1}\left\{\mathbf{u}^{(n)} + \gamma_{n+1}(\mathbf{G}\mathbf{u}^{(n)} + \mathbf{k} - \mathbf{u}^{(n)})\right\} + (1 - \rho_{n+1})\mathbf{u}^{(n-1)}$$

Here the parameters $\{\rho_{n+1}\}$ and $\{\gamma_{n+1}\}$ are given by

$$(2.5) \qquad \gamma_1 = \gamma_2 = \cdots = \frac{2}{2 - M(\mathbf{G}) - m(\mathbf{G})}$$

$$\rho_{n+1} = \begin{cases} 1 & n = 0 \\ \left(1 - \frac{\sigma^2}{2}\right)^{-1} & n = 1 \\ \left(1 - \frac{\sigma^2}{4}\rho_n\right)^{-1} & n \geq 2 \end{cases}$$

where

$$(2.6) \qquad \sigma = \frac{M(\mathbf{G}) - m(\mathbf{G})}{2 - M(\mathbf{G}) - m(\mathbf{G})}$$

where $m(\mathbf{G})$ and $M(\mathbf{G})$ are the smallest and largest eigenvalues of $\mathbf{G}$, respectively. In a given case one normally uses estimates m_E and M_E for $m(\mathbf{G})$ and $M(\mathbf{G})$ respectively. We refer to the parameters m_E and M_E as "acceleration parameters."

For a symmetrizable basic iterative method, the rate of convergence of the Chebyshev acceleration procedure, with exact estimates m_E and M_E, is an order-of-magnitude faster than that of the unaccelerated method. However, the rate of convergence is often very sensitive to the choice of m_E and M_E. Moreover in the case of Chebyshev acceleration applied to a basic iterative method with a splitting parameter, such as the SSOR method, one must choose the splitting parameter as well as the acceleration parameters, m_E and M_E. In the case of the SSOR method, however, it can be shown that $m(\mathbf{G}) \geq 0$; hence one can choose $m_E = 0$.

For conjugate gradient acceleration of a symmetrizable basic iterative method one must choose an auxiliary matrix $\mathbf{Z}$ such that $\mathbf{Z}$ and $\mathbf{Z}(\mathbf{I} - \mathbf{G})$ are SPD. In the case where $\mathbf{A}$ and $\mathbf{Q}$ are SPD, typical choices of $\mathbf{Z}$ are $\mathbf{Z} = \mathbf{A}$ and $\mathbf{Z} = \mathbf{Q}$. There are several equivalent forms of CG acceleration; see, e.g., the discussion of Young et al. [51]. One such form, which was considered by Concus et al. [4], see also Hageman and Young [20], is given by (2.4) where

$$(2.7) \quad \begin{aligned} \gamma_{n+1} &= \frac{\left(\mathbf{Z}\boldsymbol{\delta}^{(n)}, \boldsymbol{\delta}^{(n)}\right)}{\left(\mathbf{Z}\boldsymbol{\delta}^{(n)}, (\mathbf{I} - \mathbf{G})\boldsymbol{\delta}^{(n)}\right)} \\[2mm] \rho_{n+1} &= \left[1 - \frac{\gamma_{n+1}}{\gamma_n}\frac{\left(\mathbf{Z}\boldsymbol{\delta}^{(n)}, \boldsymbol{\delta}^{(n)}\right)}{\left(\mathbf{Z}\boldsymbol{\delta}^{(n-1)}, \boldsymbol{\delta}^{(n-1)}\right)}\frac{1}{\rho_n}\right]^{-1}, \quad n \geq 1, \quad (\rho_1 = 1) \end{aligned}$$

Here the "pseudo-residual" vector $\boldsymbol{\delta}^{(n)}$ is given by

$$(2.8) \qquad\qquad \boldsymbol{\delta}^{(n)} = \mathbf{G}u^{(n)} + \mathbf{k} - \mathbf{u}^{(n)}$$

It should be noted that there are no acceleration parameters involved. Moreover, it can be shown that, for a given basic iterative method, CG acceleration converges at least as fast (in a certain norm) as any polynomial acceleration procedure including Chebyshev acceleration, applied to the basic method. On the other hand, CG acceleration requires the computation of inner products after each iteration and thus may require more time per iteration than Chebyshev iteration. Also, for some parallel computers the communication time needed to compute the inner products may be costly.

It should be noted that, given an accurate estimate M_E of $M(\mathbf{G})$, one can obtain a good bound on the norm of the error $\|\mathbf{u}^{(n)} - \bar{\mathbf{u}}\|$, where $\bar{\mathbf{u}} = \mathbf{A}^{-1}\mathbf{b}$ is the true solution of (2.1), in terms of the norm of the pseudo-residual vector $\boldsymbol{\delta}^{(n)}$. Thus one can estimate $\|\mathbf{u}^{(n)} - \bar{\mathbf{u}}\|$ by

$$(2.9) \qquad\qquad \left\|\mathbf{u}^{(n)} - \bar{\mathbf{u}}\right\| \approx \frac{1}{1 - M_E}\left\|\boldsymbol{\delta}^{(n)}\right\|$$

(For a suitable choice of norm one can replace "$\approx$" by "$\leq$".) Thus an accurate estimate of $M(\mathbf{G})$ is useful in determining when an iterative process should be terminated. We remark that for the CG method an accurate estimate of $M(\mathbf{G})$ can be obtained by solving an eigenvalue problem for a tridiagonal matrix involving the $\{\gamma_i\}$ and the $\{\rho_i\}$ of (2.7); details are given in Hageman and Young [20]; see also Mai [28].

3   A Priori Techniques

In this section and in the next section, we review some of the techniques that can be used to estimate optimum iteration parameters. We consider two types of procedures, namely "a priori" techniques, which are usually carried out before the iteration process begins, and "adaptive techniques" which are carried out along with the iterative process. The a priori techniques described in this section include analytic techniques, spectral methods, and procedures involving the use of related differential equations.

Analytic Techniques

In some cases it is possible to relate the eigenvalues of the iteration matrix G corresponding to a relatively complicated iterative method, such as the SOR method, to those of a relatively simple method, such as the Jacobi method. Thus, for example, if A is a consistently ordered matrix then the eigenvalues λ of the SOR matrix and the eigenvalues μ of the iteration matrix B of the Jacobi method are related by

$$(3.1) \qquad\qquad \lambda + \omega - 1 = \omega\mu\sqrt{\lambda}$$

see Young [46] for details. If A is an SPD matrix the optimum value of ω for the SOR method is

$$(3.2) \qquad\qquad \omega_b = \frac{2}{1 + \sqrt{1 - S(B)^2}}$$

where $S(B)$ is the spectral radius of B. Thus for the consistently ordered case the determination of the optimum value of ω can be reduced to the determination of $S(B)$. Moreover for some linear systems derived from problems involving elliptic partial differential equations with constant coefficients one can find $S(B)$ using spectral methods as described below.

Even if A is not consistently ordered, but is an "L-matrix" (i.e., if $a_{i,i} > 0$ for all i and $a_{i,j} \leq 0$ for $i \neq j$) the use of (3.2) gives a "good" value of ω. This follows from work of Kahan [25]; see also Wachspress [44] and Young [46], Chapter 12.

Another example of the use of an analytic technique involves the SSOR method for the case where A is SPD.[1] Good estimates for the optimum value of ω and the corresponding value of $S(S_\omega)$, where S_ω is the matrix of the SSOR method, can be obtained; see for instance Habetler and Wachspress [18], Ehrlich [8], and Young [46,47,48]. The values of ω and $S(S_\omega)$ are given in terms of the largest eigenvalue $M(B)$ of the matrix B corresponding to the Jacobi method and $S(\mathbf{LU})$ where $\mathbf{L}$ and $\mathbf{U}$ are strictly lower triangular and strictly upper triangular matrices, respectively, and $B = \mathbf{L} + \mathbf{U}$. Thus it is shown in Young [48] that a good value, ω_1, of ω is given by

$$(3.3) \qquad\qquad \omega_1 = \frac{2}{1 + \sqrt{1 - 2M(B) + 4\tilde{\beta}}}$$

---

[1] Ehrlich [8] obtained estimates for the optimum ω for block SSOR for certain separable problems using spectral techniques.

where

(3.4)
$$\tilde{\beta} = \max\left\{\frac{1}{4}, S(\mathbf{LU})\right\}$$

The corresponding value of $S(S_{\omega_1})$ satisfies

(3.5) $\quad S(S_{\omega_1}) \leq \left(1 - \dfrac{1 - M(B)}{\sqrt{1 - 2M(B) + 4\tilde{\beta}}}\right) \Big/ \left(1 + \dfrac{1 - M(B)}{\sqrt{1 - 2M(B) + 4\tilde{\beta}}}\right)$

It is not difficult to see that for the SSOR method to be effective $S(\mathbf{LU})$ must not exceed $\frac{1}{4} + \epsilon$ for some small ϵ. This condition is satisfied if one uses the natural ordering of the grid points and if the coefficients of the differential equation are sufficiently smooth; see Young [48].

Spectral Methods

For linear systems corresponding to separable partial differential equations with constant coefficients over rectangular regions, the eigenvalues of $\mathbf{A}$ can often be found using "spectral methods," i.e., methods based on the use of finite Fourier series.[2] Thus, consider the "model problem" based on the Dirichlet problem in the unit square. Using the standard 5-point difference equation with a square grid of size $h = M^{-1}$ with M an integer, we have the difference equation

(3.6) $\quad u(x + h, y) + u(x - h, y) + u(x, y + h) + u(x, y - h) - 4u(x, y) = 0$

for all interior grid points (x, y). Multiplying the equations by -1 and moving the terms involving the (known) boundary values to the right hand side, we obtain a system of the form (2.1) where $\mathbf{A}$ has 4's on the main diagonal, -1's in certain off-diagonal elements, and 0's elsewhere. It can be verified directly that the eigenvectors of the difference operator corresponding to $\mathbf{A}$ are

(3.7) $\qquad\qquad v_{p,q}(x, y) = \sin p\pi x \sin q\pi y$

and the corresponding eigenvalues are

(3.8) $\qquad\qquad \mu_{p,q} = \frac{1}{2}[\cos p\pi h + \cos q\pi h]$

where $p, q = 1, 2, \ldots, M - 1$. Thus the eigenvalues μ of the Jacobi method lie in the interval

(3.9) $\qquad\qquad -\cos \pi h \leq \mu \leq \cos \pi h$

Hence Chebyshev acceleration can be used for the Jacobi method with

(3.10) $\qquad\qquad -m_E = M_E = \cos \pi h$

---

[2]For one of the earliest papers to use finite Fourier series to analyze difference equations see Phillips and Wiener [35].

A similar technique can be used for rectangular regions with an elliptic differential equation of the form

$$(3.11) \qquad\qquad (Au_x)_x + (Cu_y)_y + Fu = G$$

where A, C and F are constants; see, e.g., Young [46], Chapter 6. Again $M(B)$ and $m(B)$ can be computed. Moreover, bounds on $M(B)$ and $m(B)$ can be computed even for non-rectangular regions and for variable coefficients. The bounds involve upper and lower bounds for A, C, and F in the region. For non-rectangular regions, a monotonicity theorem is used which shows that $S(B)$ is an increasing function of the region involved.

Spectral methods can sometimes be used to determine the eigenvalues of the iteration matrix L_ω for the SOR method even when no relation, such as (3.1), is available between the eigenvalues of L_ω and the eigenvalues of B. Thus, if one uses the standard 9-point difference equation instead of the 5-point difference equation for the model problem, the matrix $\mathbf{A}$ is not consistently ordered and (3.1) does not hold. (Of course such a relation would hold if line SOR were used.) Nevertheless, the difference equation is separable and the eigenvalues of the SOR method can be determined, implicitly, as described by Adams, LeVeque and Young [1] (see also Van de Vooren and Vliegenthart [40]), by the following polynominal equation

$$
\begin{aligned}
(3.11) \quad & \alpha^4 - \left[\frac{4}{5}\omega c_q + \frac{2}{25}\omega^2 c_p^2 c_q\right]\alpha^3 - \left[2(1-\omega) - \frac{4}{25}\omega^2 c_q^2 + \frac{1}{25}\omega^2 c_p^2(c_q^2 + 4)\right]\alpha^2 \\
& + \left[\frac{4}{5}\omega(1-\omega)c_q - \frac{2}{25}\omega^2 c_p^2 c_q\right]\alpha + (1-\omega)^2 = 0
\end{aligned}
$$

Here $\lambda = \alpha^2$ is an eigenvalue of L_ω and

$$(3.12) \qquad\qquad \begin{cases} c_p = \cos p\pi h \\ c_q = \cos q\pi h \end{cases}$$

In order to find $S(L_\omega)$, one must solve the quartic equation (3.11) numerically where $p, q = 1, 2, \ldots, M-1$. One can then determine $S(L_\omega)$. Next, one can find the value of ω which minimizes $S(L_\omega)$ by a direct search procedure.

Using asymptotic analysis it can be shown that for small h

$$(3.13) \qquad\qquad S(L_\omega) \approx 1 - 1.79\pi h$$

We remark that an approximate asymptotic result, namely,

$$(3.14) \qquad\qquad S(L_\omega) \approx 1 - 2.35\pi h$$

can be obtained using the much less laborious analysis of Garabedian [13], which is based on the use of a related hyperbolic differential equation.

Use of Differential Equations

A technique that has sometimes been used for analyzing the behavior of an iterative method for solving an elliptic problem is to consider a related time-dependent problem. Essentially the iteration number, n, is replaced by the time variable t. Thus, suppose one is trying to solve the model problem using the "extrapolated Jacobi method" defined by

$$(3.15) \qquad \mathbf{u}^{(n+1)} = \gamma(B\mathbf{u}^{(n)} + \mathbf{c}) + (1 - \gamma)\mathbf{u}^{(n)}$$

where

$$(3.16) \qquad \begin{cases} B = \mathbf{I} - \mathbf{D}^{-1}\mathbf{A} \\ \mathbf{c} = \mathbf{D}^{-1}\mathbf{b} \end{cases}$$

Here $\mathbf{D}$ is defined in Sec. 2. The number γ is the extrapolation factor. The extrapolated Jacobi method corresponds to the forward difference method for solving the time-dependent equation

$$(3.17) \qquad u_t = u_{xx} + u_{yy}$$

The extrapolation factor γ depends on the time step. If one uses variable time steps, this correspond to the use of variable extrapolation with the Jacobi method, as defined by

$$(3.18) \qquad \mathbf{u}^{(n+1)} = \gamma_{n+1}(B\mathbf{u}^{(n)} + \mathbf{c}) + (1 - \gamma_{n+1})\mathbf{u}^{(n)}$$

Here again the $\{\gamma_{n+1}\}$ are determined by the time steps. We remark that, with a suitable choice of the $\{\gamma_{n+1}\}$, the convergence of the Jacobi method can be speeded up by an amount comparable to the speed-up achieved using the Chebyshev acceleration. The use of variable extrapolation with the Jacobi method was studied by Huang [22] in his doctoral thesis.

Another example of the use of a related differential equation to analyze the convergence of an iterative method for solving a linear system is the analysis of Garabedian [13] for the SOR method to which we have previously referred.

4 Adaptive Techniques

In many cases, it is difficult or impractical to determine the iteration parameters in advance to sufficient accuracy. For example, suppose one is using the SOR method and that the coefficient matrix $\mathbf{A}$ is consistently ordered. If the true value of $S(B)$ is .9999 but if one uses instead an estimate of .99 to determine omega, the rate of convergence may be reduced by a factor of nearly 20. This illustrates the fact that it is seldom possible to get a sufficiently good estimate of $S(B)$ using a priori methods. In order to get $S(B)$ to sufficient accuracy, it is often necessary to use adaptive techniques.

The basic idea of an adaptive procedure is as follows: One starts the iterative process with initial estimates of the optimum parameters. Periodically the convergence of the procedure is tested and compared with the expected convergence rate (which one would obtain if the exact optimum parameters were used). If the convergence rate is satisfactory the process is continued; otherwise new parameter estimates are determined according to some procedures. The development of such procedures is not simple in general, since one must develop procedures to tell how rapidly the method, with the given choice of parameters, is converging and, if the method is found to be converging too slowly, how the parameters should be changed to speed up the convergence.

In this section, we discuss the following: the search for omega for the SOR method; adaptive procedures for Chebyshev acceleration; and the choice of omega for the SSOR method with Chebyshev and CG acceleration. We also consider variational-based adaptive procedures, including a composite adaptive procedure developed by Mai [28].

The Search for Omega for the SOR Method

Early work on the search for omega for the SOR method was based on the use of (3.1) and (3.2) as well as other related formulas. One such approach was considered by Young [45]. Use was made of the fact that if $\mathbf{A}$ is consistently ordered then the eigenvalues of the matrix L for the Gauss-Seidel method are the squares of the eigenvalues of the matrix B for the Jacobi method. Thus by (3.2) the optimum value of ω is given by

$$(4.1) \qquad \omega_b = \frac{2}{1 + \sqrt{1 - S(L)}}$$

The eigenvalues of L are real, nonnegative and less than one. Hence, one can try to determine $S(L)$ using the ordinary power method. Then, having estimated $S(L)$, one can determine ω_b by (4.1).

A drawback of the above scheme is that many iterations of the Gauss-Seidel method are required to determine $S(L)$ and that the iterations are essentially wasted. A much more sophisticated scheme was used by Hageman and Kellogg [19], where the power method is speeded up by the use of Chebyshev polynomials. Other procedures are described by Varga [41] and by Bilodeau et al. [3].

Ideally, one would like to avoid carrying out "wasted iterations." Instead it would be desirable if every iteration could be useful, to some extent, even if perhaps not as useful as though the optimum value of the ω is used. Several schemes have been proposed for doing this including schemes of Kulsrud [27], Carré [5], Rigler [37], and Reid [36]. Each scheme is based on the following ideas. If one knew the spectral radius the SOR matrix, L_ω, corresponding to a given ω, one could, using (3.1), obtain a good estimate of $S(B)$ by

$$(4.2) \qquad S(B) = \frac{S(L_\omega) + \omega - 1}{\omega \sqrt{S(L_\omega)}}$$

Having found $S(B)$, one could get an improved value of ω using (3.2).

The basic procedure is as follows. Choose a value of ω not greater than ω_b. (For example, let $\omega = 1$.) Iterate for several iterations and estimate $S(L_\omega)$. Then compute $S(B)$ from (4.2) and then a new value of ω by (3.2).

The problem of adaptively determining the optimum value of ω, for the SOR method for the case where the coefficient matrix is SPD and consistently ordered, is complicated by the fact that the matrix L_ω of the SOR method may have principal vectors of grade 2. This tends to slow the convergence and, together with the fact that the eigenvalues of L_ω are often complex, makes the convergence erratic and also makes it hard to estimate $S(L_\omega)$. A rather elaborate program, which is based on the ideas given above and which takes into account a great many of the potential difficulties, has been developed by L. Hageman. The program is described in Chapter 9 of Hageman and Young [20]. The program is also included in the ITPACK 2C software package; see Kincaid et al. [26].

Adaptive Chebyshev Acceleration

For Chebyshev acceleration of a basic iterative method, two acceleration parameters, m_E and M_E, are normally required in addition to the splitting parameters, if any. However, in many cases a good lower bound, say $\underline{m}$, is available for $m(\mathbf{G})$. An example is the SSOR method where it is known that $0 \leq m(\mathbf{G})$. In such cases, one can let m_E be fixed and equal to $\underline{m}$. There remains one acceleration parameter, M_E, to be determined. For a given value of M_E, one can determine how fast the method should be converging if M_E were equal to $m(\mathbf{G})$. Moreover, one can measure the actual rate of convergence. Also procedures are available to improve M_E whenever it is found that the convergence is too slow. An adaptive procedure based on these components is described in Hageman and Young [20]. An important property of the procedure is that the values of M_E are nondecreasing and are never greater than $m(\mathbf{G})$.

Extensive numerical experiments over a wide range of problems indicate that the "overhead" associated with the adaptive process is seldom more than 25%–30%; see Hageman and Young [20] and Jea and Young [24].

Several approaches have been used for the case where m_E and M_E must both be determined adaptively, see, e.g., Diamond [6], Hageman and Young [20], and Mai [28]. Recently, Mai and Young [29] have considered the use of a "dual adaptive procedure," based on a suggestion of Tom Manteuffel. Numerical experiments based on this procedure indicate that it is somewhat more effective in many cases then the Hageman and Young procedure and the Mai procedure.

An alternative approach for finding $m(\mathbf{G})$ and $m(\mathbf{G})$ based on the use of modified moments has been used by Golub and Kent [14].

Adaptive SSOR with Chebyshev and CG Acceleration[3]

Since the SSOR method is symmetrizable it is usually applied in conjunction with an acceleration procedure—either Chebyshev acceleration or CG acceleration. In either case the relaxation factor ω must be estimated. If Chebyshev acceleration is used $M(S_\omega)$ must also be determined. In Hayes and Young [21] and Grimes et al. [16] an adaptive procedure is described which is focused on estimations of $S(B)$ and $S(\mathbf{LU})$. The procedure is based on analytic formulas given in Sec. 3 relating the good value of ω and the corresponding value of $M(S_\omega)$ to $S(B)$ and $S(\mathbf{LU})$. The procedure is used in the ITPACK 2C software package; see Kincaid et al. [26]. Another adaptive procedure for finding ω was given by Benokraitis [2]. An alternative procedure based on a composite adaptive procedure corresponding to a variational method is described below.

Variational-Based Adaptive Methods: The Composite Adaptive Procedure

Let us again consider the linear system (2.1) where $\mathbf{A}$ is SPD. It is easy to show that the problem of solving (2.1) is equivalent to that of minimizing the quadratic form

$$(4.3) \qquad F(\mathbf{u}) = \frac{1}{2}(\mathbf{u}, \mathbf{Au}) - (\mathbf{b}, \mathbf{u})$$

Suppose that we have a basic iterative method (2.2) for solving (2.1) where the splitting matrix $\mathbf{Q}$ depends on a (splitting) parameter, say ω. Given $\mathbf{u}^{(n)}$ it would seem reasonable to choose ω to minimize $F(\mathbf{u}^{(n+1)})$. In the case of the SOR method, for example, it can be shown that (assuming $\mathbf{D} = \mathbf{I}, \mathbf{L} = \mathbf{C}_L$, and $\mathbf{U} = \mathbf{C}_U$)

$$(4.4) \qquad F\left(\mathbf{u}^{(n+1)}\right) - F\left(\mathbf{u}^{(n)}\right) = -\frac{1}{2}\left[\frac{2-\omega}{\omega}\right]\left(\boldsymbol{\delta}^{(n)}, \boldsymbol{\delta}^{(n)}\right)$$

where $\boldsymbol{\delta}^{(n)} = \omega(I - \omega\mathbf{L})^{-1}\mathbf{r}^{(n)}$ and $\mathbf{r}^{(n)} = \mathbf{b} - \mathbf{Au}^{(n)}$. One could determine ω to minimize $F(\mathbf{u}^{(n+1)})$ at each step or periodically. This might involve a direct search procedure. However, in some cases, for example for the case where $\mathbf{A}$ is red-black the relation $(\mathbf{I} - \omega\mathbf{L})^{-1} = \mathbf{I} + \omega\mathbf{L}$ holds and the problem of minimizing $F(\mathbf{u}^{(n+1)})$ is equivalent to that of minimizing a quartic polynomial in ω.

We now describe a somewhat more general approach which is related to the method of steepest descent.

Given $u^{(n)}$ choose λ_n and ω to minimize

$$(4.5) \qquad F\left(\mathbf{u}^{(n+1)}\right) = F\left(\mathbf{u}^{(n)}\right) - \lambda_n\left(\boldsymbol{\delta}^{(n)}, \mathbf{r}^{(n)}\right) + \frac{\lambda_n^2}{2}\left(\boldsymbol{\delta}^{(n)}, \mathbf{A}\boldsymbol{\delta}^{(n)}\right)$$

Here $\boldsymbol{\delta}^{(n)} = \mathbf{Q}(\omega)^{-1}\mathbf{r}^{(n)}$.

---

[3]Evans and Forrington [11] considered an iteration procedure for finding the optimum value of SSOR using a procedure not related to an acceleration procedure.

The minimization of $F(\mathbf{u}^{(n+1)})$ with respect to λ_n gives

$$(4.6) \qquad \lambda_n = \frac{\left(\boldsymbol{\delta}^{(n)}, \mathbf{r}^{(n)}\right)}{\left(\boldsymbol{\delta}^{(n)}, \mathbf{A}\boldsymbol{\delta}^{(n)}\right)}$$

and

$$(4.7) \qquad F\left(\mathbf{u}^{(n+1)}\right) - F\left(\mathbf{u}^{(n)}\right) = -\frac{1}{2}\frac{\left(\boldsymbol{\delta}^{(n)}, \mathbf{r}^{(n)}\right)^2}{\left(\boldsymbol{\delta}^{(n)}, \mathbf{A}\boldsymbol{\delta}^{(n)}\right)}$$

One then can seek to choose ω to minimize $F(\mathbf{u}^{(n+1)})$. Again this might be done after each iteration or periodically. A direct search might be required unless an analytic formula can be found.

We now consider a more general approach which we refer to as the "composite adaptive procedure." We assume that $\mathbf{Q}(\omega)$ is SPD for each ω so that the basic iterative method is symmetrizable. It can be shown that if CG acceleration with $\mathbf{Z} = \mathbf{Q}$ is applied to (2.2) then, for each n, $F(\mathbf{u}^{(n+1)})$ is minimized with respect to any polynomial acceleration procedure based on (2.2). Here we write the CG acceleration procedure in the two-term form

$$(4.8) \qquad \begin{aligned} \mathbf{u}^{(n+1)} &= \mathbf{u}^{(n)} + \lambda_n \mathbf{p}^{(n)} \\ \mathbf{p}^{(n)} &= \boldsymbol{\delta}^{(n)} + \alpha_{n,n-1}\mathbf{p}^{(n-1)} \end{aligned}$$

where λ_n and $\alpha_{n,n-1}$ are chosen to minimize $F(\mathbf{u}^{(n+1)})$. For the composite adaptive procedure where ω varies, we choose an integer $s \geq 0$ and let

$$(4.9) \qquad \mathbf{p}^{(n)} = \boldsymbol{\delta}^{(n)} + \alpha_{n,n-1}\mathbf{p}^{(n-1)} + \cdots + \alpha_{n,n-s}\mathbf{p}^{(n-s)}$$

where again $\alpha_{n,n-1}, \alpha_{n,n-2}, \ldots, \alpha_{n,n-s}, \lambda_n$ and ω are chosen to minimize $F(\mathbf{u}^{(n+1)})$.

The case $s = 0$ reduces to the procedure defined by $(4.5)-(4.7)$ above. The case $s = 1$ has been analyzed by Mai [28]. Numerical experiments have been carried out which show that the convergence rate is improved and that the optimum value of ω is obtained to reasonable accuracy. However, the amount of work required to find ω at each step is quite high. It is expected that with the use of parallel computers the search for the optimum ω can be greatly speeded up.

5 The Nonsymmetric Case

For many problems involving partial differential equations the matrix $\mathbf{A}$ is not SPD. For example consider the convection-diffusion equation

$$(5.1) \qquad u_{xx} + u_{yy} + Du_x = 0.$$

If the standard 5-point finite difference equation based on central difference formulas is used, the matrix $\mathbf{A}$ is not symmetric and hence is not SPD. If $\mathbf{A}$ is not SPD, then most of the standard basic iterative methods are not symmetrizable; i.e., $\mathbf{I} - \mathbf{G}$ is not similar to an SPD matrix. One consequence of this is that the eigenvalues of the iteration matrix $\mathbf{G}$ are not necessarily real and less than one.

We now describe some of the approaches that can be used to solve linear systems where the matrix is not SPD. One approach is to use the ordinary SOR method. One can also use Chebyshev iteration. In each case, the formulas are the same as in the symmetrizable case, but the optimum parameters are determined in a different way. A number of generalized CG acceleration algorithms are also available; see, e.g., Hageman and Young [20]. As an alternative to generalized CG acceleration algorithms, one can use various Lanczos algorithms. These are also described in Hageman and Young [20] and in Young et al. [51].

The SOR Method

For the case where $\mathbf{A}$ is consistently ordered the relation between the eigenvalues of the SOR matrix L_ω and those of the Jacobi matrix B holds. For separable problems, it is often possible to find regions in the complex plane which contain all eigenvalues of B. Programs are available (see, e.g., Young and Eidson [49], Young and Huang [50] and Huang [23]) for finding the optimum value of ω when all eigenvalues of B are known. Usually it is sufficient to have available certain key eigenvalues of B.

For more general problems Ehrlich [9] and others have developed an "Ad-Hoc SOR" procedure. This procedure involves choosing values of ω at individual mesh points. The value of ω used at a given mesh point is that which one would obtain for a representative rectangle if the coefficients of the differential equations were constant throughout the rectangle. (In such a related problem the optimum value of ω can be obtained analytically.) This ad hoc procedure has been found to work well in practice; however, a rigorous analysis of the procedure remains to be done.

Chebyshev Acceleration

If the eigenvalues of the basic iterative method $\mathbf{G}$ are complex, but have real parts less than one, we can apply Chebyshev acceleration. The method is defined by (2.4)–(2.6). However, γ and the $\{\rho_i\}$ are given in terms of two real numbers d and σ^2 which relate to an "optimum ellipse." This ellipse contains all eigenvalues of $\mathbf{G}$ and has center at d and foci at $d \pm |\sigma|$ if σ^2 is positive, and at $d \pm |\sigma|i$ if σ^2 is negative. Here $\sigma^2 = b^2 - a^2$ where $\mathbf{A}$ and b are the lengths of the semi-major axis and the semi-minor axis, respectively, of the ellipse. (See Hageman and Young [20], Chapter 12, for more details.)

To find d and σ^2 one needs to first find all eigenvalues of $\mathbf{G}$ (or at least certain key eigenvalues). Manteuffel [30,31] developed an adaptive program which periodically determines key eigenvalues of $\mathbf{G}$ and from these determines d and σ^2. Another approach to finding the key eigenvalue of $\mathbf{G}$ is the hybrid procedure of Elman et al.

[10]. This procedure involves first carrying out several iterations using the GMRES procedure (see Saad and Schultz [38]) and then using the coefficients thus obtained to estimate the key eigenvalues. From these key eigenvalues the values of d and σ^2 are determined.

Huang [23] developed an alternate procedure for finding the optimum values of d and σ^2, given all of the eigenvalues of $\mathbf{G}$.

Generalized CG Methods and Lanczos Methods

Both the generalized CG acceleration methods (ORTHODIR, ORTHOMIN, and ORTHORES) and the Lanczos acceleration methods (including the biconjugate gradient method of Fletcher [12]) do not require any acceleration parameters. However, for the generalized CG acceleration procedures many of the terms must be discarded; otherwise, the machine time and storage needed would be prohibitive. Thus for the generalized CG procedures, one must choose a nonnegative integer s, and discard all but s terms on each iteration. All of these methods may include splitting parameters. Mai [28] has proposed a composite adaptive procedure for combining the adaptive determination of the splitting parameters with a generalized CG procedure. Numerical results indicate that the splitting parameter is determined fairly quickly and accurately; however, there is a need to improve the efficiency of the minimization procedure which must be carried out during each iteration.

Acknowledgements

This research was supported by the Department of Energy under Grant DE-FG05-87ER25048 in part by the National Science Foundation under Grant DCR-8518722, and by the U.S. Air Force Office of Scientific Research and Development under Grant AFSOR-85-0052 with The University of Texas at Austin. This paper was originally typed by the second author in TeX and retyped in LaTeX by Lisa Laguna.

References

[1] Adams, L.M., R.J. LeVeque, and D.M. Young. "Analysis of the SOR method for the 9-point Laplacian," *SIAM J. of Numer. Anal.* **25** (1988) 1156–1180.

[2] Benokraitis, V.J. "On the Adaptive Acceleration of Symmetric Successive Overrelaxation," Doctoral thesis, The University of Texas at Austin, 1974.

[3] Bilodeau, G.G., W.R. Cadwell, J.P. Dorsey, J.M. Fairey, and R.S. Varga. "PDQ–An ABM-704 Code to Solve the Two-dimensional Few-

group Neutron-diffusion Equations," Report WAPD-TM-70, Bettis Atomic Power Laboratory, Westinghouse Electric Corp., Pittsburg, Pennsylvania, 1957.

[4] Concus, P., G.H. Golub, and D.P. O'Leary. "A generalized conjugate gradient method for the numerical solution of elliptic partial differential equations," in *Sparse Computation* (Bunch and Rose. eds.), New York: Academic Press, 309–332, 1976.

[5] Carré, B.A. 'The determination of the optimum acceleration factor for successive over-relaxation," *Comput. J.* **4** (1961) 73–78.

[6] Diamond, M.A. "An Economical Algorithm for the Solution of Finite Difference Equations," Doctoral thesis, University of Illinois, Urbana, 1971.

[7] Dupont, T., R. Kendall, and H.H. Rachford, Jr. "An approximate factorization procedure for solving self-adjoint elliptic difference equations," *SIAM J. Numer. Anal.* **5** (1968) 559–573.

[8] Ehrlich, L.W. 'The block symmetric successive overrelaxation method," *J. Soc. Indust. Appl. Math.* **12** (1964), 807–826.

[9] Ehrlich, L.W. "The ad hoc SOR method: a local relaxation scheme," *Elliptic Problem Solvers II*, edited by Garrett Birkhoff and Arthur Schoenstadt, Orlando: Academic Press, 1984.

[10] Elman, H.C., Y. Saad, and P.E. Saylor. "A Hybrid Chebyshev Krylov Subspace Algorithm for Solving Nonsymmetric Systems of Linear Equations," Report TR-301, Yale University, New Haven, 1984.

[11] Evans, D.J. and C.V.D. Forrington. "An iterative processor optimizing successive overrelaxation," *Comput. J.* **6** (1963), 271–273.

[12] Fletcher, R. "Conjugate gradient methods for indefinite systems," *Proc. of the Dundee Biennial Conference on Numerical Analysis*, (G.A. Watson, ed.), Berlin and New York: Springer-Verlag, 1975, 73–89.

[13] Garabedian, P.R. "Estimation of the relaxation factor for small mesh size," *Math. Tables Aids Comput.* **10** (1956), 183–185; 228.

[14] Golub, G.H. and M.D. Kent. "Estimates of Eigenvalues for Iterative Methods," Report NA-87-02, Numerical Analysis Project, Computer Science Department, Stanford University, Stanford, 1987.

[15] Golub, G.H. and R.S. Varga. "Chebyshev semi-iterative methods, successive over-relaxation iterative methods, and second- order Richardson iterative methods," *Numer. Math.*, Parts I and II **3** (1961), 147–168.

[16] Grimes, R.G., D.R. Kincaid, and D.M. Young. "ITPACK 2.0 User's Guide," Report CNA-150, Center for Numerical Analysis, The University of Texas at Austin, 1979.

[17] Gustafsson, I. "Stability and Rate of Convergence of Modified Incomplete Cholesky Factorization Methods," Report 79.02R (Doctoral thesis), Department of Computer Sciences, Chalmers University of Technology, Goteborg, Sweden, 1979.

[18] Habetler, G.J. and E.L. Wachspress. "Symmetric successive overrelaxation in solving diffusion difference equations," *Math. Comp.* **15** (1961), 356–362.

[19] L.A. Hageman and R.B. Kellogg, "Estimating optimum overrelaxation parameters," *Math. Comp.*, **22** (1968), 60–68.

[20] Hageman, L.A. and David M. Young. *Applied Iterative Methods*, New York: Academic Press, 1981.

[21] Hayes, L. and D.M. Young. "The Accelerated SSOR Method for Solving Large Linear Systems," Report CNA-123, Center for Numerical Analysis, The University of Texas at Austin, 1977.

[22] Huang, C.Y. "Optimization of Explicit Time-Stepping Algorithms and Stream-Function-Coordinate (SFC) Concept for Fluid Dynamic Problems," Doctoral thesis, The University of Texas at Austin, 1987.

[23] Huang, R. "On the Determination of Iterative Parameters for Complex SOR and Chebyshev Methods," Report CNA-187, Center for Numerical Analysis, The University of Texas at Austin, 1983. (Also, Masters thesis.)

[24] Jea, K.C. and D.M. Young. "On the effectiveness of adaptive Chebyshev acceleration for solving systems of linear equations," *J. Comp. Appl. Math.* **24** (1988) 33–54.

[25] Kahan, W. "Gauss-Seidel Methods of Solving Large Systems of Linear Equations," Doctoral thesis, University of Toronto, Canada, 1958.

[26] Kincaid, D.R., J.R. Respess, R.G. Grimes and D.M. Young. "ITPACK 2C: a fortran package for solving large sparse linear systems by adaptive accelerated iterative methods," *ACM Trans. on Math. Software* **8** (1982), 302–322.

[27] Kulsrud, H.E. "A practical technique for the determination of the optimum relaxation factor of the successive over-relaxation method," *Comm. Assoc. Comput. Mach.* **4** (1961), 184–187.

[28] Mai, T.-Z. "Adaptive Iterative Algorithms for Large Sparse Linear Systems," Report CNA-203, Center for Numerical Analysis, The University of Texas, Austin, Texas, (1986). (Also, Doctoral dissertation.)

[29] Mai, T.-Z. and D.M. Young. "A Dual Adaptive Procedure for the Automatic Determination of Iteration Parameters for Chebyshev Acceleration." Report CNA-224, Center for Numerical Analysis, The University of Texas at Austin, 1988.

[30] Manteuffel, T.A. "The Tchebyshev iteration for nonsymmetric linear systems," *Numer. Math.* **28** (1977), 307–327.

[31] Manteuffel, T.A. "Adaptive procedure for estimating parameters for the nonsymmetric Tchebyshev iteration," *Numer. Math.* **28** (1978), 183–208.

[32] Manteuffel, T.A. "An incomplete Cholesky factorization technique for positive definite linear systems," *Math. Comp.* **34** (1980), 473–497.

[33] Meijerink, J.A. and H.A. Van der Vorst. "An iterative solution method for linear systems of which the coefficient matrix is a symmetric *M*-matrix," *Math. Comp.* **31** (1977), 148–162.

[34] Peaceman, D.W. and H.H. Rachford. "The numerical solution of parabolic and elliptic differential equations," *J. Soc. Indus. Appl. Math.* **3** (1955), 28–41.

[35] Phillips, H.B. and N. Wiener. "Nets and the Dirichlet problem," *J. Math. and Phys.* **58** (1923), 105–124.

[36] Reid, J.K. "A method for finding the optimum successive over-relaxation parameter," *Comput. J.* **9** (1966), 200–204.

[37] Rigler, A.K. "Estimation of the successive over-relaxation factor," *Math. Comp.* **19** (1965), 302–307.

[38] Saad, Y. and M.H. Schultz. "GMRES: A Generalized Minimal Residual Algorithm for Solving Nonsymmetric Linear Systems," Technical Report 254, Department of Computer Science, Yale University, New Haven, 1983.

[39] Stone, H.L. "Iterative solutions of implicit approximations of multi-dimensional partial differential equations," *SIAM J. Numer. Analy.* **5** (1968), 530–538.

[40] Van de Vooren, A.I., and A.C. Vliegenthart. "The 9-point difference formula for Laplace's equation," *J. Eng. Math.* **1** (1967), 187–202.

[41] Varga, R.S. "A comparison of the successive overrelaxation method and semi-iterative methods using Chebyshev polynomials," *J. Soc. Indus. Appl. Math.* **5** (1957), 39–46.

[42] Varga, R.S. "Factorization and normalized iterative methods," *Boundary Problems in Differential Equations* (R.E. Langer, ed.), 121–142, University of Wisconsin Press, Madison, 1960.

[43] Varga, R.S. *Matrix Iterative Analysis*, Englewood Cliffs: Prentice-Hall, 1962.

[44] Wachspress, E.L. *Iterative Solution of Elliptic Systems and Application to the Neutron Diffusion Equations of Reactor Physics*, Englewood Cliffs: Prentice-Hall, 1966.

[45] Young, D.M. "Ordvac solutions of the Dirichlet problem," *J. Assoc. Comput. Math.* **2** (1955), 137–161.

[46] Young, D.M. *Iterative Solution of Large Linear Systems*, New York: Academic Press, 1971.

[47] Young, D.M. "Second-degree iterative methods for the solution of large linear systems," *J. of Approx. Theory* **5** (1972), 137–148.

[48] Young, D.M. "On the accelerated SSOR method for solving large linear systems," *Adv. in Mathematics* **23** (1977), 215–271.

[49] Young, D.M. and H.D. Eidson. "On the Determination of the Optimum Relaxation Factor for the SOR Method When the Eigenvalues of the Jacobi Method are Complex," Report CNA-1, Center for Numerical Analysis , The University of Texas at Austin, 1970.

[50] Young, D.M. and R. Huang. "Some Notes on Complex Successive Overrelaxation," Report CNA-185, Center for Numerical Analysis, The University of Texas at Austin, 1983.

[51] Young, D.M., K.C. Jea and T.-Z. Mai. "Preconditioned conjugate gradient algorithms and software for solving large sparse linear systems," in *Linear Algebra in Signals, Systems, and Control* (Datta, Johnson, Kaashoek, Plemmons, and Sontag, eds.), SIAM, Philadelphia, 1988, 260–283.

Index

acceleration parameters, 294, 296, 297, 303, 307
accuracy, 53
ad hoc
    SOR procedure, 81, 84, 306
    method, 86
adaptive, 215, 217, 228, 229
    Chebyshev, 217
        acceleration, 68, 303
    method, 230
    parameter, 228
    procedures, 294, 301, 302, 304
ADI
    convergence, 263
    equations, 252
    iteration, 253, 268
    method, 252, 294
    minimax problem, 251–253
    nonsymmetric systems, 268
    parameters, 254
adjoint, 158
    left, 158
    right, 158
algorithm, 216, 222, 229, 227, 230
alternating direction implicit,
    See ADI
alternating method, 275
analytic techniques, 293, 294, 298
approximate factorization methods, 295
approximation, 226
arc, 223
arithmetic-geometric mean, 261
Arnoldi
    iteration, 268
    method, 219
asynchronous algorithms, 194
    iteration, 193

auxiliary matrix, 297

B-normal, 159
backward spectrum analysis, 252
balanced projection methods, 153
basic iterative method, 20, 293, 294
Bendixson's Theorem, 262
biconjugate gradient (BCG) algorithm, 163, 307
block Arnoldi method, 269
    conjugate gradient method, 167
    diagonal scaling, 1
    directed graph of type 2, 240
    Gauss-Seidel, 101
    Gaussian elimination, 278
    iterative methods, 88, 91
    Jacobi, 92, 98, 99, 101
    partitioned matrix, 239
    partitioning, 238
    tridiagonal, 217, 222
boundary, 229
boundary conditions, 135, 136
    Dirichlet, 53
    general, 44
    mixed, 55, 141
    Neumann, 141
bounded iterates, 154, 157
breakdown of iterative methods, 152, 156, 157, 159

cell Reynolds numbers, 91
centered differences, 93, 97, 101
chaining, 217, 228
Chebyshev
    acceleration, 68, 293, 294, 296
    iteration, 219, 225
    -like method, 150
    polynomials, 219

313